T0290340

Introduction to General Relativity and Cosmology (Second Edition)

Online at: https://doi.org/10.1088/2514-3433/acc3ff

AMERICAN ASTRONOMICAL SOCIETY 

AAS Editor in Chief

Ethan Vishniac, Johns Hopkins University, Maryland, USA

About the program:

AAS-IOP Astronomy ebooks is the official book program of the American Astronomical Society (AAS) and aims to share in depth the most fascinating areas of astronomy, astrophysics, solar physics, and planetary science. The program includes publications in the following topics:

GALAXIES AND COSMOLOGY

INTERSTELLAR MATTER AND THE LOCAL UNIVERSE

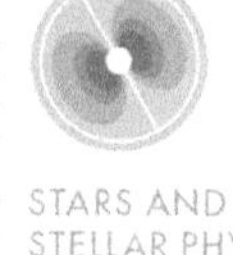

STARS AND STELLAR PHYSICS

EDUCATION, OUTREACH, AND HERITAGE

HIGH-ENERGY PHENOMENA AND FUNDAMENTAL PHYSICS

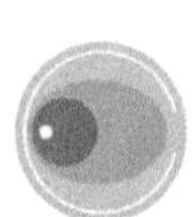

THE SUN AND THE HELIOSPHERE

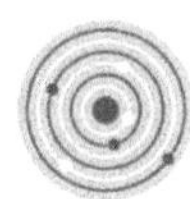

THE SOLAR SYSTEM, EXOPLANETS, AND ASTROBIOLOGY

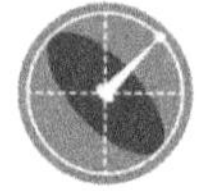

LABORATORY ASTROPHYSICS, INSTRUMENTATION, SOFTWARE, AND DATA

Books in the program range in level from short introductory texts on fast-moving areas, graduate and upper-level undergraduate textbooks, research monographs, and practical handbooks.

For a complete list of published and forthcoming titles, please visit iopscience.org/books/aas.

About the American Astronomical Society

The American Astronomical Society (aas.org), established 1899, is the major organization of professional astronomers in North America. The membership (~7,000) also includes physicists, mathematicians, geologists, engineers, and others whose research interests lie within the broad spectrum of subjects now comprising the contemporary astronomical sciences. The mission of the Society is to enhance and share humanity's scientific understanding of the universe.

Editorial Advisory Board

Introduction to General Relativity and Cosmology (Second Edition)

Ian R Kenyon
School of Physics and Astronomy, University of Birmingham, Birmingham, UK

IOP Publishing, Bristol, UK

ISBN 978-0-7503-3763-2 (ebook)
ISBN 978-0-7503-3761-8 (print)
ISBN 978-0-7503-3764-9 (myPrint)
ISBN 978-0-7503-3762-5 (mobi)

DOI 10.1088/2514-3433/acc3ff

Version: 20230801

AAS–IOP Astronomy
ISSN 2514-3433 (online)
ISSN 2515-141X (print)

British Library Cataloguing-in-Publication Data: A catalogue record for this book is available from the British Library.

Published by IOP Publishing, wholly owned by The Institute of Physics, London

IOP Publishing, No.2 The Distillery, Glassfields, Avon Street, Bristol, BS2 0GR, UK

US Office: IOP Publishing, Inc., 190 North Independence Mall West, Suite 601, Philadelphia, PA 19106, USA

To Valerie

Contents

Preface

The intention is to present to the student a modern, compact and digestible account of general relativity and modern cosmology. In recent decades there have been significant changes. The LIGO/Virgo collaboration has detected gravitational waves from the merger of black holes and then from neutron star mergers, with ~100 mergers observed by early 2020. This is making the properties of black holes, and tests of GR in the strong regime, accessible as never before. In 2019 the galactic black hole M87* was imaged for the first time by the Event Horizon Telescope collaboration. The study using SN1a supernovae as *standard candles* led to the discovery of acceleration of the expansion of the universe, a process started around 5 Gyr ago, or "recently" as cosmologists say. This acceleration is due to the existence of some field, most likely a scalar field; a field whose properties in most respects match the properties of Einstein's cosmological constant, and which appears to account for the bulk of the energy in the universe. In the common view, another scalar field was likely responsible for the earlier inflation certainly by a factor more than 10^{30}, terminating when the universe was only of order 10^{-35} s old.

Taken together these observations indicate that the aim of bringing to the student a unified picture is appropriate and timely.

Acknowledgments

The author is grateful for the continued support and warm encouragement of Professor Paul Newman, Head of the Particle Physics Research Group, and of Professor Bill Chaplin, the Head of the School of Physics and Astronomy at the University of Birmingham. This has given me access to the essential facilities needed for carrying out the project.

Thanks to Professor Sean McGee, from the School of Physics and Astronomy for reading and commenting on most of the cosmological component of this book; and also for producing essential figures with CMBfast and for patiently answering many questions. Thanks to Dr Geraint Pratten, Royal Society University Research Fellow, also of the School of Physics and Astronomy, who was kind enough to read and comment on two cosmology chapters and took the trouble to help me through an area of particular difficulty. Dr Geoff Brooker of Wadham once again took on the task of knocking my text into shape: my thanks to him for reading and commenting on the whole text with his usual insight. Thanks also to Professor Frank van den Bosch of Yale, for taking time to answer specific questions that puzzled me in his area of expertise. All this help was illuminating and invaluable. Errors are the author's alone.

Special thanks to Dr Mark Slater of the Particle Physics Research Group in the School of Physics and Astronomy for his invaluable, patient, and cheerful help with software, and for installing Linux and suites of software on a sequence of laptops. This gave me the necessary stable and reliable computing environment in which to operate.

My thanks to people at IOP Publishing: the Senior Editor John Navas, Leigh Jenkins, Associate Director David McDade, and his production team. Their help with preparation of figures, acquisition of permissions, and the process of e-publication, was throughout courteous and efficient. My thanks too, to the American Astronomical Society for jointly publishing the book.

Thanks to the publishers and authors who permitted re-use of material under copyright, or creative commons, etc. Each case is fully detailed in the text as it occurs.

I am indebted to Oxford University Press and their Senior Science Editor, Sonke Adlung, for ceding to me the copyright of *General Relativity* published in 1990 by Oxford. Sonke also forwarded to me originals of the figures. Oxford also allowed me to adapt a portion of Chapter 11 (Quantum measurement) from *Quantum 20/20: Fundamentals, Entanglement, Gauge Fields, Condensates and Topology* published by Oxford University Press in 2019. These generous acts are warmly appreciated.

Last, but not least, my heartfelt thanks to Dr Yoshinari Mikami for supplying me with a Latex file of the equations from his Japanese translation of *General Relativity*, and for his careful checking of the formulae. His kind act gave a timely boost toward getting this text launched and saved several weeks of additional keypunching and checking.

About the Author

Ian R. Kenyon

Ian R. Kenyon is an elementary particle physicist in the School of Physics and Astronomy at the University of Birmingham, UK. He took part in the discovery of the carriers of the weak force, working at CERN for three years on the design, construction, data-taking and analysis of the UA1 experiment. Earlier he designed and built the optics for the Northwestern University 50 cm liquid helium bubble chamber. More recently he worked on the H1 experiment at the HERA electron–proton collider and on optoelectronics: the design and construction with CERN and Hewlett-Packard of the then fastest link between computers, commercialized by HP, and on the CERN-funded RD23 programme for optoelectronic readout from LHC detectors. He is the author of four advanced textbooks for physics undergraduates: *Elementary Particle Physics*, *General Relativity* (also translated into Japanese), *The Light Fantastic: A Modern Introduction to Classical and Quantum Optics* and *Quantum 20/20: Fundamentals, Entanglement, Gauge Fields, Condensates and Topology*. This present text builds on the earlier *General Relativity* to cover the development together of general relativity and cosmology.

Author Tribute

Prof Paul Newman, Particle Physics Group Leader, University of Birmingham

During the last stages of the preparation of this book, Prof Ian Kenyon sadly passed away. Ian was stalwart of Birmingham's School of Physics and Astronomy and a cornerstone of its particle physics group for over half a century. Fundamental science has advanced enormously during Ian's research career; his significant role in the discovery of the W and Z bosons by the CERN UA1 experiment stands out among his many contributions. Ian remained full of energy up to the very end of his life, attending the university daily, writing prolifically, organising group seminars and enjoying conversations on a wide range of topics with an eclectic selection of scientists. His curiosity about nature and his tenacity in seeking to understand its basic mechanisms were as remarkable as the breadth of his knowledge. This book and his other titles form part of a rich and lasting legacy. He was an outstanding physicist who will be fondly remembered and greatly missed.

Physical Constants and Parameters

Gravitational constant	$G = 6.673 \times 10^{-11}\ \mathrm{kg^{-1}\ m^{-3}\ s^{-2}}$
Speed of light	$c = 2.998 \times 10^{8}\ \mathrm{m\ s^{-1}}$
Planck's constant/2π	$\hbar = 1.055 \times 10^{-34}\ \mathrm{J\ s} = 6.582 \times 10^{-16}\ \mathrm{eV\ s}$
Electron charge	$e = 1.602 \times 10^{-19}\ \mathrm{C}$
Electron mass	$m_{\mathrm{e}} = 0.511\ \mathrm{MeV}\ c^{-2}$
Proton mass	$m_{\mathrm{p}} = 938.3\ \mathrm{MeV}\ c^{-2} = 1.672 \times 10^{-27}\ \mathrm{kg}$
Avogadro's number	$N_{\mathrm{A}} = 6.022 \times 10^{23}\ \mathrm{mol^{-1}}$
Boltzmann's constant	$k_{\mathrm{B}} = 1.381 \times 10^{-23}\ \mathrm{J\ K^{-1}} = 8.617 \times 10^{-5}\ \mathrm{eV\ K^{-1}}$
Stefan's constant	$\sigma = 5.67 \times 10^{-8}\ \mathrm{W\ m^{-2}\ K^{-4}}$
Fine structure constant	$\alpha = 7.297 \times 10^{-3}$
Bohr radius	$a_0 = 0.529 \times 10^{-10}\ \mathrm{m}$
Solar mass	$M_{\odot} = 1.99 \times 10^{30}\ \mathrm{kg}$
Solar luminosity	$L_{\odot} = 3.83 \times 10^{26}\ \mathrm{J\ s^{-1}}$
Sun–Earth distance	$\mathrm{AU} = 1.50 \times 10^{11}\ \mathrm{m}$
Earth's radius	$R_{\oplus} = 6.378 \times 10^{6}\ \mathrm{m}$
Luminosity factor	$c^5/G = 3.63 \times 10^{52}\ \mathrm{W}$
Thomson cross-section	$\sigma_{\mathrm{T}} = 6.652 \times 10^{-29}\ \mathrm{m^2}$
One year	$3.156 \times 10^{7}\ \mathrm{s}$
Megaparsec	$3.09 \times 10^{22}\ \mathrm{m}$ = 3.26 megalight years
Planck mass	$M_{\mathrm{P}} = 2.18 \times 10^{-8}\ \mathrm{kg}$
Planck length	$\ell_{\mathrm{P}} = 1.62 \times 10^{-35}\ \mathrm{m}$
Planck time	$t_{\mathrm{P}} = 5.39 \times 10^{-44}\ \mathrm{s}$
Neutron mean life	$\tau_{\mathrm{N}} = 880\ \mathrm{s}$
CMB temperature	2.7255 K

ΛCDM model parameters used:	
Critical density of universe	$\rho_c = 9.14 \times 10^{-27}\ \mathrm{kg\ m^{-3}} = 1.35 \times 10^{11}\ M_{\odot}\ \mathrm{Mpc^{-3}}$
Matter fraction	$\Omega_{\mathrm{m0}} = 0.30$
Dark energy fraction	$\Omega_{\Lambda 0} = 0.70$
Radiation fraction	$\Omega_{\mathrm{r0}} = 0.000\ 09$
Baryon fraction	$\Omega_{\mathrm{b0}} = 0.045$
Hubble's constant	$H_0 = 70\ \mathrm{km\ s^{-1}\ Mpc^{-1}}$
Hubble distance	$c/H_0 = 4290\ \mathrm{Mpc}$
Hubble time	$1/H_0 = 14.0\ \mathrm{Gyr}$

AAS | IOP Astronomy

Introduction to General Relativity and Cosmology (Second Edition)

Ian R Kenyon

Chapter 1

Introduction

1.1 Prologue

This chapter contains a brief survey of the twin topics of general relativity and cosmology, highlighting the key experimental discoveries and concepts. The body of the book fleshes out these themes. Einstein's theory is the first topic, including the tests that his theory has passed within the solar system. Two consequent topics of great current interest come next: black holes and gravitational waves. This allows a smooth transition into cosmology: to an account of our understanding of how the universe developed. The narrative shows the success of a model for the universe picturing it as a space that initially expanded violently and almost instantaneously, then more sedately and is now destined to expand ever more rapidly. This is a model whose framework was provided by Einstein. The contents of the universe in this successful model are mainly dark energy and cold dark matter, which paradoxically are not directly detectable. Ordinary matter accounts for only about 5% of the energy in the universe and the stars are just 5% of that 5%. Our understanding of the whole has been built principally on observations of this tiny residue and of radiation from the initial compressed, extremely hot phase of the universe. Our model, which combines dark energy (λ) with cold dark matter (CDM) and is called λ CDM, provides a consistent interpretation of evolution of the universe over some 13,800 million years since the Big Bang that set it off.

Chapters 2–6 are used to introduce the concepts of the general theory of relativity (GR) and Einstein's equation linking spacetime curvature and matter. Chapter 7 describes the success of GR in passing the many tests made within the solar system. Chapter 8 is devoted to the analysis of black hole properties and the observations that convince us of the existence of stellar and galactic black holes. Chapter 9 develops the theory of gravitational waves and describes their discovery and their study carried out using km-sized optical interferometers. Chapters 10–17 are used to

doi:10.1088/2514-3433/acc3ffch1

present the general relativistic dynamics of the universe, and then to describe the quantitative success of the ΛCDM model in explaining the evolution of the universe.

1.2 Einstein's Insight

The general theory of relativity proposed by Einstein in 1915 imposed a new view of the spacetime that we inhabit: instead of matter moving through a passive spacetime continuum GR asserts that the presence of matter distorts spacetime. This distortion causes the deflection of material particles and light from their classical paths. Einstein predicted on the basis of GR that starlight which passes the limb of the Sun on its way to the Earth is deflected by 1.750 arcsec—not a great deal but measurable. Soon after, in 1919, Eddington organized an expedition to the island of Princip é, which photographed the star field around the Sun during a solar eclipse. When a comparison was made with night photographs of the same star field the predicted general relativistic deflection was confirmed. The curious properties proposed for spacetime helped turn this result into a popular news item. Science fiction in all the media shows how awareness of the new concepts has penetrated popular culture. Today, a century on, data obtained with modern telescopes, detectors, and the associated computing facilities is so detailed and refined that we can say with confidence that massive black holes, predicted by GR, lie at the centers of most galaxies; while gravitational waves from distant black hole mergers are regularly being detected by the collaboration using the km-scale LIGO/Virgo interferometers.

Newton's law of gravitational attraction, the predecessor to GR, states that two masses m_1 and m_2 separated by a distance r feel a mutual gravitational attraction

$$F = G\frac{m_1 m_2}{r^2}, \tag{1.1}$$

where G is the universal gravitational constant. Despite its successes, this law is fundamentally flawed: it is time-independent, meaning that the gravitational force would act instantaneously at all distances, in flat contradiction to Einstein's special theory of relativity (SR) of 1905, which requires that no signal should travel faster than c, the speed of light in a vacuum. Coulomb's law of electrostatics shares this weakness, and in this electromagnetic case the difficulty is resolved by using Maxwell's equations, which are consistent with SR. Coulomb's law is then seen to be the limiting form of one of these equations when the charges are slow moving, that is to say, in the quasistatic limit.

Einstein therefore sought equations to describe gravitation that would be consistent with SR, and his search was crowned with success a decade later in 1915. Crucially Einstein came to appreciate a fundamental connection: that the presence of matter causes the curvature of spacetime and in turn this curvature influences the paths of matter and light, a view going beyond both classical and special relativistic mechanics. Gravitation is then built into the theory from the outset through the key equation that links spacetime curvature to the matter distribution. Further, gravitational effects propagate with the same speed as light.

In the limit of low velocities and small gravitation effects Einstein's equation reduces to Newton's law, so that all the latter's successes are explained consistently.

The bending of light near a massive body is more spectacular in *Einstein rings*, images of sources seen through galaxies located nearer to us. Figure 1.1 shows the image of a blue galaxy gravitationally lensed by the red galaxy nearer to the Earth; the axial alignment is so good that the image forms almost a complete circle.

GR predicts the fate of massive stars. Initially the gravitational self-attraction of the stellar material leads to large internal pressures and temperatures that ignite nuclear burning. Eventually the fuel is used up and our naive expectation would be that the star should contract to a size at which the pressure and gravitational self-attraction are in balance. However, in 1931 Chandrasekhar showed that for a sufficiently massive star the gravitational collapse can continue indefinitely. What is left behind is that mysterious entity, a black hole. Spacetime is so warped that even light cannot escape from within its horizon: a sphere of radius $2GM/c^2$ for a static black hole of mass M, where G is the gravitational constant. GR also predicts the existence of gravitational waves that travel at the speed of light. When they cross a region of spacetime it is spacetime itself that vibrates. Such vibrations were detected for the first time in 2016 by the advanced LIGO 4 km Michelson interferometers. The strain produced by these waves is shown in Figure 1.2: they were emitted during the final inspiral of a binary pair of 30 solar mass black holes, merging as a single black hole. About three solar masses (times c^2) were converted to $\sim 10^{48}$ J of gravitational wave energy, which, if it could be harnessed, would power our current civilization for $\sim 10^{28}$ years.

Large black holes of order 10^{6-9} solar masses lie at the centers of most galaxies. The orbits of several stars close to the center of our Galaxy, near Sgr A*, have been

Figure 1.1. Einstein ring LRG 3-757 recorded using the Hubble Telescope's Wide-Field Camera 3. Image credit: ESA/Hubble and NASA.

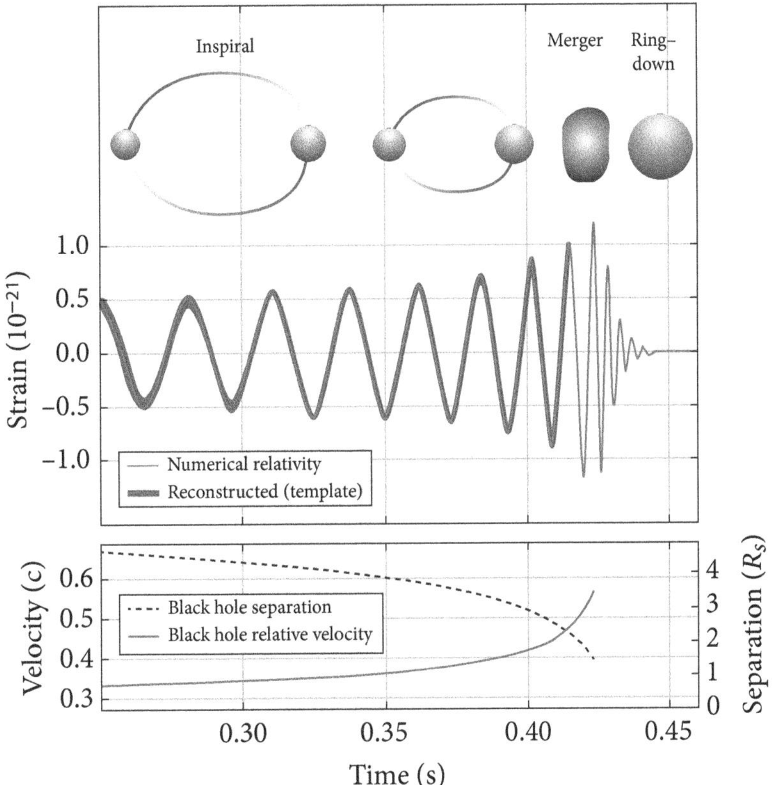

Figure 1.2. Estimated strain amplitude from GW150914 made with numerical relativity models of the black hole behavior as the holes coalesce. In the lower panel the separation of the black holes is given in units of the Schwarzschild radius $2GM/c^2$ and the relative velocity is divided by c. LIGO Open Science Center at https://losc.ligo.org/events/GW150914. The work is reported by the LIGO Scientific Collaboration and Virgo Collaboration (Abbott et al. 2016). Courtesy LIGO Collaboration. This figure is Figure 11.5 taken with permission from Kenyon (2019) published by Oxford Univ. Press in 2019.

observed over decades. In the case of the star SO2, its velocity at closest approach is fully consistent with GR but departs by 200 km s^{-1} from the Newtonian prediction (Do et al. 2019).

It is interesting to contrast gravitation with the other long-range force in nature: electromagnetism. The long range of the two forces is attested by the operation of the solar system in the first case and by the Van Allen belts and solar flares in the second. Of the two the electromagnetic force is intrinsically far stronger; the electrical repulsion between two protons is 10^{36} times stronger than their mutual gravitational attraction. The fact that gravity dominates on the large scale shows that matter in stars and galaxies must be electrically neutral to very high precision.

While quantized theories have been discovered to describe the non-gravitational forces, GR remains a classical theory. The possibility of reconciling GR and quantum mechanics, the twin achievements of twentieth century physics, has now been pursued by theoreticians for a century. There are still only tantalizing hints of a solution.

Turning to cosmology, GR has provided the solid framework within which the structure and development of the universe are described. The model of spacetime, used throughout this text, due to Friedmann, Robertson, and Walker, was derived using GR. It provides the framework for the modern coherent description of the development of the universe.

1.3 Structures Seen Today

Normal matter is composed of baryons and electrons. The baryons are protons, neutrons, and heavier nuclei, while electrons plus baryons make atoms and ions. Normal matter is called *baryonic matter* by cosmologists, with electrons included because they contribute relatively little mass, each being about one two-thousandth the mass of a proton. However the baryon number count used here is always the number of baryons, strictly defined.[1] The current mean density of baryonic matter in the universe is equivalent to 0.25 protons in each cubic meter.

Baryonic matter in the visible universe is concentrated in galaxies of which our own is an average example. It contains about 2×10^{11} stars of which our Sun is also average, having a mass 2×10^{30} kg, a mass used as the mass unit on the cosmic scale, $M_{\odot}$. We, on Earth, are 150 million km away from the Sun, a measure called one *astronomical unit (AU)*. Another measure is related to this: a *parsec* is defined as the distance at which an AU would subtend one second of arc (arcsec). Our Galaxy forms a pancake-shaped bright region of diameter ~34 kpc with a central bulge. When you go far enough from city lights, our Galaxy is seen edge-on as a diffuse band of light across the sky. The Sun lies 8 kpc from the central black hole.

Galaxies form clusters and superclusters. Our local cluster contains about 80 galaxies including the Andromeda galaxy, and our cluster forms part of the Virgo supercluster. This supercluster is about 33 Mpc across and contains around 100 clusters. Galactic clusters themselves form a vast web-like structure, which encloses voids, where galaxies are rare, that can be 100 Mpc across. Above this scale the universe is, broadly speaking, isotropic and homogeneous; that is to say, it looks the same in any direction, and at a given moment in cosmic time it would appear the same when seen from any other location. Distant stars and galaxies are seen as they were at an earlier time when the light we now receive left them. Figure 1.3 shows an area spanning 1000 square degrees and extending back 7 Gyr (7,000,000,000 years): how times are inferred from the observed redshifts is discussed below.

A working hypothesis used by cosmologists is that wherever you are in the universe it would, on the large scale, look the same. Taking this to be universally valid is known as the *cosmological principle*. As we shall see, this principle is very helpful in constraining the possible models of the universe.

[1] Particle properties are detailed in Appendix A.

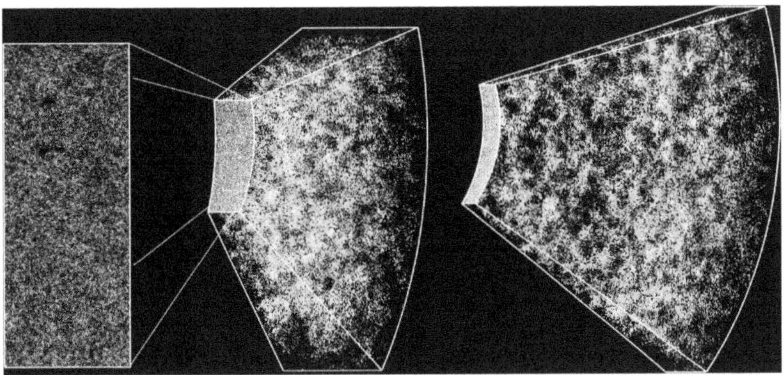

Figure 1.3. The left panel shows a 1000 square degree image taken by the Sloan Digital Sky Survey III. The other panels are three-dimensional presentations going back 7 Gyrs based on the galaxy redshifts. Image prepared from SDSS-III data by Jeremy Tinker. Courtesy Jeremy Tinker and the Sloan Digital Sky Survey.

1.4 Hubble's Law

Around the end of the 1920s Edwin Hubble (Hubble 1929; Hubble & Humason 1931) discovered that the redshift of a star's spectral lines is proportional to the star's distance d from the Earth. The redshift is given by

$$z = (\lambda_0 - \lambda)/\lambda, \tag{1.2}$$

where λ and λ_0 are, respectively, the wavelengths of the radiation on emission at some distant point and as observed on Earth. Hubble's law is then

$$z = Hd/c, \tag{1.3}$$

where the measured value of Hubble's constant at present, H_0, is around 70 km s^{-1} Mpc^{-1} (megaparsec). If the redshift is interpreted as a Doppler shift then the velocity of the source with respect to the Earth, v, is given by

$$v = Hd. \tag{1.4}$$

The redshifts of the most distant galaxies extend far above unity, which makes their interpretation in terms of a Doppler shift appear inconsistent with special relativity. This paradox is resolved once it is grasped that spacetime itself is expanding at velocities greater than c, and carries stars and galaxies with it. Special relativity only requires that the velocity locally in space can never exceed c. Thus the ratio between the emitted and observed wavelength is the factor by which spacetime has expanded between the emission of the radiation and its arrival on Earth.

Hubble's law has the property that it applies to any observer. If A observes that B has a redshift z_B and C has a redshift z_C then, using vectors to give the direction of the redshift:

$$\mathbf{z}_B = H[\mathbf{d}_B - \mathbf{d}_A]/c \text{ and } \mathbf{z}_C = H[\mathbf{d}_C - \mathbf{d}_A]/c. \tag{1.5}$$

From which we deduce the redshift of C seen by B

$$\mathbf{z} = \mathbf{z}_{\mathrm{C}} - \mathbf{z}_{\mathrm{B}} = H[\mathbf{d}_{\mathrm{C}} - \mathbf{d}_{\mathrm{B}}]/c, \tag{1.6}$$

consistent with the cosmological principle that all locations are equivalent. Any relation other than linear between z and d would give an inconsistent result, making Hubble's law the only acceptable relation connecting distance and redshift.

Hubble's constant is determined by the expansion rate of the universe, which itself changes. This makes Hubble's *constant* time dependent, and thus different at different times in the life of the universe. We define a scale factor a for the universe so that the wavelength of light $\lambda \propto a$, and set the current value to be unity: $a_0 = 1$. Consider a particular time t when the wavelength was λ and changed by $\Delta\lambda$ in a short time Δt due to the expansion of the universe. During the time Δt light traveled a distance $\Delta d = c\Delta t$. Then using Equations (1.2) and (1.3) Hubble's constant at time t is

$$H = \frac{c\,\Delta\lambda}{\lambda\,\Delta d} = \frac{\Delta\lambda}{\lambda\,\Delta t} = \frac{\Delta a}{a\,\Delta t} = \frac{\dot{a}}{a}. \tag{1.7}$$

This convenient equation will be used a great deal.

It is important to grasp that light itself travels in a vacuum at velocity c with respect to the local space; this velocity is unaffected by the expansion. As we have just discovered, the relative velocity of a source relative to the Earth can exceed c thanks to the continuous expansion of the universe. From the moment the relative velocity first exceeds c, no further light from the source will ever reach us.[2] Any motion in addition to motion due to the expansion of the universe is known as *secular motion*. Our secular motion is the vector sum of the motion of the Earth around the Sun, that of the Sun round our Galaxy, and of the Galaxy falling into the Virgo cluster.

A prerequisite for measuring Hubble's constant is that there exist classes of stars or other sources whose absolute luminosity can be reliably predicted. The first such *standard candles* to be identified were the Cepheid variable stars: Henrietta Leavitt (Leavitt 1908) made the crucial discovery that the frequency of their oscillation in intensity is correlated with their absolute luminosity. Edwin Hubble measured the flux of radiation on Earth (apparent luminosity) from a Cepheid and compared this to the flux leaving the Cepheid surface (absolute luminosity) as estimated from its frequency. Comparing these values gives the distance from Earth; in a flat static universe the flux falls off as the inverse square of the distance.

More recently the *tip of the red giant branch (TRGB)* method has provided another standard candle. Stars in the range of a few solar masses have a distinct history. While fusing hydrogen to helium they grow cooler (as red giants) and more luminous, then a thermonuclear reaction is ignited in the helium core converting three helium nuclei to a carbon nucleus. There is a *helium flash*, the temperature rises

[2] We will continue to receive the light previously emitted, which gets progressively more redshifted and less intense. This effect is explained in Chapter 8 when, in an equivalent way, a source falling into a black hole is extinguished.

relatively rapidly and the luminosity falls gradually. The result is that the luminosity distribution of the red giants is continuous up to a maximum luminosity, and then cuts off sharply. This produces the TRGB shown in Figure 1.4. Following the helium ignition the stars trajectory is to move rapidly to the left. Observation shows that the absolute luminosity at the tip is independent of the era of star formation, or other variables, making the TRGB luminosity an excellent standard candle. Cepheids are often located in densely populated and dusty regions making for

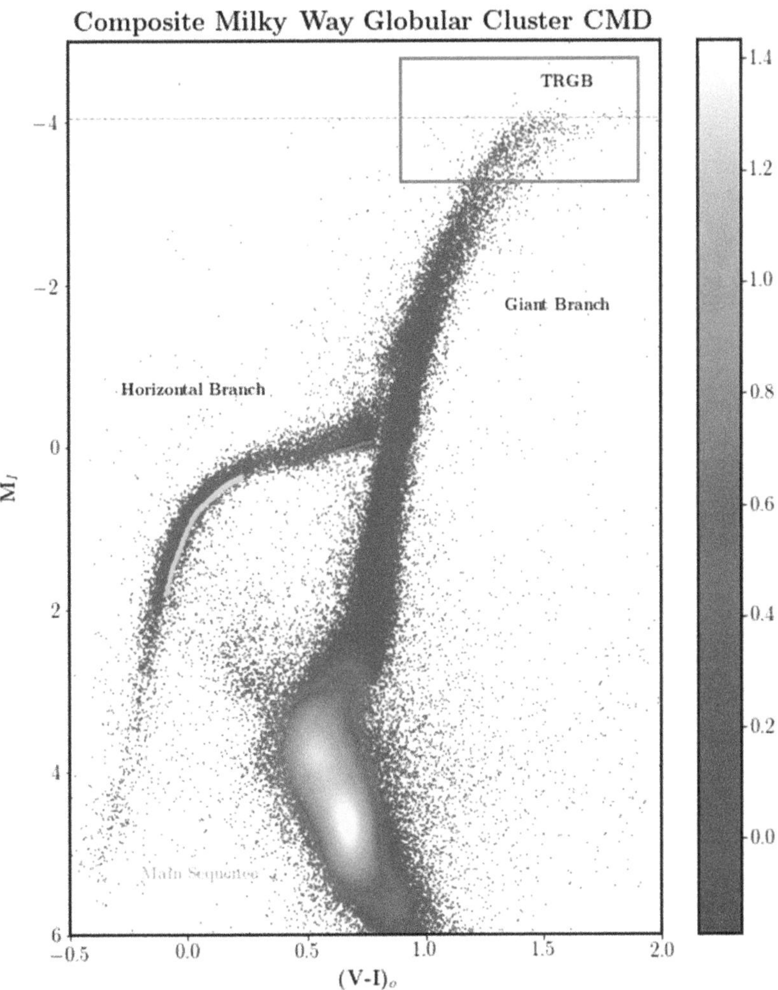

Figure 1.4. The Hertzsprung–Russell plot for stars in a globular cluster in our Galaxy. M_I is the absolute magnitude of luminosity measured with an infrared filter. The precise definition of magnitude is given in Section 1.13. (V − I) is the difference between the brightness using filters for yellow-green light (V) and infrared light (I) and measures the temperature of the star. This uses the fact that as temperature rises the spectrum of a black body source moves to shorter wavelengths. The color scale indicates the population density. The sharp luminosity cutoff is at magnitude −4.05. Figure 1 from Freedman (2021). Published by the American Astronomical Society under Creative Commons Attribution 4.0 licence.

uncertainty in estimating their luminosity. On the other hand stars at the TRGB are usually well isolated.

Nowadays distances to stars within our Galaxy are obtained from the parallax observed over the solar year by space telescopes such as *Hipparcos* and *Gaia*, which are located in orbits close to the Earth. The angular resolution of the observations made are μas (micro arcsecond). Cepheid and TRGB measurements carry the distance scale to nearby galaxies. Much brighter standard candles are used further off, the *supernovae type Ia (SNe Ia)*. These are events that may occur when a white dwarf star and a larger star are gravitationally bound (a binary pair). Material from the larger partner flows onto the white dwarf and eventually this ignites a thermonuclear explosion that is, for several hours, more than a billion times brighter than the Sun. The peak luminosity and rate of luminosity decay have characteristic correlations so that such explosions provide standard candles that can be observed at remote distances. Observations discussed in Chapter 17 have revealed that the expansion of the universe is accelerating due to a mysterious component of the universe called *dark matter*.

1.5 Olbers' Paradox

When you look up into the night sky, well away from city lights, you see that there are dark regions between the stars. In 1826 Heinrich Olbers found this surprising if, as was then supposed, the universe were infinite in time and space. Consider what you would then see within a narrow cone subtending a small solid angle α at your eye. At a distance R from the Earth the number of stars in that cone would be proportional to αR^2, and the light received from each star would be proportional to R^{-2}. In other words the overall light received from a layer of the universe at distance R would be independent of R; so the total intensity would steadily build up as R grows indefinitely. No matter what line you draw in the cone it would inevitably hit a star, so the whole cone (and sky) would be brightly illuminated. The resolution of Olbers' paradox is that the universe has only existed for a finite time. On the simple assumption that the universe has expanded at a constant rate we can infer that the present age $1/H_0$ would be about 14 Gyr. As we shall see this is not far from the current best estimate.

1.6 The Big Bang and the Cosmic Microwave Background

At some moment, around 14 Gyr ago, the universe would have been confined to a submicroscopic region, which was unimaginably hot, and it then expanded explosively: the *Big Bang*. Alpher, Bethe and Gamow (Alpher et al. 1948), among others, inferred that the radiation from the Big Bang would have a black body spectrum. They predicted that as a result of traveling through the expanding universe the wavelengths would be stretched so that today they lie in the microwave part of the spectrum. This radiation was first detected by Penzias and Wilson (Penzias & Wilson 1965) in 1965, and has been studied using satellites with increasing temperature precision, angular resolution over the sky, and with polarization determination, most recently by the Planck Collaboration. This radiation is called the cosmic microwave background (CMB).

Figure 1.5 shows the measured spectrum of the CMB plotted against $2\pi/\lambda$. The data agrees very precisely with the black body spectrum predicted with Planck's formula at a temperature of 2.7255 K. This is the most perfect example that exists of a black body spectrum! The spectrum peaks at $2\pi/\lambda = 5.92$ cm^{-1}, making the wavelength 1.06 mm (frequency $\nu = 282$ GHz, energy $\hbar\nu = 1.17$ meV per photon). Planck's formula for the number of photons with energy in the range E to $E + \mathrm{d}E$ per unit volume in black body radiation is

$$n(E)\mathrm{d}E = \frac{8\pi}{c^3h^3}\frac{E^2\mathrm{d}E}{\exp(E/[k_\mathrm{B}T]) - 1}, \tag{1.8}$$

where k_B is Boltzmann's constant. We introduce a parameter $x = E/[k_\mathrm{B}T]$ so the distribution becomes

$$n(x)\mathrm{d}x = \frac{8\pi}{c^3h^3}[k_\mathrm{B}T]^3\frac{x^2\mathrm{d}x}{\exp x - 1}, \tag{1.9}$$

showing that the shape of the distribution is unchanged during the expansion of the universe and scales like T^3. Similarly the energy density in black body radiation scales as T^4:

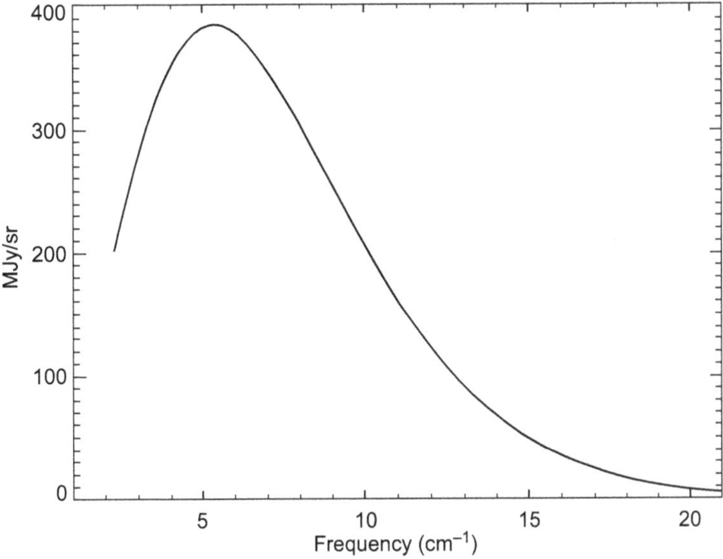

Figure 1.5. The spectrum of the CMB measured with FIRAS, the Far Infrared Absolute Spectrometer on the COBE satellite. Figure from Fixsen et al. (1996). Courtesy of Professor Fixsen on behalf of the copyright holders. The agreement with the line displaying the prediction of the black body spectrum at 2.7255 K is such that the error bars on the data points would only emerge if the line width were reduced one hundredfold. 1 MJy is 10^{-20} W m^{-2} Hz^{-1}.

$$W(x)\mathrm{d}x = \frac{8\pi}{c^3h^3}[k_\mathrm{B}T]^4 \frac{x^3\mathrm{d}x}{\exp x - 1}.$$

Integration[3] of the above expressions over all energies yields:

$$\text{Number density} = 2.026 \times 10^7\, T^3\ \mathrm{m}^{-3}, \tag{1.10}$$

$$\text{Energy density} = 7.558 \times 10^{-16}\, T^4\ \mathrm{J\,m}^{-3}, \tag{1.11}$$

$$\text{also Energy flux} = 5.670 \times 10^{-8}\, T^4\ \mathrm{W\,m}^{-2}, \tag{1.12}$$

the latter equation is known as Stefan's law. There are thus 4.11×10^8 photons per m^3 in the CMB at present. The number of protons is around 0.25 per m^3, so the ratio of protons to photons in the universe today is

$$\eta \approx 6 \times 10^{-10}. \tag{1.13}$$

What is more, a fact for use later, this ratio remained constant back to the time that the CMB was emitted. Also for use later we have

$$\lambda \propto [a = 1/(1 + z)] \propto 1/T, \tag{1.14}$$

with a set to unity at present.

Departures of the CMB from a uniform temperature across the sky are only parts in 10^5, as illustrated by color-coding in Figure 1.6. The fluctuations seen were the outcome of quantum fluctuations in spacetime at a time close to the Big Bang. These fluctuations provided the template for the large scale structures seen in the universe today: the galaxies and clusters of galaxies. Refining what has been stated: apart from the inherited quantum fluctuations, of only parts in 10^5, the CMB is isotropic and homogeneous in the frame in which the distant galaxies appear at rest. In this frame the Sun itself is traveling at 400 km s^{-1} under the gravitational attraction of local matter and the Earth travels at 30 km s^{-1} relative to the Sun. Because of this motion the CMB has a corresponding dipole anisotropy

$$T(\theta) = T(0)[1 + \cos\theta]. \tag{1.15}$$

$T(\theta)$ is the CMB temperature in a direction making an angle θ with the axis provided by the Earth's motion with respect to the distant galaxies; the Earth's secular motion. Once the CMB spectrum is transformed to the frame of the distant galaxies it becomes isotropic, apart from the fluctuations that are displayed in Figure 1.6. The spectrum displayed in Figure 1.5 is the average after correcting for the secular motion. The frame preference just noted illustrates a fundamental concept, known as Mach's Principle: namely that matter in the universe provides a unique reference frame. Such a preferred frame is anathema to SR, but is perfectly consistent with GR. A simple illustration of Mach's principle apparent on the Earth is the following. If a pendulum were suspended at the North Pole, the plane of swing would rotate

[3] $\int_0^\infty x^2\,\mathrm{d}x/[\exp x - 1] = 2.404$ and $\int_0^\infty x^3\,\mathrm{d}x/[\exp x - 1] = \pi^4/15$.

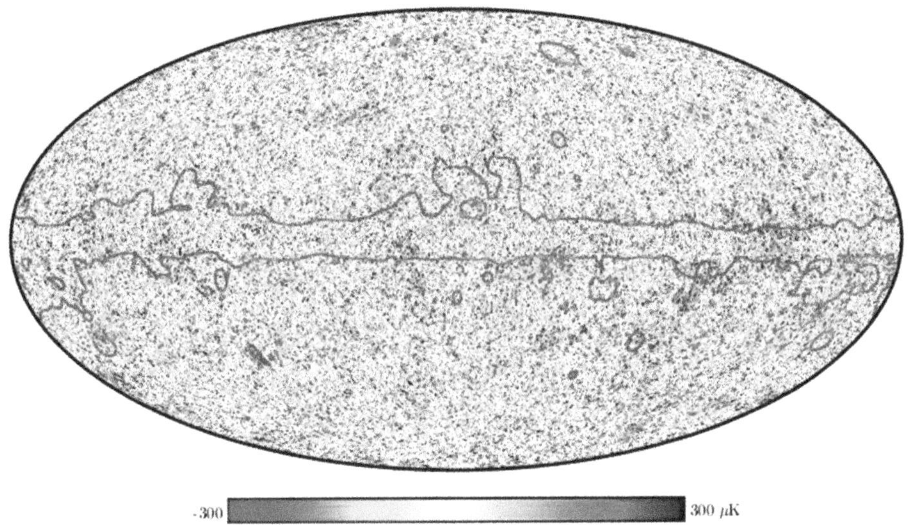

Figure 1.6. Temperature variation of the cosmic microwave background recorded by the Planck Collaboration: Planck 2018 results: I Overview and the cosmological legacy of Planck (Planck Collaboration et al. 2020). Courtesy EDP Sciences and the European Southern Observatory. This standard display is an equal area (Mollweide) projection of both hemispheres of the sky with the galactic mid-plane lying along the horizontal center line. The direction toward the center of the Galaxy lies at the center of the plot. As in an atlas, the left-hand edge wraps round onto the right-hand edge, this time enclosing the viewer. The gray boundary encloses our Galaxy, whose foreground emissions have been removed.

once every 24 hours relative to the Earth; but the plane would not rotate in the frame defined by the distant galaxies.

1.7 Inflation

The CMB is close to being both homogeneous and isotropic: if we look north and south we view regions of the CMB which were never in causal contact, so how could they be so similar? The unexpected solution to this problem is that the steady expansion of the universe was punctuated, early on, at around 10^{-36} s after the Big Bang, by an almost instantaneous expansion by a factor greater than $\sim 10^{30}$. What we now see, across the full 4π angle of the sky, *would* have been in causal contact before inflation. This violent *inflation* is believed to be the effect of the universe dropping from one equilibrium state to another of lower energy. It was an essentially quantum mechanical transition, after which the universe continued expanding more leisurely to the present time. This expansion would also be enough to account for why the spacetime we inhabit is to the limit of observational precision a flat spacetime.

1.8 Dark Matter

In 1964 Vera Rubin and colleagues (Rubin et al. 1980) brought into focus another productive puzzle, first spotted by Zwicky (1937) in the 1930s; namely that the motion of stars in galaxies, and of galaxies in clusters of galaxies, seemed to require much more mass than that of the visible stars and interstellar gas. Figure 1.7 shows the distribution of tangential velocities determined for matter orbiting galaxies from emission line spectra. Beyond the volume populated by visible stars the orbital velocities are obtained from the spectra of diffuse gas clouds in orbit around a galaxy. The striking feature in the figure is that orbital velocities plateau in regions extending well beyond the visible stars. Suppose a galaxy's mass, M, was concentrated in its stars. Then at a radius r beyond the stars the tangential velocity v of a

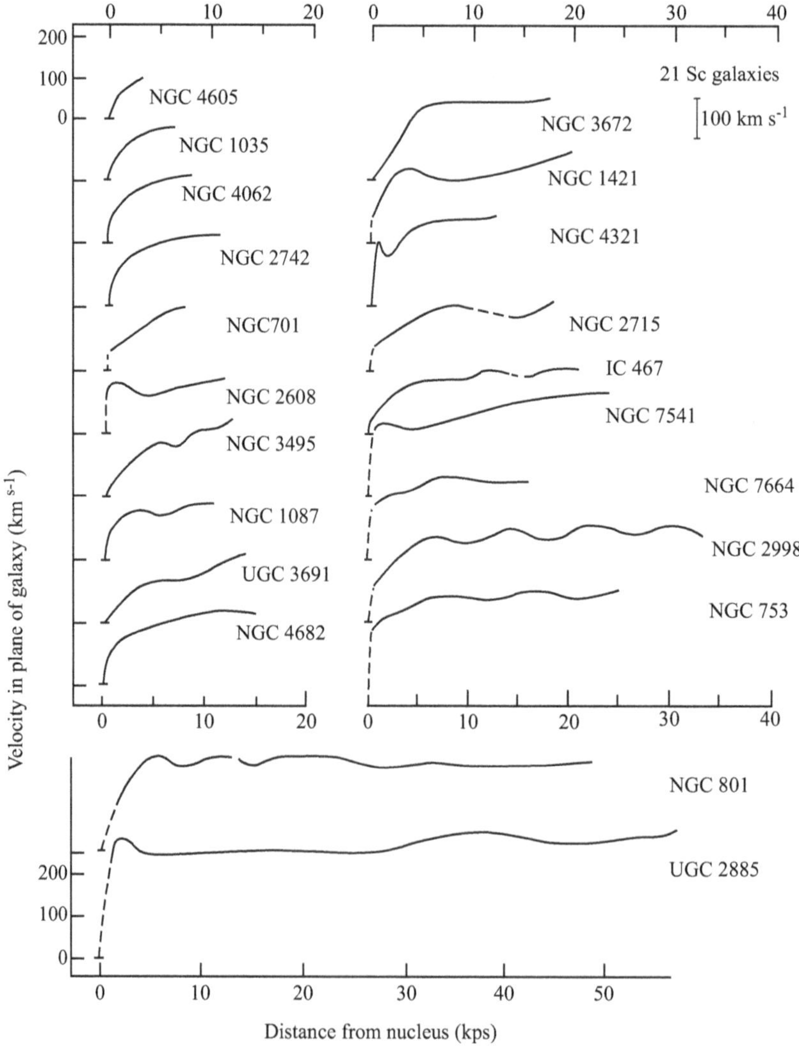

Figure 1.7. Rotational velocities, determined from emission line spectra, as a function of the distance from the center of the galaxy. Adapted from Figure 5 in Rubin et al. (1980). Copied from Figure 11.4 in Kenyon (1990), published by Oxford Univ. Press, courtesy of the late Professor Rubin, the American Astronomical Society and Oxford Univ. Press.

gas cloud of mass m in a stable orbit maintained by the gravitational attraction of the galaxy would be given by

$$mv^2/r = GMm/r^2, \text{ whence } v = GM/r, \tag{1.16}$$

which is wildly at odds with the plateau observed at large radii! There has to be invisible matter extending to the radii in the plateau and this matter must act gravitationally. Alternative possibilities such as the variation of the gravitational strength with distance, and hence a modification of GR, have been closely examined but none provide an explanation consistent with the full range of the observed kinematics. As we shall see, this *dark matter* outweighs normal matter by a factor of over 6. Dark matter is then presumed to distort spacetime and to respond to spacetime curvature in the same way as baryonic matter. Its invisibility shows that it does not interact with electromagnetic radiation, neither emitting nor absorbing this radiation.

1.9 Structure Formation

Only 5% of the matter in the universe is in the form of baryonic matter, from which you and I are made. The rest is mysterious dark matter, which only interacts with baryonic matter, and itself, through gravity. Both types of matter were involved in the steps by which the stars, galaxies, and galaxy clusters form, but in their different ways. As the universe cooled dark matter accumulated in the gravitational potential wells formed by quantum fluctuations during inflation. Not so the baryons, they formed a plasma with radiation: hydrogen atoms continually forming from protons and electrons, and being continually ionized by the radiation. Eventually the universe and plasma cooled enough so that the photon energy was no longer sufficient to ionize hydrogen atoms. That change decoupled the baryons from radiation, which traveled freely to us as the CMB. The baryons instead fell into the potential wells already loaded with, and deepened by, dark matter. These accumulations of matter became the scaffolding within which baryonic structures would form. The ability of baryonic particles to radiate energy was crucial. Baryons in a gravitational potential well could radiate photons and these would exit the potential well carrying off energy. In this way baryons descended deeper into the gravitational wells and formed bound structures: stars, galaxies and galaxy clusters.

If dark matter had been hot, that is to say the components traveled at speeds approaching c, it would have streamed out of gravitational potential wells and washed out smaller structures lighter than galaxy clusters. This is not what we see: it follows that dark matter particles have been non-relativistic for all the period of structure formation. Dark matter is therefore termed *cold dark matter*. Intensive searches in particle physics experiments and in astronomical data have thus far failed to detect any dark matter candidates.

At decoupling, around 365,000 years after the Big Bang, baryons became bound in neutral atoms, and therefore became transparent to radiation. Since then the CMB has traveled freely to reach us. It carries, unchanged, the pattern imposed by the quantum fluctuations of spacetime during inflation. This pattern is in the form of temperature and polarization fluctuations. For example, a photon emerging just after decoupling from a gravitational potential well would lose energy (and hence cool) by an amount equal to the well depth. The temperature fluctuations detected are only parts in 10^5 and the density of matter at decoupling would have shown similar small fluctuations. While the CMB fluctuations froze, these matter fluctuations have grown from parts in 10^5, through the action of gravity and electromagnetic interactions, to become the dense structures we see today.

For a long period after decoupling there were no structures that could radiate. This era, called the dark ages, lasted until roughly 1 Gyr later, when the first stars had had time to form. The first stars were upwards of one hundred times more massive than the Sun (100 $M_\odot$). Gravitational self-attraction compressed these stars so that their cores became hot and dense enough to ignite thermonuclear fusion. Then the radiation emitted by these early stars began to reionize the enveloping, mainly atomic hydrogen gas. When after lives of around 10^6 yr these first stars exploded as supernovae this produced heavy elements that seeded the environment. Consequently later generations of stars and their planets contained the crucial elements that organic life requires: oxygen, carbon, nitrogen, and phosphorus. For reference the elements beyond hydrogen and helium in the periodic table are called *metals* by cosmologists. The structure of the universe has been built hierarchically, from the bottom up. First stars, then galaxies, and finally clusters of galaxies that are, even now, still aggregating.

1.10 Dark Energy

A big surprise awaited experimenters when measurements were made to extend the cosmic distance scale to extract Hubble's constant at early times. The intention had been to observe the expected deceleration of the expansion of the universe brought about by the gravitational pull of the matter it contains. Brighter standard candles were introduced in order to probe further than is possible with Cepheid variables. These newly introduced standard candles are the SNe Ia. Currently they have proved useful at redshifts up to 2.3 (2.8 Gyr after Big Bang). SNe Ia originate from white dwarf stars, closely bound to a giant star in a binary pair. The gravitational attraction of the white dwarf drags matter from the partner until a critical mass is attained, and then the white dwarf undergoes a thermonuclear explosion. For many hours it becomes several billion times more luminous than the Sun, outshining its parent galaxy. The researchers found that the most distant examples of such supernovae to be unexpectedly dim. Dimming due to intervening dust can be ruled out because it would also redden the spectra of the parent galaxy. The unavoidable explanation is that over the last 5 gigayears (recently!), the expansion of the universe has been accelerating.

The most plausible explanation of this acceleration is the presence of the *dark energy*, first mooted by Einstein to account for the then-supposed static nature of the universe. It will be introduced in Chapter 6 and is identified by the symbol Λ. The total dark energy would increase in proportion to the size of the universe, and, unlike matter or radiation, it would be uniformly distributed. Again, unlike matter or radiation, its gravitational effect is repulsive; and it is this repulsion that is accelerating the expansion of the universe.

1.11 The Model of the Universe

The explanation of the features and history of the universe given above and throughout the book follows what is called the ΛCDM model. In this model the material components of the universe are baryonic matter, cold dark matter (CDM) and dark energy (Λ), hence the name. A primary Big Bang is essential to explain the CMB: an early nearly instantaneous inflation in the size of the universe is needed to explain why the universe is nearly if not quite flat and also so homogeneous. The model uses the framework of GR, while the laws of physics as they are applied on Earth are presumed to apply across the universe. A consistent description of the evolution of the universe has been obtained, which works for all that we currently know about its 13.8 Gyr life.

1.12 The Telescopes

The comprehensive knowledge we have of the universe owes everything to the sophisticated telescopes, detectors, and ancillary computing hardware that came to maturity in the late twentieth century. There are telescopes to image sources emitting electromagnetic radiation of wavelengths differing by a factor of 10^{20}, from gamma-rays to radiowaves. Here a very brief survey is made of these scientific and technological marvels, which may whet the reader's appetite to learn more about them.

Observations with telescopes are necessarily snapshots of processes at work that have taken millions, or more, years to complete. It is by having snapshots spread over a billion years that an understanding of the evolution of structures is achieved.

The access provided by radio, microwave, and optical plus near-infrared detection is illustrated in Figure 1.8. The evolution of the cosmic microwave background from decoupling to the present is shown. The Lyα line and the 21 cm line are characteristic emissions from hydrogen. These lines are emitted by a variety of sources that were present over long periods, so the result is a band rather the single line required for the evolution of the CMB.

The angular resolution achievable with a telescope is determined by several factors. Diffraction imposes an absolute minimum of $\Delta\theta = \lambda/d$ where d is the objective diameter and λ the wavelength used. The Hubble space telescope with a 2.4 m aperture has a resolution at 600 nm of 2.5×10^{-7} or 50 mas (milli-arcsec). In the case of viewing a galaxy of similar diameter to the Milky Way, 30 kpc, at a distance of 10 Mpc Hubble's resolution would be better than 2.5 pc, allowing 12,000 independent pixels across the image. This resolution at the lunar surface is 90 m. Hubble is equipped with detectors to record images in the infrared, visible, and ultraviolet spectra. For example, it is used to detect and follow the decay of distant supernovae in the early universe, with the important results introduced above.

In 2022 the *James Webb Space Telescope* (JWST) was deployed. It has a 6.5 m diameter primary mirror with instruments for imaging and spectroscopy optimized

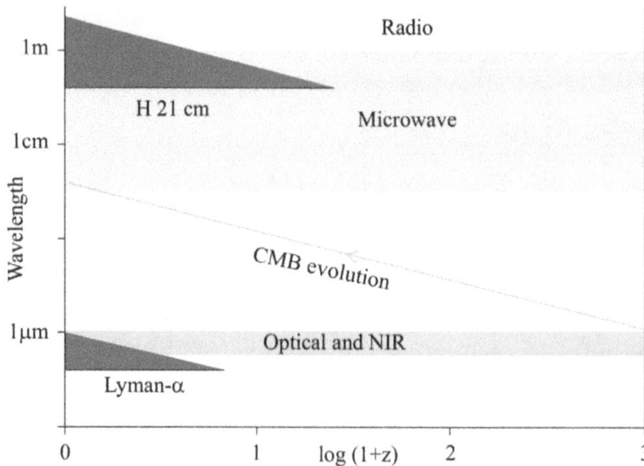

Figure 1.8. Evolution of radiation in the expanding universe as a function of $\log_{10}(1 + z)$, z being the redshift. The spectral regions accessible with specific detector types are indicated by horizontal bands. The CMB was emitted once, while various sources have over extended periods emitted the hydrogen 21 cm line and 121 nm Lyα-line. The lower edge of the blue band at 400 nm is the wavelength below which absorption on metals in stellar atmospheres becomes very strong. This results in the *4000 Ångstrom break* in spectra. Adapted, with permission, from the ICHEP2020 talk by David Kirkby, University of California, Irvine. https://faculty.sites.uci.edu/dkirkby/.

for detection in the infrared: from 0.6 to 28.5 μ m. By optimizing on the near-infrared higher redshift galaxies and stars become detectable compared to, for example, the HST. Therefore, the JWST will view further back in time to the appearance of the earliest stars: an era that is poorly understood at present. Studies using the JWST will for the first time access events at redshift to between 8 and 10, well into the first billion years after the Big Bang. The resolution, limited by the detectors, is 100 mas, comparable to the HST. A complementary advantage is that infrared light penetrates the dust clouds that obscure star formation, better than visible light. This is because dust particles scatter, very efficiently, radiation whose wavelength is less than or equal to the dust particles' size.

Many earthbound telescopes already have larger objectives: the two KEK telescopes have 10 m diameter objectives each made up of 36 hexagonal segments that work as one. The turbulence in the atmosphere brings distortions that alter over times of order 10 ms and would vitiate the advantage of objective size. The problem is reduced by locating telescopes on high mountains in dry climates, such as the KEK telescopes on the Maunakea extinct volcano in Hawaii, and the four 8.2 m diameter European Southern Observatory Very Large Telescopes (VLTs) at 2600 m altitude at Cerro Paranal in the Atacama desert in Chile. Compensation for turbulence is still needed, and is provided by a flexible mirror in the light path through the telescope after the objective, a mirror that can be deformed rapidly in a controlled manner. The mirror is warped so as to retain a point image in the image plane of the telescope of a bright point source in the field of view; this target may be a star or it can be generated by a laser beam incident on sodium atoms in the upper atmosphere and causing these to emit in turn. Three giant telescopes in the 30 m diameter class are under development, with corresponding increased resolution and light-gathering capability.

More recently it has proved possible to interfere light from more than one telescope by making underground piped connections that are maintained constant in length to a small fraction of a wavelength. This gives a resolution improvement in the case of the KEK pair from 40 mas to 4 mas; and similarly for the VLTs.

More modest telescopes have performed important surveys. The 2.5 m diameter wide-angle telescope at Apache Point, New Mexico has carried out a program called the Sloan Digital Sky Survey since 2000. For example it was used to observe galaxies and quasars (active galactic nuclei) at redshifts from 0.6 to 1.0 and from 0.8 to 3.5 respectively over one third of the sky. As a result a three-dimensional slice of the universe is recorded and accessible; reaching back to when the universe was a quarter its present age. This and similar data has been used as input in making a comparison between the structures seen today and the overdense regions signaled by the CMB.

At millimeter and microwave wavelengths the telescope primary mirror, *dish*, is a metal mesh mirror: the mesh spacing is made smaller than the wavelengths of interest so that it appears solid to such radiation. The radio telescope at Arecibo in Puerto Rico had a 300 m diameter dish. Atmospheric turbulence is no problem but absorption by air means that only certain wavelengths are usable from Earth. X-rays are almost totally absorbed, so that satellite mounted detectors are essential.

The detectors that have produced the detailed data on the CMB have been satellite mounted: the latest, Planck, was active from 2009 to 2013. Its parking orbit was around 1.5 Mkm from the Earth in the opposite direction to the Sun (L_2 Lagrangian point) so as to be well-shielded enough to detect the universal 2.7 K CMB black body radiation. The bulk of the spectrum is concentrated between 30 GHz and 850 GHz (10 mm and 0.3 mm wavelength) and, serependitiously, this lies in a gap between higher frequency radiation from dust and lower frequency synchrotron radiation from the Galaxy. The primary mirror has a 1.5 m aperture giving a resolution between $5'$ and $30'$ across the accessible spectrum. The detectors from 30 to 100 GHz are high electron mobility transistors like those in satellite dishes; from 100 to 850 GHz the detectors are bolometers. A typical device is a roughly 1 cm diameter mesh of radial and azimuthal gold coated 1 μm wires resembling an ideal spider's web. At the center sits a transition edge sensor at 0.1 K that conducts when warmed by the radiation absorbed on the mesh. The mesh gaps are shorter than the wavelength so that radiation is efficiently absorbed but cosmic rays that would warm the wires almost all pass between them. Signals from the detectors are first amplified by electronics cooled at 40 K, and then processed, digitized, and transmitted to Earth. After 18 months the helium coolant was exhausted and later the satellite was passivated. Over its lifetime the temperature precision achieved was 2 μ K at the lowest frequency: this and the angular resolution were adequate to examine the quantum fluctuations of the CMB in great detail, as described later. Some bolometers had a rectangular pattern of wires, with only those in one orientation being made electrically conductive. In this way the polarization of the CMB could be measured, a property of significance in determining cosmological parameters.

There is a significant difference between short wavelength detection from the microwave region downward in wavelength, and radiowave detection. Detectors of the former respond to the radiation intensity, those for the latter to the electric field, the radiation amplitude. It means that interference between signals from detectors at a pair of remote millimeter-wave detectors is directly achievable. The Event Horizon Telescope collaboration carried this to a logical extreme in 2019 using radio telescopes at 1.3 mm wavelength. The signals from a set of eight radio telescopes spread over an area of linear dimensions 10,000 km were recorded and accurately time stamped. Then the data from the hard disks was "interfered" much later to recover the interference pattern of the source viewed. This was M87*, the 66 million solar mass black hole at the center of the M87 galaxy. Figure 1.9 is the image obtained, showing the shadow cast by a black hole at the center of the galaxy M87 on the intense microwave radiation from material falling into it. The resolution is $\Delta\theta = 1.3$ mm/10, 000 km, that is 20 mas.

At the other end of the energy scale are water Cerenkov detectors, which record muons (μ-leptons). The muons are one of the end products of interactions of cosmic rays incident on the Earth's upper atmosphere; being weakly interacting they can survive to reach the Earth's surface at full energy. They are traveling close to c, the speed of light in a vacuum, and hence they exceed the speed of light in water. As a result they produce a shock wave of light, the Cerenkov radiation, akin to the sonic boom of something exceeding the speed of sound in air. Arrays of tanks of water instrumented with photomultipliers to detect the Cerenkov light, electronics, and Wi-Fi connections, are distributed over, in cases, areas of square km. Figure 1.10 shows the angular distribution of muon showers relative to the Crab Nebula, observed by the Tibet Air Shower Array. This array consists of 64 tanks covering

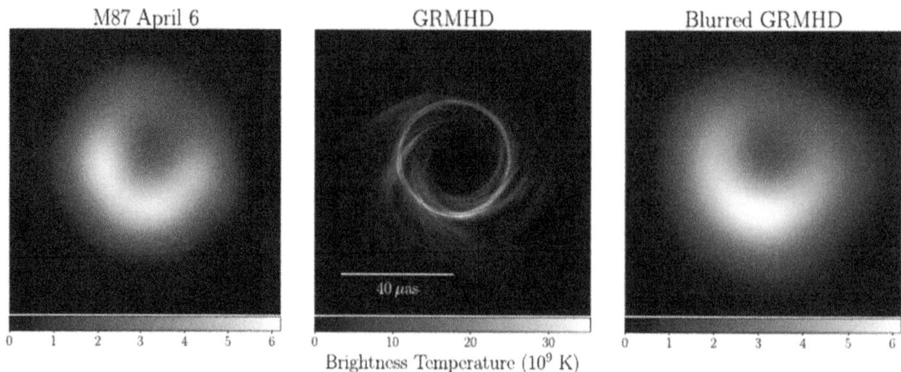

Figure 1.9. The image of the black hole M87* recorded by the Event Horizon Telescope (EHT) on the left; in the center the predicted image with perfect resolution; on the right the prediction taking account of the interferometer's intrinsic resolution. The 66 billion solar mass black hole was imaged by eight radio telescopes at 1.3 mm wavelength. The bright flare is from its accretion disk. Figure from The Event Horizon Telescope Collaboration et al. (2019), reproduced with permission under CCBY-SA-3.0.

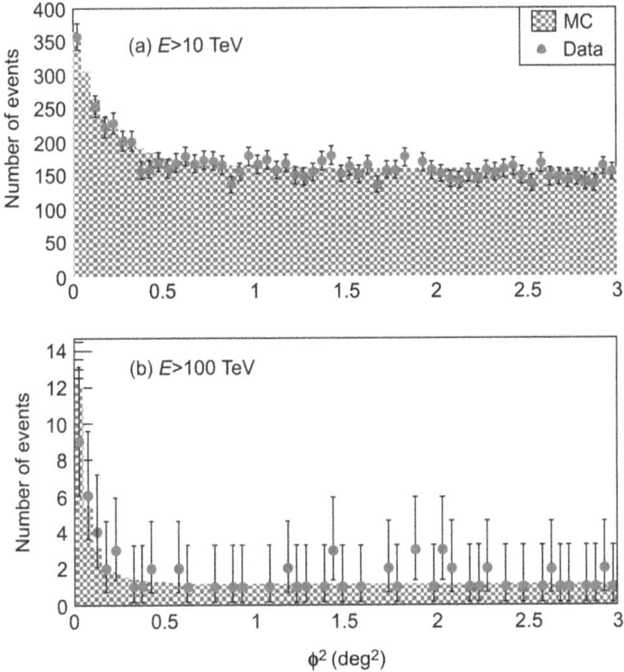

Figure 1.10. Distribution of events as a function of the square of the incident angle with respect to the Crab Nebula direction. The data points are compared with simulations shown by solid histograms. Figure from Amenomori et al. (2019). Courtesy of the American Physical Society.

3400m^2, part of a much larger array of 600 plastic scintillators covering 65,700 m^2; all at an altitude of 4300 m in Tibet. Evidently there is a signal due to high energy photons, above 100 TeV, originating in the Crab Nebula.

Telescopes detecting electromagnetic radiation are now complemented by the giant interferometers that detect gravitational waves from inspiralling black hole binaries and neutron star binaries. The aLIGO interferometer is described in detail in a later chapter.

1.13 Luminosity

Stars emit radiation strongly across the spectrum from the infrared into the ultraviolet. Filters are used with telescopes to select bands of the spectrum and so gain more information. For example, the hotter the star the bluer its spectrum. Bands covering from the ultraviolet to the near-infrared are shown in the accompanying Table 1.1. *Luminosity* is the electromagnetic energy radiated from a source per unit time, measured in watts. *Flux* is the electromagnetic energy flowing across unit area per unit time, measured in W m^{-2}. Magnitude is a measure inherited from the classical world when the brightest stars were designated magnitude 1, and those barely visible as magnitude 6; which already makes for confusion. As regards

Table 1.1. Spectral Bands Showing Nominal Wavelength Ranges

Symbol	Range	Wavelength (nm)
U	Ultraviolet	325–390
B	Blue	390–490
V	Yellow/Green	490–580
R	Red	580–730
I	Infrared	730–950

In practice the filter used determines the range.

the eye's response to changes in intensity, Fechner's law approximates quite well: for an intensity I the response is

$$S = 2.3\log_{10}(I/I_0),$$

where I_0 is a constant. Nowadays the agreed way to define magnitudes is to interpolate between magnitudes 1 and 6, and extend beyond, with a logarithmic scale that makes use of Fechner's law. The *apparent* magnitude is determined from the measured incident flux f as

$$m = -2.5\log_{10}(f/f_0), . \tag{1.17}$$

The negative sign is needed to take care of the inversion of the magnitude scale, with brighter sources having the lower magnitude. The constant flux f_0 is set to 2.53×10^{-8} W m^{-2}, so that the ancient reference magnitudes are matched quite well. With this choice a difference in m of 5 conveniently gives a ratio of 100 between the luminosities being compared. The relevant quantity for direct comparison between sources is the *absolute* magnitude. This is defined to be the magnitude that a source would have if viewed from a distance of 10 pc:

$$M = -2.5\log_{10}\left[\frac{L}{L_0}\right], \tag{1.18}$$

where L is the luminosity of the source and L_0 is the zero-point luminosity giving zero magnitude, namely 3.0128×10^{28} W. Flux falls off with the square of the (luminosity) distance d_L, so that

$$M = m - 5\log_{10}\left[\frac{d_L}{10\text{pc}}\right]. \tag{1.19}$$

Flux, luminosity, and magnitude summed across the whole spectrum are known as *bolometric* flux, *bolometric* luminosity and *bolometric* magnitude. Quantities for a single spectral band have a subscript attached, as M_B. The absolute bolometric magnitude of the Sun is 4.75: that of Rigel, the brightest (bottom right), star in Orion is 0.12. The brightest sources are distant quasars with absolute magnitudes reaching −30. Most stars have absolute magnitudes in the range of +20 to −10. The Sun's bolometric luminosity $L_\odot$ is 3.86×10^{26} W: the brightest stars have luminosities up to 10^6 $L_\odot$, and stars with luminosities down to 10^{-4} $L_\odot$ are detectable.

As the temperature of a perfect black body rises so the ratio of blue light emitted to red light emitted increases in a predictable manner. Stars and galaxies are approximately black body emitters, so it follows that the difference between the B-band and V-band magnitudes is expected to change in a well-defined way with temperature. Taking the difference between magnitudes in the B- and V-bands gives

$$M_B - M_V = -0.865 + 8540/T. \tag{1.20}$$

This is useful at temperatures T for which the wavelength of the spectral peak λ lies in the visible spectrum. Wien's law for black body radiation relating the temperature to the peak wavelength is $\lambda T = 2.898$ mm K, so the range of usefulness is from 4,000 K to 10,000 K.

1.14 Summary of Results in Special Relativity

The two basic postulates of SR are these: first, the laws of physics take the same form in all inertial frames, that is in all frames in free fall; second, the velocity of light is a constant written as c. Take a point in spacetime (which we shall call an event) having coordinates x, y, z in a rectangular Cartesian system, at time t. The four-space plus time coordinates form a four-vector with components

$$x_0 = ct, \ x_1 = x, \ x_2 = y, \ x_3 = z. \tag{1.21}$$

We also write $\mathbf{r}$ for the spatial three vector. In another inertial frame moving with relative velocity $v = \beta c$ parallel to the x-axis the new (primed) coordinates of the same event are given by the Lorentz transformation

$$\begin{aligned} x_0' &= \gamma(x_0 - \beta x_1), \\ x_1' &= \gamma(x_1 - \beta x_0), \\ x_2' &= x_2, \quad x_3' = x_3. \end{aligned} \tag{1.22}$$

where $\gamma = 1/(1 - \beta^2)^{1/2}$, with $0 \leqslant \beta < 1$. The interval between two events P_1 and P_2 with coordinate separations $(\Delta x_0, \Delta x_1, \Delta x_2, \Delta x_3)$ is defined to be

$$\Delta s^2 = (\Delta x_0)^2 - (\Delta x_1)^2 - (\Delta x_2)^2 - (\Delta x_3)^2 = (\Delta x_0)^2 - \Delta \mathbf{r}^2, \tag{1.23}$$

which can readily be shown to be invariant under Lorentz transformations. If a clock travels from P_1 to P_2 then the time interval it measures between these two events is called the *proper time* $\Delta\tau$ with $c\Delta\tau = \Delta s$. Note also that the components of $\Delta\mathbf{r}$ perpendicular to the velocity vector are unaltered. There are other important four-vectors; for example, the energy-momentum four-vector with components

$$p_0 = E/c, \ p_1 = p_x, \ p_2 = p_y, \ p_3 = p_z, \tag{1.24}$$

where E is the total energy and $\mathbf{p}$ is the relativistic momentum. The Lorentz transformation takes the same form for all four-vectors: in the case of the four-momentum we can take Equation (1.22) and replace x everywhere by p.

Each four-vector has an invariant, formed in the same way as Equation (1.23). The energy-momentum invariant is

$$p_0^2 - \mathbf{p}^2 = E^2/c^2 - \mathbf{p}^2 = m^2c^4, \tag{1.25}$$

where m would be the rest mass in the case of a particle. The four-momentum is related to the four-velocity v whose components are

$$v_0 = c\gamma, \ v_1 = \gamma v_x, \ v_2 = \gamma v_y, \ v_3 = \gamma v_z. \tag{1.26}$$

Here v_x, v_y, and v_z are the standard components of the velocity, meaning that in time dt the distance traveled along the x-direction will be $v_x dt$. Also $\gamma = 1/(1 - \beta^2)^{1/2}$, where $\beta c = (v_x^2 + v_y^2 + v_z^2)^{1/2}$. The invariant for the velocity four-vector is c.

The invariants can be positive, negative or zero. Let us cite P_1 at the origin in Figure 1.11. If the spacetime invariant interval Δs^2 from event P_1 to event P_2 is zero then $c\Delta t$ exactly equals $\Delta \mathbf{r}$ and a light ray can travel from event P_1 to event P_2: such an interval is called light-like. P_2 lies upward along a diagonal line. When $c\Delta t$ exceeds $\Delta \mathbf{r}$ the interval is positive and P_2 can be reached from P_1 by traveling slower that the speed of light. P_2 lies in the upward gray region. Such an interval is called time-like because we can choose an inertial frame such that $\mathbf{r}_2 = \mathbf{r}_2$, leaving only a separation in time. Finally if the interval Δs^2 is negative $c\Delta t$ is less than $\Delta \mathbf{r}$ so that no information can pass from P_1 to P_2. P_2 then lies in the white region. Such a separation is called space-like, and in this case an inertial frame can be found in which $t_2 = t_1$, leaving a spatial separation. In this case whatever happens at P_1 can have no influence on what happens at P_2 and vice versa. Figure 1.11 illustrates these different cases shown with P_1 located at the origin with a single spatial axis. With three dimensions the accessible future forms a three-dimensional cone in spacetime around the time axis.

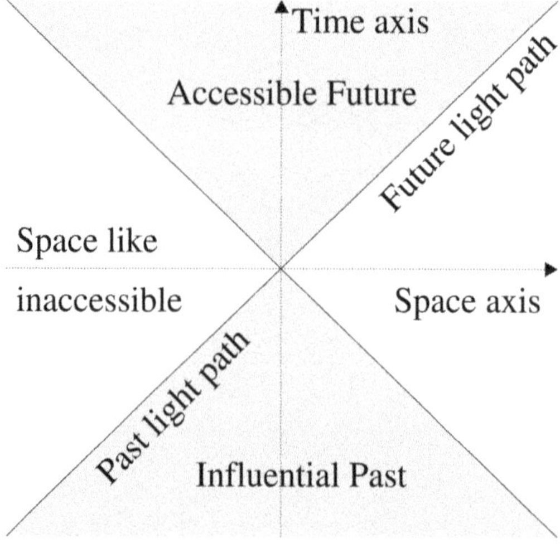

Figure 1.11. Section through the light cones at one point in spacetime showing one spatial dimension.

1.15 Exercises

1. Calculate whether the following spacetime intervals are space-like, time-like or light-like: (1.0, 3.0, 0.0, 0.0); (3.0, 3.0, 0.0, 0.0); (3.0, −3.0, 0.0, 0.0); (−3.0, 3.0, 0.0, 0.0); (0.0, 3.0, 0.0, 0.0); (3.0, 1.0, 0.0, 0.0).
2. Use Stefan's formula for the energy density, $\rho(E)$, in black body radiation $\rho(E) = \alpha T^4$, where α is 7.565×10^{-16} J m^{-3} K^{-4}. What is the energy density of the CMB at the current temperature of 2.7255 K? Take the mean energy per photon to be $2.7k_B$T. Make an estimate of the number of CMB photons per cubic centimeter.
3. The average density of baryonic matter (nucleons, either protons or neutrons, and electrons) in the universe is 4.2 10^{-28} kg m^{-3}. How many nucleons are there per cubic meter? What is the ratio of CMB photons to nucleons today?
4. Suppose a star made of antimatter annihilates on a star made of matter. Calculate the energy release. The most energetic long duration cosmic sources are quasars emitting up to 10^{40} W. Is it likely there is much antimatter is the universe?
5. The transition between the ground and first excited state in atomic hydrogen produces radiation of wavelength 121.6 nm, called the Lyα line. The continuous spectrum from quasars is marked by the absorption lines due to absorption by intergalactic clouds of hydrogen encountered on the way to the Earth. If one such Lyα line is redshifted to 150 nm how far is the absorbing hydrogen cloud from the Earth?
6. What is the thermal balance, heat in, heat out, of an astronaut during a free spacewalk? Stefan's law for radiation from a black body at temperature T gives an energy flux per square meter $E = 5.67 \times 10^{-8} T^4$.
7. Sirius, the brightest star of all in the night sky, has an apparent magnitude -1.46 and is 2.67 pc away. What is its absolute magnitude? The TRGB stars have an absolute magnitudes close to −4.02: what is their luminosity?

Further Reading

Two excellent parallel texts on cosmology are: Rich J 2009 *Fundamentals of Cosmology* (2nd ed.; Berlin: Springer) and Ryden B 2017 *Introduction to Cosmology* (2nd ed.; Cambridge: Cambridge Univ. Press). Both were prepared before the discovery of gravitational waves.

Susskind L 2006 *The Cosmic Landscape* (Boston, MA: Little, Brown and Company). This is a more popular book on cosmology by a noted theorist.

Narlikar J V 2002 *An Introduction to Cosmology* (3rd ed.; Cambridge: Cambridge Univ. Press). This gives an extended and fuller account of similar material, written by a practiced author. It is somewhat dated.

Cottrell G 2016 *Telescopes: A Very Short Introduction* (Oxford: Oxford Univ. Press). This text provides a compact introduction to modern telescopes used across the electromagnetic spectrum.

References

Abbott, B. P., Abbott, R., Abbott, T. D., et al. 2016, PhRvL, 116, 061102
Alpher, R. A., Bethe, H., & Gamow, G. 1948, PhRv, 73, 803
Amenomori, M., Bao, Y. W., Bi, X. J., et al. 2019, PhRvL, 123, 051101
Do, T., Hees, A., Ghez, A., et al. 2019, Science, 365, 664
Fixsen, D. J., Cheng, E. S., Gales, J. M., et al. 1996, ApJ, 473, 576
Freedman, W. L. 2021, ApJ, 919, 16
Hubble, E. 1929, PNAS, 15, 168
Hubble, E., & Humason, M. L. 1931, ApJ, 74, 43
Kenyon, I. R. 1990, General Relativity (Oxford: Oxford Univ. Press)
Kenyon, I. R. 2019, Quantum 20/20: Fundamentals, Entanglement, Gauge Fields, Condensates and Topology (Oxford: Oxford Univ. Press)
Leavitt, H. S. 1908, AnHar, 60, 87
Penzias, A. A., & Wilson, R. W. 1965, ApJ, 142, 419
Planck Collaboration, Aghanim, N., Akrami, Y., et al. 2020, A&A, 641, A1
Rubin, V. C., Ford, W. K., & Thonnard, N. 1980, ApJ, 238, 471
The Event Horizon Telescope Collaboration, Akiyama, K., Alberdi, A., et al. 2019, ApJ, 875, L5
Zwicky, F. 1937, ApJ, 86, 217

Ian R Kenyon

Chapter 2

The Equivalence Principle

A number of experiences familiar to modern man indicate that there is a close resemblance between the gravitational force and the effects of acceleration. High-speed centrifuges generating large inertial forces are used to separate materials from liquid suspensions that would sediment only slowly if at all under gravity. Pilots of jet aircraft making tight turns feel forces that are labeled *g-forces*. These are examples of centrifugal acceleration. Linear acceleration is less easily sustained, for example, during lift-off of a space probe, and then only for seconds. Einstein realized that the parallel between constant acceleration and a constant gravitational force is a principle of nature, the *equivalence principle*. This states that a region of uniform gravitational field and a uniform accelerating frame are equivalent (Section 2.1). That is to say, there is no way to distinguish between them provided that measurements do not extend beyond the region of uniformity. One conclusion that will be drawn from this principle is that gravitational fields affect electromagnetic radiation: light leaving a star is redshifted, and light passing a star is deflected from a classical straight line path (Section 2.4). These effects and their experimental verification are described below.

Frames in free fall become special: gravitational effects vanish and experiments will always give results consistent with special relativity (SR). A frame in free fall is thus an inertial frame according to the usage of SR.

2.1 The Equivalence Principle

The origin of the equivalence principle (EP) goes back to the experiment ascribed to Galileo. When Galileo compared the rate of falling of different materials he was attempting to answer a fundamental question: he wanted to know whether the gravitational attraction on different materials was the same. Even now it is not at all obvious that it should be so. Matter is made up from different particle species and

doi:10.1088/2514-3433/acc3ffch2

the proportions of these vary from material to material. Each atom contains a nucleus, which is made from nucleons (i.e., neutrons and protons) with electrons circulating around the nucleus. Nucleons feel the strong force whereas electrons do not: thus it is reasonable to ask whether nucleons and electrons feel the same gravitational force. The nucleon-to-electron ratio varies from unity in hydrogen to about 2.5 for elements with high atomic number, so that any difference in the gravitational force felt by nucleons and electrons would appear as a difference in the gravitational acceleration of elements of high and low atomic number. In addition nuclei are lighter than the sum of the constituent nucleon masses by the nuclear binding energy. This nuclear binding energy varies from zero for hydrogen to 0.7% of the mass $\times\, c^2$ for iron. Hence if the gravitational force depended, like the strong force, on the number of nucleons rather than the mass there would be a difference of 0.7% between the gravitational acceleration of iron and hydrogen. Finally there is the gravitational binding energy of matter, which is a fraction 4.64×10^{-10} (0.19×10^{-10}) of the total mass in the case of the Earth (Moon). Again, it is reasonable to ask whether the Earth and Moon fall toward the Sun with the same acceleration: how does gravity act on gravity?

The EP was already implicit in the Newtonian analysis of Galileo's experiment. The force acting on a mass $m_{\rm g}$ in a gravitational field g is

$$F = m_{\rm g}g. \tag{2.1}$$

Applying Newton's second law of motion gives the acceleration a of the mass

$$F = m_{\rm i}a. \tag{2.2}$$

A distinction is being made here between the gravitational mass $m_{\rm g}$ and the inertial mass $m_{\rm i}$. Inertial mass appears in expressions for kinematic energy $m_{\rm i}v^2$ and momentum $m_{\rm i}v$, so that its definition is made independent of any weighing process. Eliminating F from the last two equations gives the acceleration

$$a = (m_{\rm g}/m_{\rm i})g. \tag{2.3}$$

Tests from Galileo's time up to the present reveal no variation in the rate of fall from material to material. Therefore $m_{\rm g}/m_{\rm i}$ has the same value for all materials, and by choosing units appropriately we make this ratio equal to unity. Einstein interpreted this result as follows: the motion of a neutral test body released at a given point in spacetime is independent of its composition, which is the *weak equivalence principle (WEP)*. A test body is by definition small. Massive objects such as the Earth and the stars are, unlike test bodies, bound gravitationally and the question arises whether the gravitational mass remains the same as the inertial (bound) mass in such cases. If it is postulated that they are the same then the equivalence principle becomes the *strong equivalence principle (SEP)*.

Einstein next considered the implications of the equivalence principle for motion in free fall. One example is the International Space Station (ISS), another is a capsule falling radially toward the Earth, and a final example is a capsule drifting through intergalactic space. In all three cases we need to ignore any drag forces from the surrounding gas. Einstein posed a searching question for such systems: can an astronaut inside a closed capsule determine his state of motion without looking out of the capsule, whether it is in a uniform gravitational field or uniformly accelerating? If the astronaut drops a ball it accelerates at the same rate as the capsule and will remain at rest relative to the capsule, whatever their shared acceleration. The occupants of the ISS find this property sometimes useful, sometimes not so useful! However, if the capsule is in a region where the gravitational field is *not* uniform then the motion is in principle detectable. To give a concrete example, consider the capsule to be falling radially toward the Earth. Then dropping not one but two balls will be an effective strategy because the gravitational forces on them converge at the center of the Earth as in Figure 2.1. The astronaut could measure the resultant movement of the balls toward each other, given a large enough capsule and a long enough time interval. Such effects owe their origin to gradients in the field and are called tidal effects.

The weak equivalence principle can now be restated as follows so as to exclude tidal effects: the results of *local* experiments in a state of free fall are independent of the motion. *Local* is used here to express the restriction to a region sufficiently small that the gravitational field is effectively uniform. Einstein then generalized the equivalence principle to this form:

> The results of local experiments in all freely-falling frames are independent of both the location and the time.

In a nutshell, physics is the same in all free falling frames. Notice that one frame in free fall can be accelerating with respect to another such frame. For example, we can compare satellites in free fall around different stars. The EP can be viewed as an extension of the first postulate of SR. SR requires that the result of an experiment is

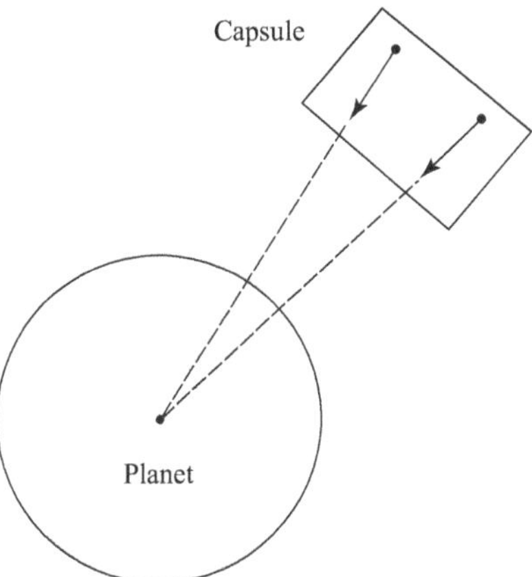

Figure 2.1. The trajectories of objects in free fall within a space capsule.

the same for all inertial frames, but has nothing to say about the effect of gravitational acceleration; the EP requires that the result of local experiments be the same in all freely-falling frames. In the special theory of relativity it is assumed that a single inertial frame can be applied to the whole universe, but at the cost of neglecting acceleration! In the general theory of relativity the natural frame anywhere is chosen to be a frame in free fall, but we cannot cover the whole universe with one such frame.

The vast body of experimental data that has been accumulated in support of SR can be reconciled with the EP by making the inference, as Einstein did, that physics in free fall must be consistent with SR. Thus, we should add a rider to the EP:

> The results of local experiments in free fall are consistent with SR.

2.2 Experimental Tests of the Equivalence Principle

The equivalence principle is such an essential component of GR that very varied and refined tests have been made. The WEP has been tested with test bodies in torsion balances and the SEP by lunar laser ranging (LLR). The first modern experiment was that of Roll, Krotkov, and Dicke in 1964 (Roll et al. 1964) and illustrates the principles of the torsion balance method. Equal-mass test bodies, 30 mg of gold and aluminum, are hung as shown in Figure 2.2 from an equal arm balance. If the acceleration of gold toward the Sun is larger than for aluminum then at 0600 the horizontal arm will rotate clockwise and at 1800 counterclockwise. The torque can be calculated as follows.

First, the balance of forces on the Earth in orbit is

$$\frac{GMm_g(e)}{R^2} = \frac{m_i(e)v^2}{R},$$

where M is the Sun's mass, $m_g(e)$ and $m_i(e)$ the Earth's gravitational and inertial mass, respectively, v its velocity, and R the Earth–Sun distance. Whence

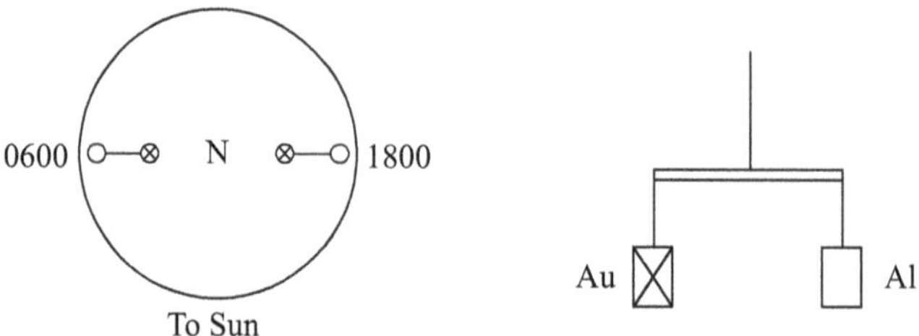

Figure 2.2. The experiment of Roll, Krotkov and Dicke (Roll et al. 1964): on the left viewed from above the North Pole; on the right a side view.

$$v^2 = \frac{GM}{R} r(\mathrm{e})$$

where $r(\mathrm{e})$ is the ratio $m_g(\mathrm{e})/m_i(\mathrm{e})$. The balance of forces on the gold mass is

$$F(\mathrm{Au}) + \frac{GMm_{\mathrm{g}}(\mathrm{Au})}{R^2} = \frac{m_{\mathrm{i}}(\mathrm{Au})v^2}{R},$$

where F is the pull from the balance arm. Rearranging this equation gives

$$F(\mathrm{Au}) = \frac{GMm_{\mathrm{i}}(\mathrm{Au})}{R^2}[r(\mathrm{e}) - r(\mathrm{Au})].$$

There is a similar equation for the aluminum. Taking both inertial masses to be equal to m, the torque on the balance at 0600 is

$$\Gamma = [F(\mathrm{Au}) - F(\mathrm{Al})]\, l = \frac{GMml}{R^2}\eta,$$

where $\eta = r(\mathrm{Au}) - r(\mathrm{Al})$ is called the Eötvös parameter. If η were non-zero there would be a net torque that reverses every 12 hours. The authors looked for the 24 hour oscillations that such a torque would produce and saw none. Their upper limit was

$$\eta_{\mathrm{Au/Al}} \leqslant 3 \times 10^{-11}.$$

Adelberger and colleagues have made more refined tests (Schlamminger et al. 2008). Their torsion balance was mounted on a turntable rotating at a constant rate, which replaces the Earth's rotation. Then, as seen by someone sitting on the turntable, an oscillation of the balance arm at the rotation frequency would signal a violation of the WEP. The rotation frequency was around 1 mHz, a hundred times that of the Earth's rotation. Oscillations of the balance arm were detected by an optical system rotating with the turntable. At the latitude of the experiment in Washington state the horizontal component of the Earth's gravitational field is 1.68 cm s^{-2}, three times that of the Sun, so that the gravitational attraction of the Earth was used to test the WEP.

This approach has several advantages over the experiments relying on the Earth's rotation. It insulates the experiment from 24 hour cycles in temperature and electrical power; it reduces noise in the apparatus, which generally has a $1/f$ or $1/f^2$ dependence, where f is the signal frequency. The sensitivity improves as the wire's thickness is reduced and the mass correspondingly reduced so as not to break the wire: 4.84 gm was the practical choice for each test body. The container was evacuated, thermally insulated, shielded by mu-metal from ambient magnetic fields and located in a temperature stabilized underground room. Data were recorded over 75 days giving

$$\eta_{\mathrm{Be/Ti}} \leqslant (0.3 \pm 1.8) \times 10^{-13}.$$

The absence of differential acceleration, at this level of precision, toward the Earth further validates the WEP.

The WEP and Newton's inverse square law of gravitation work well from the scale of the solar system down to centimeter distances. Adelberger and colleagues took one step further: they adapted the torsion balance and ruled out any detectable effect due to additional forces of ranges above around 50 μm (Schlamminger et al. 2008).

The MICROSCOPE experiment reported a test of the WEP in the quiet environment of a satellite orbiting at an altitude of 700 km (Bergé et al. 2017; Touboul et al. 2022). The test bodies were coaxial hollow cylinders of different alloys, Pt/Rh and Ti/Al/Va, chosen both to have widely different nucleon/electron ratios and to be easy to machine. The position of each was monitored capacitively, and each could be moved by applying a voltage to electrodes on a silica frame enclosing both test masses. Figure 2.3 illustrates the principle of the measurement. If the WEP holds the accelerations toward the Earth, in the direction of the red arrows, should be equal. Their relative acceleration in the axial direction, along the black arrows, was monitored by comparing the electrostatic forces needed to hold them at rest with respect to each other. If the WEP is violated the force difference would oscillate at the orbital frequency. In order to increase the frequency of such oscillations, and hence of data taking, the satellite could be spun around an axis

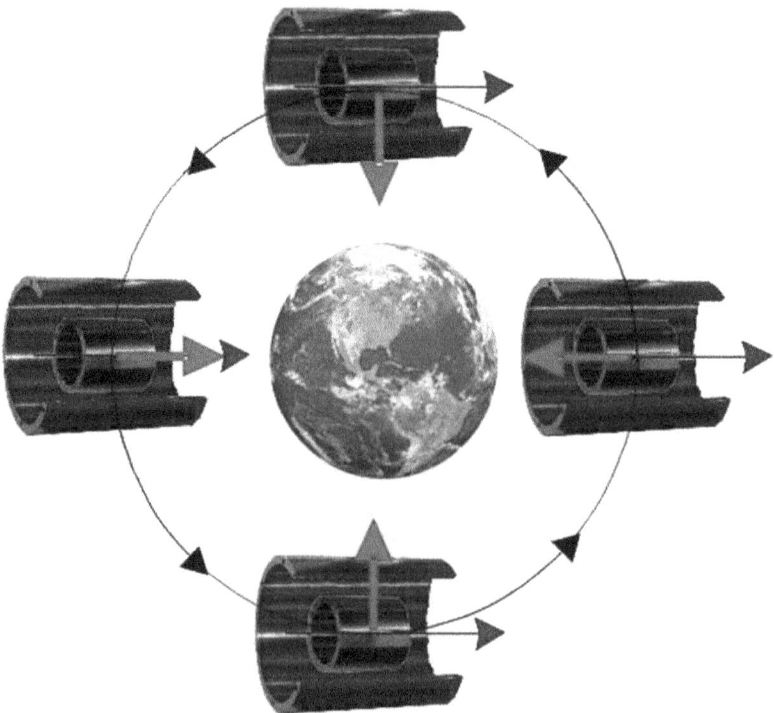

Figure 2.3. MICROSCOPE experiment to test the SEP. Figure from Bergé et al. (2017). Courtesy of Professor Berg é and the Institute of Physics.

perpendicular to the orbital plane. The latest limit on the Eö tv ös parameter reported in 2022 (Touboul et al. 2022) is

$$\eta_{\mathrm{Pt/Ti}} = (-1.5 \pm 2.3 \pm 1.5)10^{-15} \tag{2.4}$$

showing the statistical and systematic errors separately. A second, near-duplicate device is carried in the satellite, in which the test bodies are identical. With this, a confirmatory null result was obtained. The limit improves by over a factor ten on the torsion balance limit.

2.3 Lunar Laser Ranging

The inertial masses of the Earth and the Moon decreased during their formation by the gravitational self-binding energy. It is reasonable to ask whether the gravitational attraction exerted by the Sun on these two satellites is correspondingly reduced. If the gravitational attraction is not reduced in this way then the orbit of the Moon around the Earth changes from a near circular orbit of radius 3.84×10^8 m to one stretched toward the Sun by a distance

$$\delta R = C \cos \phi, \tag{2.5}$$

where C parametrizes the size of the effect and ϕ is the lunar phase. In the case that gravity acts differently on self-energy the ratio of gravitational to inertial mass would become

$$r = 1 + \xi \frac{U}{Mc^2}, \tag{2.6}$$

where U is the gravitational self-energy and Mc^2 is the mass energy. If ξ is unity then C would be 13 m. Comparing the Earth and the Moon

$$r_{\mathrm{E}} - r_{\mathrm{M}} = -4.45 \times 10^{-10} \xi. \tag{2.7}$$

The experiment performed is to measure the return time of laser pulses reflected from the Moon. These pulses are reflected from arrays of corner cubes deployed either by Apollo astronauts or by lunar landers. A corner cube is made up of three orthogonal mirrors making the corner of a cube, several cm in size. They have the property, in common with rear reflectors on cycles, that incident light is reflected in turn off each mirror and leaves finally in the reverse of its direction at incidence. Diffraction of the laser beam leaving the laser and at the corner cube leads to reduction in power between outgoing and return beam of 10^{21}. Using pulses of nanosecond duration the return time of 2.56 s can be measured with high precision. The data taking has been continuous over decades, and measurements with mm precision have been achieved. The result reported (Williams et al. 2004) is

$$\xi = (4.4 \pm 4.5)10^{-4}. \tag{2.8}$$

In conclusion we can say that there is no experimental evidence for a violation of the equivalence principle.

2.4 The Gravitational Spectral Shift and the Deflection of Light

The first important consequence of the SEP pointed out by Einstein is the prediction that gravitational fields affect radiation. Consider a capsule of height D in free fall, from rest, toward the Earth with acceleration g. At the instant of release a source on the roof of the capsule emits a pulse of light at frequency ν. The pulse reaches the floor after a time D/c. Throughout this time an occupant of the capsule will observe that the light has frequency ν. At the moment the pulse reaches the capsule floor the capsule will have attained a velocity downwards

$$V = gD/c.$$

Hence an external observer at rest observing the flash, at the same moment, sees it Doppler shifted to a higher frequency. The effect of the relative velocity V of the source toward the external observer is to give a fractional change in frequency of

$$\frac{\Delta\nu}{\nu} = V/c = \frac{gD}{c^2}.$$

Note that the relevant velocity is that of the source frame relative to the external observer at the instant the light is detected. Now choosing r to be the distance from the center of the Earth, $g = GM/r^2$ where M is the mass of the Earth. Also the change in distance from the center of the Earth is $\Delta r = -D$. Making these substitutions in the above equation we get

$$\frac{\Delta\nu}{\nu} = -\frac{GM\Delta r}{r^2c^2}. \tag{2.9}$$

The crests of the waves emitted by the source can be regarded as the ticks of an atomic clock, and if the rate of ticking ν increases, the same time interval τ would contain more ticks. Hence the frequency shift implies that time intervals measured in the two frames by identical clocks would differ. The interval between ticks emitted by the source is given by $1/\tau = \nu$. Thus

$$\frac{\Delta\tau}{\tau} = -\frac{\Delta\nu}{\nu} = \frac{GM\Delta r}{r^2c^2}. \tag{2.10}$$

Integration from a point where the potential is negligible (infinity) gives

$$\tau(r) = \tau(\infty)\exp\left[-\frac{GM}{rc^2}\right] = \tau(\infty)\left[1 - \frac{GM}{rc^2}\right]:$$

to an approximation that is valid on Earth where GM/rc^2 is 10^{-10}. Then

$$\tau(r) = \tau(\infty)\left(1 + \frac{\varphi}{c^2}\right), \tag{2.11}$$

where φ is the gravitational potential at r. Notice that because the gravitational potential is always negative $\tau(\varphi = 0)$ is always larger than $\tau(\varphi)$.

The time interval $\tau(\infty)$ measured remotely is called the *coordinate time* and the time $\tau(\varphi)$ measured where the gravitational potential is φ is called the *proper time*.

The remote observer measures the time intervals to be dilated and light to have undergone a gravitational redshift. Later we shall need to relate the squares of time intervals in different frames, so we write this for later use

$$d\tau^2 = dt^2\left(1 - \frac{2GM}{rc^2}\right) \tag{2.12}$$

where $d\tau$ is the proper time and dt is the coordinate time. The derivation leading to Equation (2.12) is heuristic, but the equation itself is rigorously correct in GR.

A precise measurement of the gravitational redshift was made by Pound and Rebka in 1960 (Pound & Rebka 1960) using photons that dropped down inside a 22.6 m tower at Harvard. The predicted spectral shift is only

$$\Delta\nu/\nu = 2.46 \times 10^{-15}.$$

Pound and Rebka exploited the contemporary discovery by Mö ssbauer of decays by gamma emission in which the whole crystal recoils rather than the parent nucleus. This means that the spectral line width is unusually narrow. In the case of

$$^{57}\mathrm{Fe}^{*} \rightarrow \gamma + {}^{57}\mathrm{Fe}$$

the 14.4 keV line has a fractional width of 10^{-12}, still 500 times the gravitational spectral shift to be measured.

Pound and Rebka placed the $^{57}Fe^{*}$ source at the top of the tower and a thin ^{57}Fe absorber at the foot of the tower. This absorber covered a scintillator viewed by a photomultiplier. It was arranged that the source could be driven slowly up or down using a transducer, so producing a Doppler shift to compensate the gravitational spectral shift. With exact compensation the absorption of photons in the ^{57}Fe absorber was maximized and the photomultiplier count rate minimized. Pound and Rebka scanned across the narrow line profile by varying the drive velocity. This gave a large gain in sensitivity in locating the line center, and a measured gravitational spectral line shift

$$\Delta\nu/\nu = (2.57 \pm 0.26) \times 10^{-15},$$

which agrees, within the small quoted error, with the prediction obtained from the SEP. Other more recent tests involve direct comparison of the time-keeping of atomic clocks or of masers. Vessot and colleagues in 1980 (Vessot et al. 1980) compared the rate of a hydrogen maser launched to a height of 10,000 km in a rocket with the rate of an identical maser kept in the laboratory. Two-way telemetry was used to compensate both for atmospheric effects and for the first-order Doppler shift due to the relative velocity of the masers. The comparison gave a rate difference that agreed with the prediction from the EP to parts in 10^4.

An extension of the above arguments based on the EP leads to the conclusion that the path of light is bent in a gravitational field. Using Equation (2.9) we can infer that a photon climbing a distance d against the Earth's gravitational pull loses energy

$$\Delta E = h\Delta\nu = gd\frac{h\nu}{c^2},$$

where h is Planck's constant. The kinetic energy lost by a body of mass m rising through the same distance is remarkably similar:

$$\Delta E = gmd.$$

Therefore it emerges that a photon of energy $E = h\nu$ behaves in a gravitational field as if it possessed an inertial mass E/c^2! We shall see later that in Einstein's theory of general relativity all forms of energy couple to the gravitational field. Consequently a photon feels the gravitational force and it follows that a photon follows a curved path in a gravitational field. This behavior can be pictured in the following Gedanken (thought) experiment.

Imagine a space capsule in free fall near the Earth: inside it an astronaut strapped to one wall shines a beam of light horizontally at the opposite wall. Figure 2.4(a) shows the light path as seen by the astronaut; his frame is in free fall so that the light travels in a straight line to the opposite wall. An external observer at rest sees things quite differently. At emission the lamp is at one height, but by the time the light reaches the other wall the capsule has fallen a little. This view is shown in Figure 2.4(b) where the light path is seen to curve. Einstein calculated the deviation of starlight passing near the Sun's surface on its way to the Earth; his result was 1.750 arcsec. We should note that a calculation using just the EP gives exactly half this value: time and frame distortion contribute equally.[1] The confirmation from the measurements made in 1919 has already been discussed. Higher precision is obtained by studying the apparent motion of radio sources that pass near the Sun's disk, a technique that is not restricted to times of solar eclipse. In Figure 2.5 a widely spaced pair of radio telescopes are shown receiving signals from the same source. If the source direction makes an angle θ with the baseline, which is of length

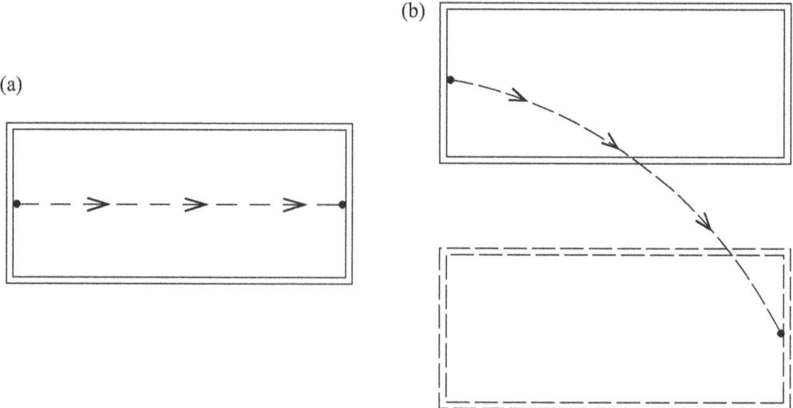

Figure 2.4. Light path in a space capsule in free fall near the Earth as seen by (a) an occupant of the capsule and (b) an external observer at rest with respect to the Earth's surface.

[1] Appendix G: Kenyon I R 1990 *General Relativity* Oxford Univ. Press (Kenyon 1990).

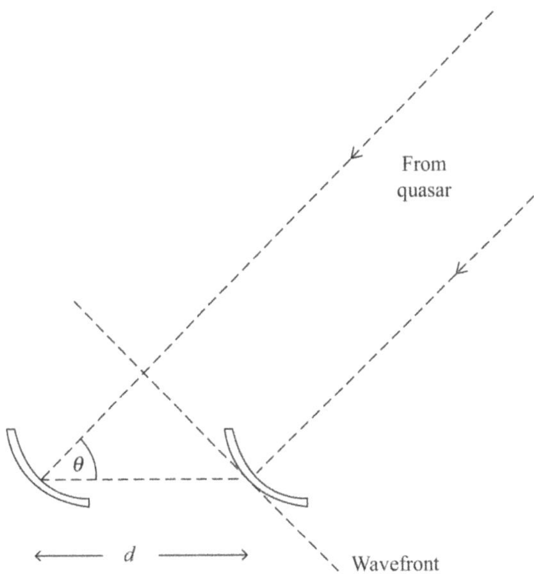

Figure 2.5. The arrangement of a pair of radio antennae used to determine the angular position of radio sources.

d, then the path difference to the two dishes is $d \cos \theta$ and the phase lag between their signals is

$$\Delta = \frac{2\pi}{\lambda} d \cos \theta$$

at a wavelength λ, and in terms of frequency ν this becomes

$$\Delta = \frac{2\pi \nu d \cos \theta}{c}.$$

Measurement of the phase difference leads to a determination of θ. Lebach et al. (1995) used 30 m diameter radio telescopes on a 4000km baseline, across the USA, to study the radio sources close to the Sun at frequencies of 2.3, 8.4, and 22.7 GHz. This is an example of very long baseline interferometry (VLBI). They found a mean deviation due the Sun's gravitational field of 0.9998 ± 0.0008 times Einstein's prediction.

2.5 Exercises

1. Show that in the Pound–Rebka experiment the expected gravitational redshift was 2.46×10^{-15}.
2. Calculate the redshift for the 768.9 nm potassium line emitted by an atom on the Sun's surface.
3. One recent theoretic prediction for the mass of pseudoscalar (spin $0\hbar$, negative parity) particles known as axions is around 20 μeV c^{-2}. Exchange of such particles would give rise to a modification of Newton's law of gravitation. At what range would such a modification be expected?

4. If some of the protons on the Sun and Earth were not matched by electrons these two bodies would be electrically charged. What fraction of the electrons would need to be removed from both bodies in order to change the attractive force between the Sun and the Earth by 1 part in a million?
5. It seems plausible that neutrinos have a mass of around 0.1 eVc^{-2}. Neutrinos are emitted from supernovae with energies around 1 MeV. How much later will neutrinos arrive on Earth than photons from a supernova distant 2 Mpc from the Earth?

Further Reading

Will C M 2018 *Theory and Experiment in Gravitational Physics* (2nd ed.; Cambridge: Cambridge Univ. Press). This is a thorough presentation by a world expert on general relativity. It includes accounts of versions of the equivalence principle, of the post-Newtonian parametrization of metric theories of gravity, and of experimental tests.

References

Bergé, J., Touboul, P., Rodrigues, M., & Liorzou, F. 2017, JPCS, 840, 012028

Kenyon, I. R. 1990, General Relativity (Oxford: Oxford Univ. Press)

Lebach, D. E., Corey, B. E., Shapiro, I. I., et al. 1995, PhRvL, 75, 1439

Pound, R. V., & Rebka, G. A. 1960, PhRvL, 4, 337

Roll, P. G., Krotkov, R., & Dicke, R. H. 1964, AnPhy, 26, 442

Schlamminger, S., Choi, K.-Y., Wagner, T. A., Gundlach, J. H., & Adelberger, E. G. 2008, PhRvL, 100, 041101

Touboul, P., Métris, G., Rodrigues, M., et al. 2022, CQGra, 39, 204009 arXiv:2209.15488

Vessot, R. F. C., Levine, M. W., Mattison, E. M., et al. 1980, PhRvL, 45, 2081

Williams, J. G., Turyshev, S. G., & Boggs, D. H. 2004, PhRvL, 93, 261101

AAS | IOP Astronomy

Introduction to General Relativity and Cosmology (Second Edition)

Ian R Kenyon

Chapter 3

Space and Spacetime Curvature

The observation of the gravitational spectral shift and of the bending of light passing close to the Sun proved that spacetime is warped by the presence of matter. This connection brings out a fundamental point: the curvature of a space can be determined from intrinsic measurements, i.e., from measurements confined to the space itself. For example, just by throwing a ball we can measure the curvature of *spacetime* near the Earth. In Figure 3.1 the maximum height reached above the launch height is h, t is the time taken to return to the launch height. ct greatly exceeds x, the horizontal distance traveled, so the path length in spacetime is close to ct. The path though parabolic is sufficiently flat that it can be can be treated as circular. From the geometry of the circle the radius of curvature of the path in spacetime is

$$R = c^2t^2/8h.$$

Taking the acceleration due to gravity to be g, and ignoring air resistance, the height reached is

$$h = gt^2/8.$$

Combining these two equations gives R, the radius of curvature in spacetime near the Earth:

$$R = c^2/g \approx \text{ one light year}.$$

For reference later we rewrite this as

$$1/R = \frac{GM_{\oplus}}{r^2c^2}, \tag{3.1}$$

where r is the distance to the center of the Earth and $M_{\oplus}$ is the Earth's mass.

In Section 3.1 the basic concepts of curvature are introduced, using the easily visualized example of two-dimensional spaces.

doi:10.1088/2514-3433/acc3ffch3

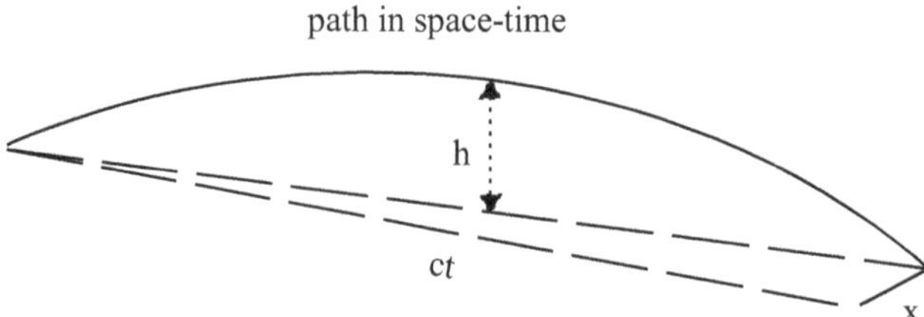

Figure 3.1. Curvature of spacetime from the flight of a ball.

> All the geometrical information about a space is quantified by the equivalent of Pythagoras' theorem, which relates the distance between nearby points in space to their coordinate separations, known as the *metric equation*. Straight lines, in the usual sense, cannot be drawn on curved surfaces or in curved spaces. Their place is taken by *geodesics*, which are the straightest lines that can be drawn between a pair of points in a curved space: that is to say they have no component of curvature in that space.

The path shown in Figure 3.1 is one such geodesic. The general definition of curvature and ways to measure it are described in Section 3.2. How you compare local vectors at different places in a curved space is described in Section 3.3. The relationship between curvature and the metric equation is described in Section 3.4, and the metric equation for Minkowski space is considered in Section 3.5. This moves the discussion forward to spacetime. Section 3.6 introduces geodesics as the paths of bodies in free fall, and tidal acceleration, which is the divergence between nearby geodesics due to the change in the gravitational field with location. Our final topic is the Schwarzschild metric, describing spacetime around a spherically symmetric mass. This is introduced here heuristically; a rigorous proof of its validity in GR is carried through in Appendix C. In Chapter 7 predictions of GR effects in the solar system will be made with the Schwarzschild metric and their successful tests described. The Schwarzschild metric will be used again to analyze the properties of black holes in Chapter 8.

3.1 Two-dimensional Surfaces

There are intrinsic properties of any curved space that reveal its curvature, a connection first appreciated by Gauss. We start by considering the surface of the sphere in Figure 3.2, which is easy to visualize from our three-dimensional standpoint. Any attempt to construct a set of rectangular Cartesian coordinates to cover the sphere runs into irretrievable difficulties. One set of generalized, or Gaussian, coordinates that can be used to cover the whole surface of the sphere are the polar angles (θ, ϕ). With the origin at the center of the Earth θ runs from 0° at the North Pole to 180° at the South Pole, and is related to the latitude. ϕ is essentially the longitude and runs from −180° to +180°. Locally, that is to say on a scale small compared to the radius of curvature r of the surface, the distance between two points

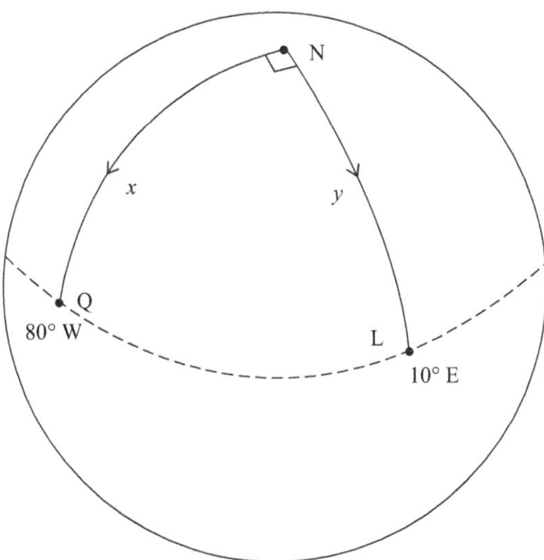

Figure 3.2. A two-dimensional spherical surface, such as the surface of the Earth.

at (θ,ϕ) and $(\theta + \mathrm{d}\theta, \phi + \mathrm{d}\phi)$ is $r\mathrm{d}\theta$ in latitude and $r \sin\theta \mathrm{d}\phi$ in longitude. Then the total separation is given by the quadratic equation

$$\mathrm{d}s^2 = r^2\mathrm{d}\theta^2 + r^2 \sin^2\theta \, \mathrm{d}\phi^2. \tag{3.2}$$

This is a *metric equation*. In order to reconstruct the geometry of the Earth from the latitude and longitude a distance scale is needed and this is what the metric equation provides. For example, the metric equation shows that near the poles ($\theta = 0°$ or $180°$) changes in longitude ϕ produce only small displacements.

The properties of a spherical curved surface, highlighted above, are common to other curved two- and higher-dimensional spaces of interest. First of all, in order to cover the whole of a curved surface or space it is necessary to use *generalized* or *Gaussian* coordinates; rectangular Cartesian coordinates are inadequate. A second property is that the local distances are given by a metric equation in the Gaussian coordinate separations.

A sphere is one example of a *Riemann* space: in such spaces it is always possible to match any arbitrary region by a flat or *Euclidean* space provided that the region taken is small enough, simplifying a surveyor's life. This is like being able to draw a straight line tangent at any point on a smooth curve. Consider a general two-dimensional Riemann surface with metric equation at a point P

$$\mathrm{d}s^2 = a \, \mathrm{d}u^2 + 2b \, \mathrm{d}u \, \mathrm{d}v + c \, \mathrm{d}v^2 \tag{3.3}$$

where (u, v) are some Gaussian coordinates. The coefficients a, b, and c are functions of position and contain all the information about the geometry of the surface. This can be converted to obviously Euclidean form by completing the squares

$$ds^2 = \left[\sqrt{a}\mathrm{d}u + \frac{b}{\sqrt{a}}\mathrm{d}v\right]^2 + \left[c - \frac{b^2}{a}\right]\mathrm{d}v^2 = \mathrm{d}x^2 + \mathrm{d}y^2,$$

where

$$x = \sqrt{a}u + \frac{b}{\sqrt{a}}v \quad \text{and} \quad y = \sqrt{c - \frac{b^2}{a}}v,$$

provided that $a > 0$ and $ac > b^2$. In this case it follows that a Euclidean surface will match the surface locally at P. In other words a plane can always be drawn at any arbitrary point on a two-dimensional Riemann surface so that it is locally tangential to the surface. A similar procedure can be followed in higher-dimensional Riemann spaces. Some coordinate transformation can always be found which converts the metric equation to a sum of squares. The Cartesian coordinates which result from this transformation describe the space tangential to the curved space at the point selected. To reiterate: Riemann spaces are locally flat (locally Euclidean).

If instead of having $ac > b^2$, it is the case that $b^2 > ac$ then

$$ds^2 = \mathrm{d}x^2 - \mathrm{d}y^2. \tag{3.4}$$

The space involved is still locally flat because it has a metric equation that reduces to a difference of squares. Its tangent space is called *pseudo-Euclidean* and the space itself is called *pseudo-Riemannian*. For most purposes we can ignore the distinction and include the pseudo-Riemannian spaces with the Riemannian spaces. Referring back to the first chapter we see that the spacetime of SR is pseudo-Euclidean, which explains our interest in such spaces.

The shortest paths joining points a finite distance apart on a curved surface are generally not simple straight lines. However they are the straightest lines that can be drawn on the surfaces between points under consideration. Such paths are called *geodesics* and naturally on a flat surface a geodesic is a straight line. Geodesics on a spherical surface are the well-known great circle routes which are used by aircraft on intercontinental flights.

Figure 3.3(a) shows a great circle path drawn to touch a line of latitude θ at P. Figure 3.3(b) shows a diametral plane section of the sphere through P and the North Pole. D is the center of curvature of the line of latitude, while C is the center of curvature of the surface. A geodesic has curvature $1/r$ and the line of latitude has curvature $1/r \sin\theta$. Resolving the curvature of the line of latitude perpendicular and parallel to the surface gives $(1/r \sin\theta)\sin\theta = 1/r$ and $\cos\theta/(r\sin\theta)$; while for the geodesic great circle the corresponding components are $1/r$ and zero. This illustrates a very important property of the set of lines through a point on a surface that are tangential (share the same direction) at that point. The component of curvature perpendicular to the surface is the same for all of them, and in the case illustrated this is $1/r$. Among these lines one, the geodesic, has only this component of curvature. It has no component of curvature lying in the surface, which makes it the straightest path possible over the surface. This characteristic is quite general and holds for geodesics in spaces that are less symmetric than that of a sphere, and in

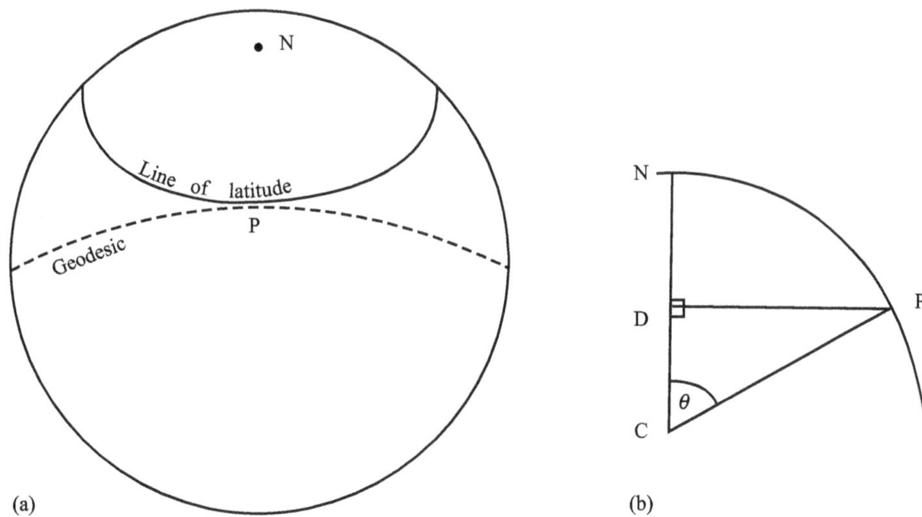

Figure 3.3. (a) A line of latitude and a geodesic (great circle) on a spherical surface which are tangential to one another at P: (b) a diametral plane section through the sphere which contains the North Pole N and the center of the Earth C. D is the center of curvature of the line of latitude at P.

spaces of higher dimension. *Intrinsic* methods by which a two-dimensional observer in two dimensions can make qualitative and quantitative measurements on curvature of the surface inhabited will be described next.

3.2 Measurement of Curvature

A quantity called the *Gaussian curvature* quantifies the local curvature of any two-dimensional surface and can be generalized in the case of higher dimensions. One simple measurement reveals whether a two-dimensional surface is curved and determines the sign of curvature. The observer pegs one end of a string of length r to a fixed point O and then makes a circuit around O so that all the while the string is fully stretched, but of course confined to lie on the surface. The observer then measures the circumference C of one complete circuit around O. The result for the flat surface drawn in Figure 3.4 is

$$C_{\mathrm{F}} = 2\pi r. \tag{3.5}$$

However if the surface is dome-shaped like the surface P in Figure 3.4 the circumference is shorter and

$$C_{\mathrm{P}} < 2\pi r. \tag{3.6}$$

In the case of the saddle-shaped surface N in Figure 3.4

$$C_{\mathrm{N}} > 2\pi r. \tag{3.7}$$

Now P is a surface of positive curvature like a spherical surface, while N has negative curvature, and so the difference $2\pi r - C$ fixes the *sign* of curvature. The

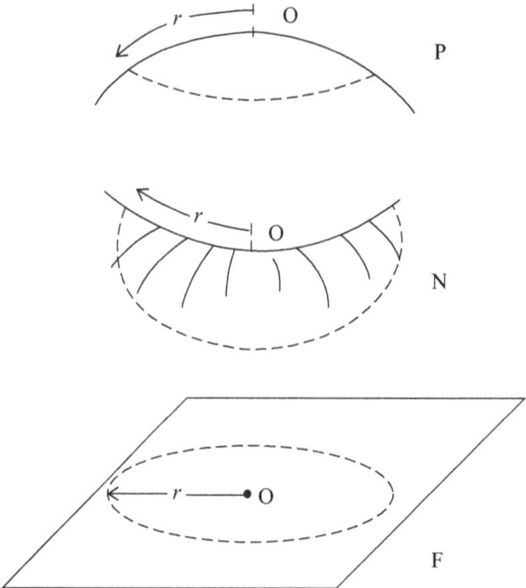

Figure 3.4. Three two-dimensional surfaces: P is dome shaped, N is saddle shaped, and F is flat. In each case the broken line marks out a path which stays exactly a distance r from O.

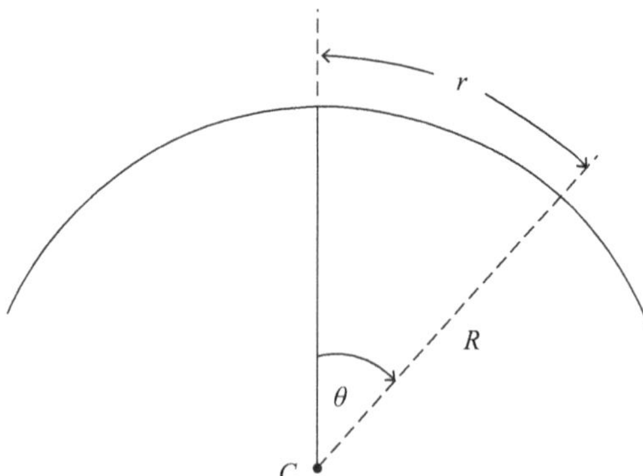

Figure 3.5. A section through a spherical surface, with center of curvature at C and radius R.

reader will find it hard to picture extending the surface N continuing everywhere with negative curvature. The reason is that surfaces of negative curvature cannot be embedded in our (nearly) flat and Euclidean three-space. The inhabitant of two-dimensional space could also refine the use of his measurement to make a quantitative determination of the curvature.

We specialize to the case of a spherical surface whose cross-section is shown in Figure 3.5. C is the center of curvature and R is the radius of curvature. The angle subtended by a length of string r at the center of the sphere is θ, and so

$$\theta = r/R \tag{3.8}$$

where R is the radius of the sphere. Then

$$C_P = 2\pi R \sin\theta = 2\pi r\left(1 - \frac{r^2}{6R^2} + \cdots\right). \tag{3.9}$$

In the limit as r tends to zero we obtain

$$R^{-2} = \frac{3}{\pi}\lim_{r\to 0}\left(\frac{2\pi r - C_P}{r^3}\right). \tag{3.10}$$

This procedure is certainly an intrinsic method for measuring the radius of a spherical surface. It does more: it gives the curvature K of any two-dimensional surface, which for a sphere is R^{-2}. In order to extend the discussion beyond the sphere we must introduce a definition of curvature valid for any such surface.

A portion of a more general two-dimensional surface is portrayed in Figure 3.6. G_1OG_1 and G_2OG_2 are geodesics across the surface which intersect at right angles; ON is the local normal to the surface. Rectangular Cartesian coordinates can be assigned at O using the tangents to G_1OG_1 and G_2OG_2 plus ON itself. These will be called the v, w, and z axes, respectively. An expression for z, which is valid for a point (v, w, z) on the surface close to O such that both v and w are small, can be obtained by a Taylor series expansion

$$z = \frac{\partial z}{\partial v}v + \frac{\partial z}{\partial w}w + \frac{1}{2}\left(\frac{\partial^2 z}{\partial v^2}v^2 + 2\frac{\partial^2 z}{\partial v\partial w}wv + \frac{\partial^2 z}{\partial w^2}w^2\right) + \cdots. \tag{3.11}$$

The first two terms vanish because the v, w plane is parallel to the surface at O. Therefore to second order in v and w we have

$$z = \frac{1}{2}\left(Lv^2 + 2Mvw + Nw^2\right). \tag{3.12}$$

This expression can be converted to a sum of squares simply by rotating the Cartesian frame around ON through an angle

$$\frac{1}{2}\tan^{-1}\left(\frac{2M}{L - N}\right). \tag{3.13}$$

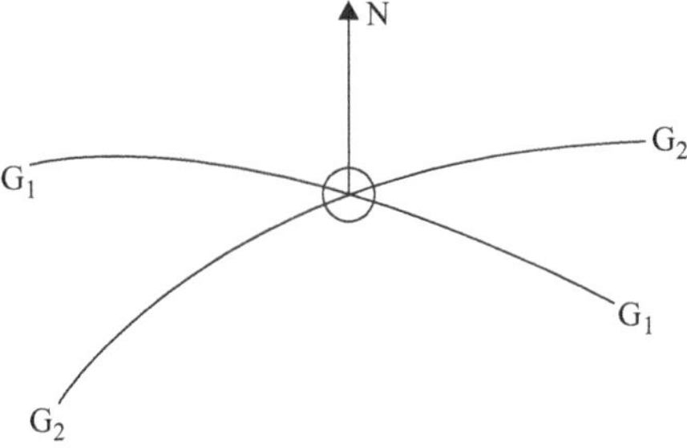

Figure 3.6. G_1OG_1 and G_2OG_2 are geodesics on a two-dimensional surface which intersect at right angles. ON is the normal to the surface at O.

In the new Cartesian frame with coordinates (x, y, z) the surface has the equation

$$z = (K_1x^2 + K_2y^2)/2. \tag{3.14}$$

The planes $x\mathrm{O}z$ and $y\mathrm{O}z$ are called the *principle planes*. Now consider the line along which this surface intersects the plane $x\mathrm{O}z$. It has the equation

$$z = K_1x^2/2, \tag{3.15}$$

and its radius of curvature R_1 is given by the formula relating the sagitta z to the chord length $2x$ for a circular arc

$$2R_1z = x^2. \tag{3.16}$$

Thus the curvature of the line of intersection is

$$1/R_1 = K_1. \tag{3.17}$$

Similarly in the orthogonal section with $y\mathrm{O}z$ the curvature is K_2. It is not too difficult to prove that K_1 and K_2 are the minimum and maximum curvatures for any plane section through the surface at O containing ON. K_1 and K_2 are called the *principal curvatures* of the surface at O. Their product is an invariant for the surface at O called the *Gaussian curvature* K; thus

$$K = K_1K_2. \tag{3.18}$$

A sphere has Gaussian curvature R^{-2} at any point on its surface. In the case of a cylinder one principal plane bisects its length along a straight line; hence one principal curvature is zero, and the Gaussian curvature is then also zero. Finally, for the saddle-shaped surface of Figure 3.4 one of the principal planes lies along the length of the saddle in the direction of the horse's spine and the other lies transverse to the saddle in the direction of the horse's ribs. The center of curvature of the first section lies above the saddle, while the center of curvature of the second section lies below the saddle. Therefore K_1 and K_2 have opposite signs and the Gaussian curvature is negative. The method already described for measuring the curvature of the sphere generalizes so that for any other two-dimensional surface the curvature is given by

$$K = \frac{3}{\pi}\lim_{r\to 0}\left(\frac{2\pi r - C}{r^3}\right). \tag{3.19}$$

Two other intrinsic methods for measuring K are worth discussing as they are methods that carry over to the analysis of spacetime curvature. The first method is based on the way the separation of geodesics grows with distance. In the case of a plane the separation between a pair of geodesics (straight lines) through a point increases linearly with the distance from this point. In contrast the separation of a pair of lines of longitude shown diverging from the North Pole N in Figure 3.7 does not vary linearly with the distance s measured from the pole. The difference in longitude is ϕ and so the separation after a distance s is

$$\eta = (R\sin\theta)\phi \tag{3.20}$$

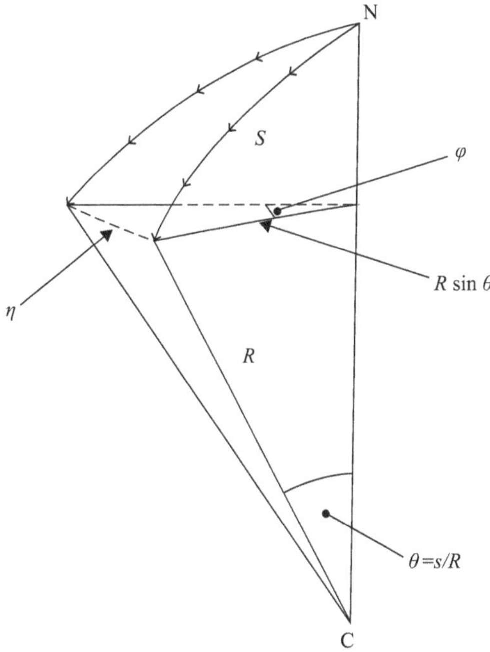

Figure 3.7. A diagram to show how the separation of a pair of geodesics on a sphere depends on the distance from their intersection point.

where R is the radius of the sphere and $\theta = s/R$. Thus

$$\eta = (R\phi)\sin(s/R). \tag{3.21}$$

Differentiating this expression gives

$$\frac{\mathrm{d}^2\eta}{\mathrm{d}s^2} = \frac{-\eta}{R^2}. \tag{3.22}$$

This relation generalizes for all two-dimensional surfaces to become

$$\frac{\mathrm{d}^2\eta}{\mathrm{d}s^2} = -\eta K, \tag{3.23}$$

where K is again the Gaussian curvature. Another effect of intrinsic curvature is revealed when a vector is transported round a closed path in such a way that it is moved parallel to itself at each step of the journey.

3.3 Local Vectors and Parallel Transport

Vectors are often drawn as extended lines with arrowheads. This convention requires interpretation for local vectors, i.e., vectors measured at a given point in space. Two examples will be considered. First, wind velocity is measured at a given observation station and refers solely to that point, for all that it may be convenient to show it on a chart as an arrow apparently extending over a long distance. In a like manner the force on an electron in an electric field has magnitude and direction but only exists at the coordinates of the electron. Wind velocity and force are both local vectors. A simple and most useful local vector in the present context is the tangent vector to a curve in space.

The comparison of local vectors at different places is easy in flat space. First, a Cartesian coordinate system is set up with one vector **a** at location A and the other **b** at B; **a** and **b** will be equal if their components are equal. Equivalently we can imagine picking up **b** and carrying it to A without changing its length or direction and then examining whether it fits **a** exactly. In this procedure **b** is said to be *parallel transported* from B to A. Comparison of local vectors in a curved space is less straightforward because it is no longer possible to set up a single Cartesian coordinate system to cover all space. Once more it is helpful to consider the situation of 2D, a two-dimensional being who lives on a spherical surface embedded in our three-dimensional space. Suppose that 2D starts at the pole in Figure 3.8 with the local vector **a** shown there. 2D's only strategy is to make small steps and to carry the local vector parallel to itself at each step. Without any reference frame to check the parallelism even this seems difficult. However, 2D can start by moving off in the direction of the local vector itself, and in this case parallel transport is well defined. What 2D is doing in this case is to trace out a geodesic—a great circle on the sphere, e.g., NA shown in Figure 3.8. After this 2D can carry any other vector parallel to itself by traveling along a geodesic and keeping the local vector at a constant angle to the geodesic. In cases where the route does not follow a geodesic it would be necessary to split the path up into infinitesimal steps, each step being along a geodesic; which is possible because the space is locally flat.

Emboldened by success, 2D could go on to parallel transport the vector **a** along the closed path NABN in Figure 3.8. Each path segment is a geodesic. Starting from the pole the local vector is carried along the direction it points (a line of longitude) to the equator (A). From A it is parallel transported along the equator to

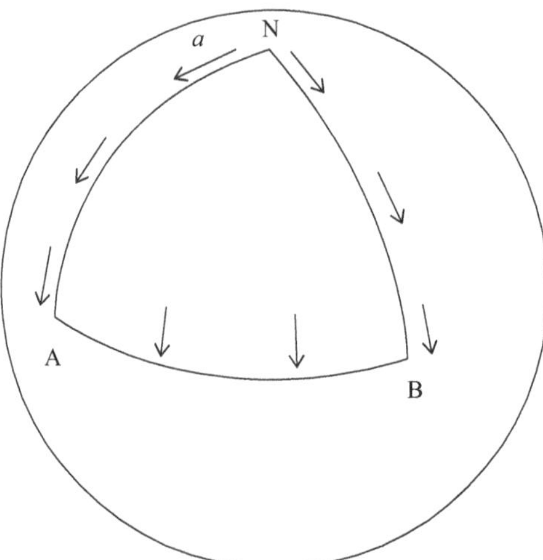

Figure 3.8. Parallel transport of a vector round the path NABN over a spherical surface.

B and then returned along another line of longitude to the pole. As a result the vector is seen to rotate through an angle ϕ, which is the separation in longitude between A and B. In general the rotation of a local vector when carried around a closed path on any two-dimensional surface is given by the expression

$$\phi = K(\text{area enclosed by path}), \tag{3.24}$$

which is easily checked for the route discussed.

An equivalent view of this effect is that the result of parallel transport in a curved space depends on the path taken.

3.4 Curvature and the Metric Equation

It has been stated that the metric equation contains the *full information on the geometry of the space.* One consequence of the primacy of the metric equation is that the Gaussian curvature is a function of the coefficients a, b, and c in Equation (3.3). This relationship was discovered by Gauss who recognized its importance by referring to it as the *Excellent theorem.*[1] Here we simply obtain the relation for an azimuthally symmetric two-dimensional surface, a result needed later when introducing the Schwarzschild metric. The selected surface has the metric equation

$$ds^2 = g_{rr}\, dr^2 + r^2 d\theta^2, \tag{3.25}$$

where g_{rr} is a function of r only. Figure 3.9 shows a section through the surface embedded in a three-dimensional space, with the axis of rotation lying in the paper, situated off to the left and parallel to the left edge of the paper. The curvature of this surface is

$$K = \frac{\partial g_{rr}/\partial r}{2\, r\, g_{rr}^2}. \tag{3.26}$$

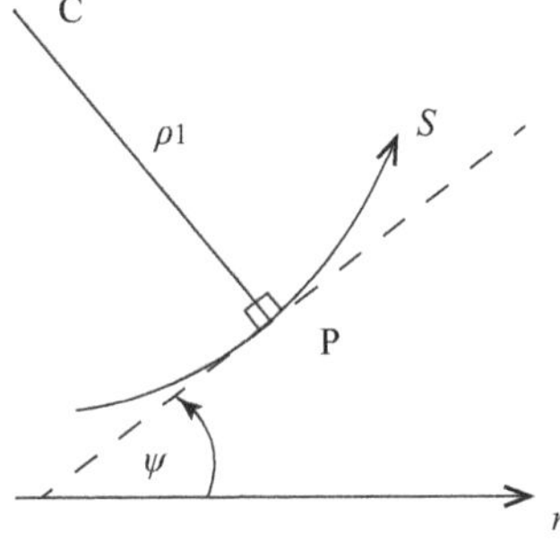

Figure 3.9. A section through a two-dimensional cylindrically symmetric curved surface. r is the coordinate distance from the axis of rotation: this axis is off to the left and lies in the plane of the paper, parallel to the left edge of the paper. s is measured along the curve drawn.

[1] A general proof is given in Appendix B: Kenyon I R (1990) *General Relativity* Oxford Univ. Press (Kenyon 1990).

Spaces with more than two dimensions require more than a single parameter to describe the Gaussian curvature at a given point: $n(n-1)/2$ independent Gaussian curvatures are required for a space of n dimensions. A more complete and compact description of curvature in n dimensions is embodied in the Riemann tensor, introduced in Chapter 6.

3.5 The Metric Equation of Special Relativity

The spacetime of SR is flat and one Cartesian coordinate system (ct, x, y, z) can be used throughout all spacetime. Intervals of proper time between events with coordinate separations $(c\mathrm{d}t, \mathrm{d}x, \mathrm{d}y, \mathrm{d}z)$ are given by

$$\mathrm{d}s^2 = c^2\mathrm{d}\tau^2 = c^2\mathrm{d}t^2 - \mathrm{d}x^2 - \mathrm{d}y^2 - \mathrm{d}z^2, \tag{3.27}$$

which is the metric equation for this spacetime (Minkowski space). This metric equation differs from a metric equation valid for Euclidean space in having both positive and negative terms on the right-hand side. The set of coefficients on the right-hand side (in this case $+1, -1, -1, -1$) is called the signature of the space and labels the properties of the space. A space with some negative coefficients and some positive is called a pseudo-Euclidean space. The coefficients can always be reduced to $+1$ or -1 by appropriate choice of the time and distance units. With the notation of Chapter 1 the metric equation can be written

$$\mathrm{d}s^2 = (\mathrm{d}x^0)^2 - (\mathrm{d}x^1)^2 - (\mathrm{d}x^2)^2 - (\mathrm{d}x^3)^2, \tag{3.28}$$

where $x^0 = ct$, $x^1 = x$, $x^2 = y$ and $x^3 = z$. A more compact way of writing this is

$$\mathrm{d}s^2 = \sum_{\mu=0}^{3}\sum_{\nu=0}^{3}\eta_{\mu\nu}\mathrm{d}x^\mu\mathrm{d}x^\nu, \tag{3.29}$$

where the components $\eta_{\mu\nu}$ form a square matrix:

$$\eta_{\mu\nu} = \begin{bmatrix} 1 & 0 & 0 & 0 \\ 0 & -1 & 0 & 0 \\ 0 & 0 & -1 & 0 \\ 0 & 0 & 0 & -1 \end{bmatrix}. \tag{3.30}$$

$\eta_{\mu\nu}$ is called a metric tensor. Subscripts and superscripts are introduced in Equation (3.29) so that the notation is consistent with that used from Chapter 4 onward. For the present the reader can take subscripts and superscripts to be equivalent. In cases for which the system has spherical symmetry, such as spacetime around a spherical star or planet, it may often be better to use polar coordinates; then

$$\begin{aligned} \mathrm{d}s^2 &= c^2\mathrm{d}t^2 - \mathrm{d}r^2 - r^2\mathrm{d}\theta^2 - r^2\sin^2\theta\mathrm{d}\phi^2 \\ &= \sum_{\mu=0}^{3}\sum_{\nu=0}^{3}\eta_{\mu\nu}\mathrm{d}w^\mu\mathrm{d}w^\nu, \end{aligned} \tag{3.31}$$

where $w^0 = ct$, $w^1 = r$, $w^2 = \theta$ and $w^3 = \phi$. This equation has a different, but still diagonal, metric tensor

$$\eta_{\mu\nu} = \begin{bmatrix} 1 & 0 & 0 & 0 \\ 0 & -1 & 0 & 0 \\ 0 & 0 & -r^2 & 0 \\ 0 & 0 & 0 & -r^2 \sin^2\theta \end{bmatrix}. \tag{3.32}$$

The notation can be greatly simplified by adopting the Einstein summation convention in which we sum over repeated indices. With this convention the metric equation simplifies to

$$\mathrm{d}s^2 = \eta_{\mu\nu}\mathrm{d}x^\mu \mathrm{d}x^\nu. \tag{3.33}$$

It is useful to distinguish between Roman and Greek suffixes. When Greek letters are used, as above, the summation is to be made over time and space coordinates. With this interpretation the right-hand side of Equation (3.33) is identical to that of Equation (3.29). However if Roman letters are used the summation is only made over space coordinates (1,2,3). The metric equation for the three-dimensional Euclidean space with Cartesian coordinates is then

$$\mathrm{d}s^2 = a_{ij}\mathrm{d}x^i \mathrm{d}x^j \tag{3.34}$$

where

$$a_{ij} = \begin{bmatrix} 1 & 0 & 0 \\ 0 & 1 & 0 \\ 0 & 0 & 1 \end{bmatrix}. \tag{3.35}$$

The observation of gravitational redshift and the deviation of electromagnetic waves passing the Sun shows that real spacetime is curved. Therefore flat Minkowski spacetime is inadequate and our analysis of curved spacetime will need to proceed along the lines mapped out above for analyzing curved space. Gaussian (generalized) coordinates x^μ can be used to cover curved spacetime. The interval $\mathrm{d}s$ between events in spacetime with coordinate separations $\mathrm{d}x^\mu$ is then given by the quadratic metric equation

$$\mathrm{d}s^2 = c^2\mathrm{d}\tau^2 = g_{\mu\nu}\mathrm{d}x^\mu \mathrm{d}x^\nu \tag{3.36}$$

where the components of the metric tensor $g_{\mu\nu}$ are functions of the position and time. At this point the strong equivalence principle supplies a key ingredient to understanding curved spacetime. This principle requires that on transforming to a freely falling frame all local experimental measurements give results in accord with SR. What this means in geometric terms is that a Minkowski frame matches the structure of real spacetime locally, but not globally. Thus we can at any event in spacetime, y, always find some frame for which

$$g_{\mu\nu}(y) = \eta_{\mu\nu}$$
$$\left.\frac{\partial g_{\mu\nu}}{\partial x^{\rho}}\right|_{y} = 0. \tag{3.37}$$

This frame is in free fall and so our spacetime belongs to the category of spaces which have a quadratic metric equation and are flat, known as pseudo-Riemann spaces.

Riemann, who was a student of Gauss, initiated the analysis of curved spaces of more than two dimensions in 1846. By the early years of the twentieth century the mathematical properties of Riemann spaces had been extensively studied and this material was available for Einstein to use. Einstein was fortunate in having a friend Marcel Grossmann who introduced him to Riemannian geometry and who worked with him until Einstein left Zurich in 1914. The formal development of GR will be outlined in Chapters 5 and 6 while in the remainder of this chapter a simpler approach will be used to infer one particularly important solution of Einstein's equation. This is the solution for empty space outside a spherically symmetric mass distribution.

3.6 Geodesics, Tidal Acceleration, and Curvature

One of the ways discussed for determining the curvature of a two-dimensional surface is to measure the deviation between nearby geodesics; in a higher-dimensional space the measurements would yield the Gaussian curvature of the two-dimensional surface containing these geodesics. This technique is equally valid in curved spacetime, starting from the physical interpretation of a geodesic.

> A geodesic in flat space (a straight line) is the path of a free body as described by Newton's first law of motion. Hence a time-like geodesic in Minkowski spacetime (also a straight line) is the path of a free body. Taking the equivalence principle as a guide we may infer that equally in curved spacetime the path of a test body in free fall follows a time-like geodesic.

Now let's consider the deviation between paths of nearby test bodies in free fall toward a spherically symmetric star. Figure 3.10 shows two such bodies A and B, both having mass m and both at a radial distance r from the star of mass M; A and B are a tangential distance ξ apart. If m is sufficiently small we can ignore the mutual attraction of the two masses. Resolving the gravitational force due to the star tangential to OA gives zero at A, while for B the force has a component

$$F_1 = \frac{GMm}{r^2}\frac{\xi}{r} \tag{3.38}$$

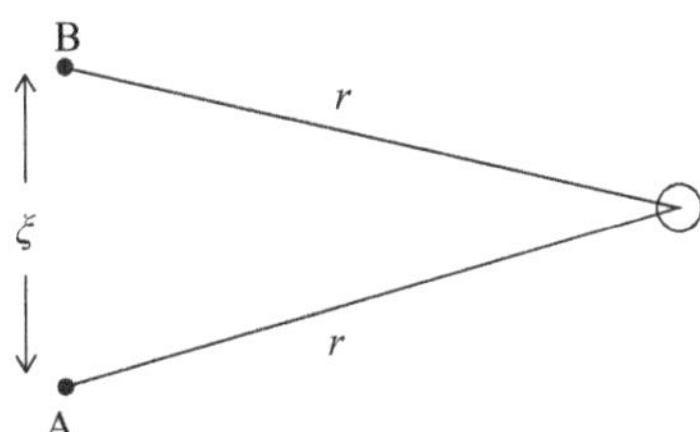

Figure 3.10. The paths of two nearby masses A and B in radial free fall toward a star whose center of mass lies at O. ξ is their tangential separation.

toward A. Therefore there is a relative acceleration between A and B given by

$$\frac{\mathrm{d}^2\xi}{\mathrm{d}t^2} = \frac{-GM\xi}{r^3}, \tag{3.39}$$

which is independent of the masses of A and B. Such an acceleration is called a differential or *tidal* acceleration; it is unaffected by the choice of reference frame. Transforming to the frame in free fall of B, for example, does not remove the relative acceleration between A and B. Tidal accelerations are non-local and are characteristic of curved spacetime. They could only be avoided if the gravitational acceleration were the same everywhere. Then a single transformation to a frame with this acceleration would enable us to impose a frame in free fall everywhere; this is the situation described by SR. Let us take the motion of A and B to be non-relativistic, so that $\mathrm{d}x^1 \ll c\ \mathrm{d}t$, and $\mathrm{d}s^2 = c\mathrm{d}\ \tau^2 \approx c^2\mathrm{d}t^2$. Then the tidal acceleration can be written

$$\frac{\mathrm{d}^2\xi}{\mathrm{d}s^2} = \frac{-GM\xi}{r^3c^2}. \tag{3.40}$$

This expression can be compared to Equation (3.23) for the geodesic deviation over a curved surface. However, we must take into account the sign change for the spatial components of the metric equation on going from three-space to spacetime, and so

$$\frac{\mathrm{d}^2\xi}{\mathrm{d}s^2} = K\xi \tag{3.41}$$

where K is the Gaussian curvature. The tidal acceleration is thus related to spacetime curvature. In the case considered the motion is transverse to OA, and so the curvature is that of the geodesic surface defined by the time direction and the spatial direction transverse to OA. Without any loss of generality we can take AOB to be in the equatorial plane of the parent star so that the acceleration is purely in the coordinate ϕ. From the comparison between the two preceding equations, the Gaussian curvature of the geodesic surface (ϕ, t) is deduced to be

$$K_{\phi t} = \frac{-GM}{r^3c^2}. \tag{3.42}$$

The above approach cannot be extended to infer spatial curvature, so for the present we infer that the curvature of the $r - \phi$ surface with $\theta = \pi/2$ has a similar form

$$K_{r\phi} = \frac{-GM}{r^3c^2}. \tag{3.43}$$

This inference is justified in Appendix C from the solution of Einstein's equation.

3.7 The Schwarzschild Metric

Empty spacetime around a spherically symmetric mass, carrying no charge or angular momentum, has a simple metric in general relativity. This, the *Schwarzschild metric*, is introduced here heuristically. A rigorous derivation is given in Appendix C. The Schwarzschild metric will be used very often in the later chapters, for example in tests of general relativity in the solar system and in deducing the properties of black holes.

The analysis of the gravitational redshift led to the relation in Equation (2.12)

$$\mathrm{d}\tau^2 = \mathrm{d}t^2\left(1 - \frac{2GM}{rc^2}\right). \tag{3.44}$$

From the present viewpoint this can be regarded as the time component of the metric equation for spacetime around a spherically symmetric mass. What is needed in order to complete the metric equation is information about the spatial terms. We can make use of the Gaussian curvature of an equatorial surface from Equation (3.26) for $K_{r\phi}$. Inserting this value in Equation (3.43) gives

$$\frac{\mathrm{d}g_{rr}}{g_{rr}^2} = -\frac{2GM}{r^2c^2}\mathrm{d}r. \tag{3.45}$$

Integrating this equation between $r = r$ and $r = \infty$ gives

$$g_{rr} = \frac{1}{1 - 2GM/rc^2}. \tag{3.46}$$

This suggests that the spatial separation of nearby points in a spherically symmetric space is

$$\mathrm{d}s^2 = \frac{\mathrm{d}r^2}{1 - 2GM/rc^2} + r^2\mathrm{d}\Omega^2 \tag{3.47}$$

with the notation $\mathrm{d}\Omega^2 = \mathrm{d}\theta^2 + \sin^2\theta\mathrm{d}\phi^2$.

Combining Equations (3.47) and (3.44) completes the metric equation for empty space outside a spherically symmetric mass distribution:

$$\mathrm{d}s^2 = c^2\mathrm{d}\tau^2 = c^2\mathrm{d}t^2\left(1 - \frac{2GM}{rc^2}\right) - \frac{\mathrm{d}r^2}{1 - 2GM/rc^2} - r^2\mathrm{d}\Omega^2. \tag{3.48}$$

Note that the relative signs of the spatial components with respect to the time component are set negative, so that in the limit of zero mass this reduces to the Minkowski metric equation of SR:

$$ds^2 = c^2 d\tau^2 = c^2 dt^2 - dr^2 - r^2 d\Omega^2. \tag{3.49}$$

Equation (3.48) is known as the Schwarzschild metric equation for which the metric tensor is

$$g_{\mu\nu} = \begin{bmatrix} \left(1 - \dfrac{2GM}{rc^2}\right) & 0 & 0 & 0 \\ 0 & -\left(1 - \dfrac{2GM}{rc^2}\right)^{-1} & 0 & 0 \\ 0 & 0 & -r^2 & 0 \\ 0 & 0 & 0 & -r^2 \sin^2\theta \end{bmatrix}. \tag{3.50}$$

Note that the components of this equation do not depend on time, and so the Schwarzschild metric is static.

Birkhoff and Langer (1923) proved that this metric equation is the unique description in GR for spacetime outside a spherically symmetric mass distribution (carrying no charge or angular momentum) and is asymptotically flat. The solution is static and there is no other non-spherically symmetric static solution. If the spherically symmetric mass distribution is contracting or expanding radially the Schwarzschild solution remains valid outside this distribution. The Schwarzschild metric will be used in calculations of the observable effects of GR in the solar system (Chapter 8) and of the properties of non-rotating neutral black holes (Chapter 9).

It is important to retain a firm grasp on what the coordinates appearing in Equation (3.50) are and what they are not. Referring back to Equation (2.2) we recall that dt is the time interval between two events measured by an observer using a clock which is in a region remote enough that spacetime is effectively flat; dt is called the *coordinate time* interval. The *proper time* interval, $d\tau$, is that measured on a clock carried by someone moving from (t, r, θ, ϕ) to $(t + dt, r + dr, \theta + d\theta, \phi + d\phi)$. The coordinate r is different from the radial distance measured from the center of mass M, which we call a. The relationship between a and r is

$$da = \left(\frac{1}{1 - 2GM/rc^2}\right)^{1/2} dr. \tag{3.51}$$

The area of the spherical surface labeled by r is $4\pi r^2$ and not $4\pi a^2$, and its circumference is $2\pi r$ and not $2\pi a$.

From the form of the Schwarzschild metric it is clear that the factor $2GM/rc^2$ is an important measure of the effect of mass on the curvature of spacetime. When $2GM/rc^2$ is small compared to unity, the curvature is small and the general relativistic effects are negligible. Conversely, if $2GM/rc^2$ approaches unity the curvature is severe and general relativistic effects dominate. Spacetime in the vicinity of a star whose radius shrinks to a value less than $2GM/c^2$ becomes so warped that the region within is effectively isolated from the rest of the universe. The phenomenon is known as a black hole and the radius of the surface of isolation, the Schwarzschild radius, is

$$r_0 = 2GM/c^2. \tag{3.52}$$

3.8 Exercises

1. Calculate the approximate rotation in the horizontal plane of a vector if it is parallel transported around the periphery of the 48 contiguous states of the USA.
2. A two-dimensional toroidal surface has these dimensions: the mean diameter is 20 m and the radius of the circular cross-section is 2 m. Calculate the Gaussian curvature of the surface at the inner and outer edges of the torus. Where does the Gaussian curvature of this surface become infinite?
3. A star with the same mass as the Sun has radius 3 km. Calculate the metric equation for spacetime near the surface.
4. Write down the equation for an equatorial line in a space described by the Schwarzschild metric at constant time ($\theta = \pi/2$, $\phi = 0$, t constant). You should obtain

$$\mathrm{d}s^2 = \frac{\mathrm{d}r^2}{1 - r_0/r},$$

   where $r_0 = 2GM_{\odot}/c^2$. Now suppose that s in this equation is used to define the length in two-dimensional flat space with Cartesian coordinates w and r. Show that the equation then defines a parabola

$$w^2 = 4r_0(r - r_0).$$

   Now consider the equatorial plane in the same space at constant time ($\theta = \pi/2$, t constant). Show that it is geometrically equivalent to a paraboloid of revolution obtained by rotating the last equation around the w-axis.
5. Consider a body in a circular equatorial orbit around a spherically symmetric mass M with angular velocity ω. Using the Schwarzschild coordinates show that the condition that the orbital period is extremal results in a radius $[GM/\omega^2]^{1/3}$.
6. The radius of the circle threading the center of the cavity of a torus is R. The torus rests on a horizontal surface. A vertical circular section of the torus has radius r. A line drawn in this plane from a point on the surface to the center of the circle makes an angle ϕ with the horizontal. Show that the Gaussian curvature at that point on the surface is $\cos\phi/[r(R + r\cos\phi)]$.

Further Reading

Berry M 1976 *Principles of Cosmology and Gravitation* (Cambridge: Cambridge Univ. Press). This old and short book contains lots of insights, and Chapter 4 is helpful in explaining curved spacetime.

References

Birkhoff, G. D., & Langer, R. E. 1923, Relativity and Modern Physics (Cambridge, MA: Harvard Univ. Press)

Kenyon, I. R. 1990, General Rellativity (Oxford: Oxford Univ. Press)

AAS | IOP Astronomy

Introduction to General Relativity and Cosmology
(Second Edition)

Ian R Kenyon

Chapter 4

Elementary Tensor Analysis

The preceding chapters provided a simple presentation of the properties of curved spacetime. Now the general theory of relativity is developed using tensors. Equations between tensors have the property that they remain valid under general transformations between Gaussian coordinate systems in curved spacetime, in particular under changes to accelerating frames. It is this property that is important for building general relativity. The elements of tensor analysis needed for assimilating general relativity are the topic of this chapter.

Scalars and vectors are the simplest tensors, and from SR it is well known that if an equality can be proved between vectors (or scalars) in one inertial frame then it is valid in other inertial frames. The new feature for curved spacetime is that the permitted transformations are quite general and not simply linear, as in the case of Lorentz transformations. For example in a constant uniform gravitational field there is a quadratic relation between position and time

$$x' = x - gt^2/2,$$

while other gravitational fields give more complicated transformations. Spacetime derivatives such as momentum, which behave as vectors under Lorentz transformations, need redefinition in order to behave as vectors under general transformations. The method for recasting all spacetime derivatives so that they are all tensors under general coordinate transformations is described in Chapter 5, giving physical laws in a form compatible with GR.

A physical law valid in SR, that is, in a frame in free fall, when recast in tensor form is automatically valid under general transformations. This technique fails for the central issue of writing a relativistic law of gravitation: Newton's law of gravitation is not compatible with SR, so we lack a starting point. SR only applies in a frame in free fall,

doi:10.1088/2514-3433/acc3ffch4

which neatly disconnects it from gravitation. What Einstein did was to discover a tensor identity between spacetime curvature and the stress–energy tensor; a tensor that quantifies appropriately the distribution of matter and associated energy. This identity is Einstein's law of gravitation and forms the keystone of GR.

These developments are treated in Chapter 6.

4.1 General Transformations

General transformations in curved spacetime are non-linear. Suppose that one set of Gaussian coordinates that covers spacetime is $(x^0, x^1, x^2, x^3) = x^\mu$, and a second set is x'^μ. The x'^μs are more or less complicated functions of the x^μ s, and equally the x^μs are functions of the x'^μ s. These relations are generally so inconvenient that it makes sense to start calculations from the differentials which transform linearly:

$$\mathrm{d}x'^\mu = \frac{\partial x'^\mu}{\partial x^0}\mathrm{d}x^0 + \frac{\partial x'^\mu}{\partial x^1}\mathrm{d}x^1 + \frac{\partial x'^\mu}{\partial x^2}\mathrm{d}x^2 + \frac{\partial x'^\mu}{\partial x^3}\mathrm{d}x^3 = \frac{\partial x'^\mu}{\partial x^\nu}\mathrm{d}x^\nu, \tag{4.1}$$

where the rightmost expression uses the Einstein summation convention. A useful shorthand is to define

$$\Lambda^\mu{}_\nu = \frac{\partial x'^\mu}{\partial x^\nu}, \quad \Lambda_\mu{}^\nu = \frac{\partial x^\nu}{\partial x'^\mu}.$$

In these definitions the order of the superscript and subscript is significant, because in general (note that the offsets between subscript and superscript change) $\Lambda^\mu{}_\nu \neq \Lambda_\nu{}^\mu$. The product

$$\Lambda^\alpha{}_\nu \Lambda_\beta{}^\nu = \delta^\alpha_\beta \tag{4.2}$$

where the Kronecker delta $\delta^\alpha{}_\beta$ is defined to be unity if α is equal to β, but zero otherwise. In the Kronecker δ the offset between subscript and superscript is superfluous.

The local vectors discussed in Section 3.3 can be expressed in terms of a set of basis vectors e_μ drawn along the local spacetime coordinate directions at an event P. Here the subscript μ on e_μ indicates that this basis vector points in the μ direction. For example if e_1 is of unit length its components would be 0, 1, 0, and 0 along the ct, x, y, and z directions, respectively; we would write these components $(e_1)^\alpha$. If P′ is separated from P by an infinitesimal distance $\mathrm{d}x^\mu$ the PP ′ is the local vector

$$\mathrm{d}x^\mu e_\mu.$$

The invariant length of PP′ is

$$\mathrm{d}s^2 = (\mathrm{d}x^\mu e_\mu) \cdot (\mathrm{d}x^\nu e_\nu) = e_\mu \cdot e_\nu \, \mathrm{d}x^\mu \, \mathrm{d}x^\nu,$$

where $\mathbf{e}_\mu \cdot \mathbf{e}_\nu$ is a more general form of a scalar product. In Minkowski spacetime with orthogonal coordinates ct, x, y and z, by comparison with Equation (3.29), we have

$$e_\mu \cdot e_\nu = \eta_{\mu\nu}.$$

With a general metric $g_{\mu\nu}$ as in Equation (3.36) we have

$$e_\mu \cdot e_\nu = g_{\mu\nu}. \tag{4.3}$$

All local four-vectors at the same event in spacetime can be expressed in terms of the same set of basis vectors: the four-momentum can be written

$$p^\mu e_\mu.$$

The components of the momentum (or any other local four-vector at the same event) will have to behave under general transformations just like the components $\mathrm{d}x^\mu$:

$$p'^\mu = \frac{\partial x'^\mu}{\partial x^\nu} p^\nu.$$

This result has implications for the invariant length of a four-vector. In the case of four-momentum this is given in a frame in free fall where SR is valid by

$$m^2c^2 = \eta_{\mu\nu} p^\mu p^\nu, \text{ and } \mathrm{d}s^2 = \eta_{\mu\nu}\, \mathrm{d}x^\mu\, \mathrm{d}x^\nu.$$

Under a general transformation in the new coordinates

$$\mathrm{d}s^2 = g_{\mu\nu}\, \mathrm{d}x^\mu\, \mathrm{d}x^\nu \text{ and } m^2c^2 = g_{\mu\nu} p^\mu p^\nu,$$

so the invariant length of *any* local four-vector is unaffected by general transformations.

Next, some examples of transformations of interest are given, using orthonormal local coordinates. The first is the rotation through θ around the z-axis. Locally it has the same form as a rotation in SR:

$$\Lambda^\mu_{\ \nu} = \begin{bmatrix} 1 & 0 & 0 & 0 \\ 0 & \cos\theta & \sin\theta & 0 \\ 0 & \sin\theta & -\cos\theta & 0 \\ 0 & 0 & 0 & 1 \end{bmatrix}.$$

A Lorentz transformation to a frame with velocity βc along the x-axis has the form

$$\Lambda^\mu_{\ \nu} = \begin{bmatrix} \cosh u & -\sinh u & 0 & 0 \\ -\sinh u & \cosh u & 0 & 0 \\ 0 & 0 & 1 & 0 \\ 0 & 0 & 0 & 1 \end{bmatrix},$$

where $\cosh u = \gamma$, $\sinh u = \beta\gamma$ and $\gamma = 1/(1-\beta^2)^{1/2}$. Finally, consider the transformation between frames momentarily coincident in velocity near a massive spherical body. The first frame is a frame in radial free fall, and the second is a frame at rest:

$$\Lambda^\mu_{\ \nu} = \begin{bmatrix} (1+2\varphi/c^2)^{1/2} & 0 & 0 & 0 \\ 0 & 1/(1+2\varphi/c^2)^{1/2} & 0 & 0 \\ 0 & 0 & 1 & 0 \\ 0 & 0 & 0 & 1 \end{bmatrix}$$

where the 1-axis is radial and φ is the gravitational potential.

4.2 Vector and Covector Components

Thus far all vectors have been written in terms of the familiar vector components, such as $dx^0 = c dt$, $dx^1 = dx$, $dx^2 = dy$ and $dx^3 = dz$. Other components, called covector components, will now be introduced. To give a physical connection we shall consider a conservative force in classical mechanics and only need to consider the spatial dimensions. The components of a conservative force are given by

$$f_i = -\frac{\partial \varphi}{\partial x^i}$$

where φ is the potential, which is a scalar function of position only. At any given location the value of φ is invariant under transformations, i.e., it remains the same. Thus the effect of a transformation on the components of f_i is given by

$$f_i' = -\frac{\partial \varphi}{\partial x'^i} = -\frac{\partial \varphi}{\partial x^j}\frac{\partial x^j}{\partial x'^i} = \frac{\partial x^j}{\partial x'^i} f_j \tag{4.4}$$

This is the inverse of the transformation given in Equation (4.1); f_i transforms like d/dx^i rather than like dx^i. Components which transform in this new way are called *covector* or *covariant* components; they are written with a subscript. Components that transform in the manner of Equation (4.1) are called *vector* or *contravariant* components. The f_i are the components of the gradient of ϕ, and so a useful physical view is that the covector components make up a *gradient*. Consider next a curve in space $x^\mu(s)$ where s measures the curve length from some reference point along it. At any location on the curve the components dx^μ/ds make up a tangent vector, and this transforms according to Equation (4.1). Therefore a physical interpretation of vector components is that they make up a *tangent*. Next we shall show that *any* vector can be expressed in terms of either its vector or covector components. Depending on the context one set or other may be of more physical importance.

The local interval PP$'$ can be taken from Equation (3.36)

$$ds^2 = g_{\mu\nu}\, dx^\mu\, dx^\nu.$$

Let us write:

$$dx_\mu = g_{\mu\nu}\, dx^\nu \tag{4.5}$$

and then

$$ds^2 = dx_\mu\, dx^\mu. \tag{4.6}$$

The quantity ds^2 is invariant under any general transformation; hence

$$ds^2 = dx'_\alpha\, dx'^\alpha = dx'_\alpha \frac{\partial x'^\alpha}{\partial x^\mu} dx^\mu. \tag{4.7}$$

Comparing Equations (4.6) and (4.7) it is seen that

$$dx'_\alpha = \frac{\partial x^\mu}{\partial x'^\alpha} dx_\mu.$$

In other words, the dx_μ are the covector components of PP ′. Thus Equation (4.5) is the general way of generating covector components from vector components.

A two-dimensional example can be used to illustrate these results. Figure 4.1 shows a local vector PP′ with the coordinate axes inclined at an angle θ. The vector component of PP ′ on the 1-axis is obtained by projecting from P′ parallel to the 2-axis, and the 2-component by projecting parallel to the 1-axis. Then in terms of unit vectors $\boldsymbol{e}_1$ and $\boldsymbol{e}_2$ along the axes:

$$\mathrm{PP}' = \mathbf{e}_1\, dx^1 + \mathbf{e}_2\, dx^2$$

The covector components given by Equation (4.5) in two dimensions:

$$dx_1 = g_{11}\, dx^1 + g_{12}\, dx^2.$$

Then using Equation (4.3) we obtain

$$dx_1 = dx^1 + dx^2 \cos\theta, \text{ and } dx_2 = dx^2 + dx^1 \cos\theta.$$

Figure 4.1 shows that these covector components dx_1 and dx_2 can be obtained by projecting perpendicularly from P′ onto the relevant axis. How this generalizes in Euclidean spaces of higher dimension is illustrated in Figure 4.2. The vector component of PP ′ is obtained by projecting from P′ onto the 1-direction over the surface through P′ that contains all the other local basis vectors: dx^1 is SP. The covector component is simply the perpendicular projection from P′ onto the 1-direction: dx_1 is S ′ P. With rectangular Cartesian coordinates in Euclidean space the distinction between vector and covector components disappears, which explains why covector components are not usually used in Newtonian mechanics. Direct visualization of vector and covector components in spacetime meets difficulties with the time components because we can only draw Euclidean spaces in our three-dimensional world. The covector components of local four-vectors are given by expressions similar to Equation (4.3). Other related quantities, the scalar products of pairs of four-vectors, are also invariant. If the lowering property of $g_{\mu\nu}$ is applied in turn to A^μ and B^ν we obtain

$$g_{\mu\nu} A^\mu B^\nu = A_\nu B^\nu = A^\mu B_\mu. \tag{4.8}$$

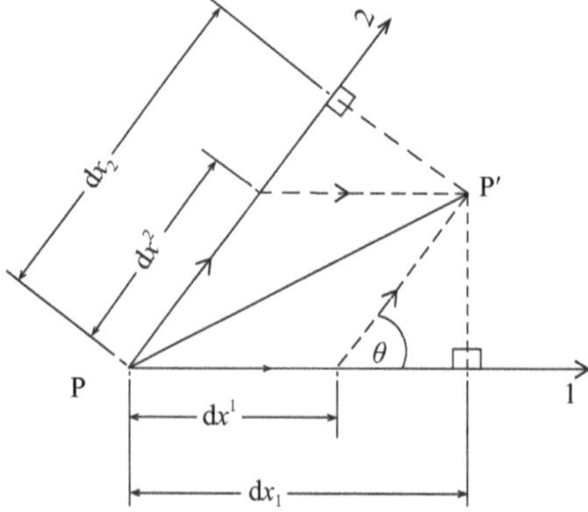

Figure 4.1. The vector and covector components of an infinitesimal vector PP ′ in a two-dimensional space.

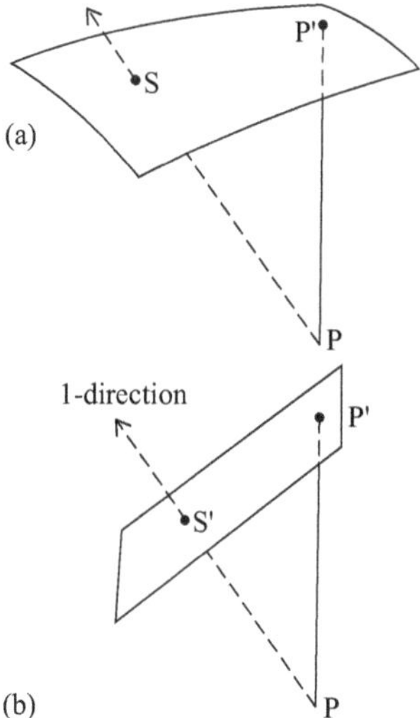

Figure 4.2. The method for constructing (a) the vector and (b) the covector 1-component of the infinitesimal vector PP $'$ in a multi-dimensional space. The surface drawn would be a normal plane where the space is flat and three-dimensional, but would be a hypersurface in higher-dimensional spaces.

4.3 Other Tensors

The metric tensor appearing in Equation (3.36) is one example of a tensor with two indices. Another such tensor which will be of great interest is the stress–energy tensor $T^{\mu\nu}$ that summarizes the energy content of matter. Tensors having n indices are called tensors of *rank n*. Vectors are rank 1 and scalars are rank 0 tensors. The simplest examples of higher rank tensors can be formed by taking products of components of vectors. Suppose that a^μ, b^μ, d_μ and e_μ are vector and covector components. Then

$$D^{\mu}{}_{\nu\sigma} = a^\mu d_\nu e_\sigma \tag{4.9}$$

has rank 3. General transformations do not alter the invariant interval ds^2. Thus

$$ds^2 = g'_{\mu\nu}\, dx'^{\mu}\, dx'^{\nu} = g'_{\mu\nu}\frac{\partial x'^{\mu}}{\partial x^\rho}\frac{\partial x'^{\nu}}{\partial x^\sigma}dx^\rho dx^\sigma. \tag{4.10}$$

In terms of the original coordinates

$$ds^2 = g_{\rho\sigma}\, dx^\rho\, dx^\sigma.$$

Comparing these last two equations gives

$$g_{\rho\sigma} = g'_{\mu\nu}\frac{\partial x'^{\mu}}{\partial x^\rho}\frac{\partial x'^{\nu}}{\partial x^\sigma}.$$

The metric tensor requires one transformation for each of its indices. This is also true for any of the tensors listed above; for example

$$D'^{\mu}_{\ \nu\sigma} = \Lambda^{\mu}_{\ \alpha}\Lambda_{\nu}^{\ \beta}\Lambda_{\sigma}^{\ \gamma}D^{\alpha}_{\ \beta\gamma}. \tag{4.11}$$

We see that each vector index brings a factor $\Lambda^{*}{}_{*}$ while each covector index brings a factor $\Lambda_{*}{}^{*}$. Valid identities between tensors only connect tensors with equal numbers of subscripts and equal numbers of superscripts (the same rank).

Furthermore, if the components of two tensors can be shown to be equal in a particular coordinate system, then the equality holds for all coordinate frames including, of course, accelerating frames. This is the vital property that makes a tensor presentation of GR so appropriate.

There are a number of other tensor properties that are needed in Chapters 5 and 6. A frequent tensor manipulation is the process of contraction. Consider a tensor $A^{\alpha\beta}_{\gamma\beta}$ where the summation is implied for all values of the repeated index β, i.e., the tensor in full is

$$A^{\alpha 0}_{\gamma 0} + A^{\alpha 1}_{\gamma 1} + A^{\alpha 2}_{\gamma 2} + A^{\alpha 3}_{\gamma 3}.$$

Applying the usual procedure to obtain this in another frame we obtain

$$A'^{\mu\nu}_{\sigma\nu} = \Lambda^{\mu}_{\ \alpha}\,\Lambda^{\nu}_{\ \beta}\,\Lambda_{\nu}^{\ \delta}\,\Lambda_{\sigma}^{\ \gamma}\,A^{\alpha\beta}_{\gamma\delta} = \Lambda^{\mu}_{\ \alpha}\,\delta^{\delta}_{\beta}\,\Lambda_{\sigma}^{\ \gamma}\,A^{\alpha\beta}_{\gamma\delta} = \Lambda^{\mu}_{\ \alpha}\,\Lambda_{\sigma}^{\ \gamma}\,A^{\alpha\beta}_{\gamma\beta}, \tag{4.12}$$

so that $A^{\alpha\beta}_{\gamma\beta}$ transforms as a rank 2 tensor. One more contraction gives $A^{\alpha\beta}_{\alpha\beta}$, which is a scalar quantity. The scalar products met in vector analysis are familiar examples of contraction. Other tensor products involving contraction are

$$\begin{aligned} &A^{\mu\nu}B_{\nu}, \text{ which is a vector} \\ &C^{\mu\nu\sigma}B^{\alpha}_{\sigma}, \text{ which is a rank 3 tensor.} \end{aligned} \tag{4.13}$$

Not all collections of numbers or functions labeled with suffixes "$F_{\mu\nu}$" constitute a tensor. Whether or not "$F_{\mu\nu}$" constitutes a tensor can be tested using the *quotient theorem*. This theorem states that if the product of $F^{\mu\nu}$ with any arbitrary tensor is also a tensor, then $F^{\mu\nu}$ is itself a tensor.

Finally we collect here some useful properties of the metric tensor. The metric tensor is symmetric under interchange of its subscripts. An antisymmetric part's contribution to ds^2 would be

$$(g_{\mu\nu} - g_{\nu\mu})\,dx^{\mu}\,dx^{\nu}/2,$$

which vanishes identically, and so it can safely be neglected. Contraction with $g_{\mu\nu}$ is used to generate *associated* tensors. For example,

$$A_\nu = g_{\nu\mu}A^\mu \text{ and } A_{\mu\nu} = g_{\mu\alpha}g_{\nu\beta}A^{\alpha\beta}.$$

The associated tensor of $g_{\mu\nu}$ itself is also of importance.

$$g_{\alpha\mu}g^{\mu\nu} = \delta^\nu_\alpha. \tag{4.14}$$

$g^{\mu\nu}$ must be symmetric if $g_{\mu\nu}$ is symmetric. Now consider the associated tensors A_μ and A^μ:

$$A_\alpha = g_{\alpha\mu}A^\mu$$

and multiply by $g^{\beta\alpha}$. This gives

$$g^{\beta\alpha}A_\alpha = g_{\alpha\mu}g^{\beta\alpha}A^\mu = g_{\alpha\mu}g^{\alpha\beta}A^\mu = \delta^\beta_\mu A^\mu = A^\beta. \tag{4.15}$$

Hence the effect of $g^{\beta\alpha}$ is to *raise* a subscript while the effect of contracting a tensor with $g_{\alpha\beta}$ is to *lower* a superscript. In the case that $g_{\alpha\beta}$ is diagonal (e.g., for the Schwarzschild metric), with no summation implied over α in this case

$$g^{\alpha\alpha} = 1/g_{\alpha\alpha}. \tag{4.16}$$

For completeness the transformation of the Kronecker delta $\delta^\nu{}_\mu$ needs to be discussed. Under the general transformation

$$(\delta')^\alpha{}_\beta = \frac{\partial x'^\alpha}{\partial x^\nu}\frac{\partial x^\mu}{\partial x'^\beta}\delta^\nu{}_\mu = \frac{\partial x'^\alpha}{\partial x^\mu}\frac{\partial x^\mu}{\partial x'^\beta} = \delta^\alpha{}_\beta, \tag{4.17}$$

which demonstrates that this tensor is the same in all frames.

4.4 Exercises

1. With the geometry of Figure 4.1 show that the distance PP $'$ is given by

$$\mathrm{d}s^2 = (\mathrm{d}x^1)^2 + (\mathrm{d}x^2)^2 + 2\,\mathrm{d}x^1\,\mathrm{d}x^2\cos\theta$$

$$= \mathrm{d}x_1\,\mathrm{d}x^1 + \mathrm{d}x_2\,\mathrm{d}x^2.$$

2. Show that $g^{\alpha\beta}g_{\alpha\beta} = n$ for an n-dimensional space.
3. Determine the transformation matrix Λ required to change from Cartesian coordinates (x,y) to polar coordinates (r,θ). Check that $\Lambda^\alpha_\nu\Lambda^\nu_\beta = \delta^\alpha_\beta$.
4. If the product of some function $F_{\alpha\beta}$ with an arbitrary tensor is a tensor show that $F_{\alpha\beta}$ is also a tensor. Try using a vector A^γ as the arbitrary tensor.

Further Reading

Laugwitz D 1965 *Differential and Riemannian Geometry* (New York: Academic). This provides a thorough review of these subjects.

AAS | IOP Astronomy

Introduction to General Relativity and Cosmology (Second Edition)

Ian R Kenyon

Chapter 5

Einstein's Theory I

The theme of this chapter is to present physical laws in a form valid under general transformations to accelerated frames. According to the principle of equivalence the physical laws in any frame in free fall are consistent with SR. Then in Chapter 3 we learnt that physical laws expressed as tensor equations automatically retain their form under general transformations.

> These two ideas are now fused using the *principle of generalized covariance*, first postulated by Einstein. This states that physical laws are expressible as tensor equations that reduce to laws consistent with SR in a frame in free fall. Hence a law valid in SR and expressed in tensors will apply in accelerating frames too. Most laws are already expressed in terms of vectors, which are tensors, so we are part way there. However the spacetime derivatives which appear in dynamical relations (e.g., the rate of change of momentum) are *not* tensors. These derivatives must be converted to *covariant derivatives* that are tensors and which reduce to the usual spacetime derivative in a frame in free fall.

Covariant derivatives are introduced here and are found to depend on entities called *metric connections* which quantify how local vectors change when they are transported through curved spacetime. These metric connections are then evaluated in terms of the derivatives of the metric coefficients $g_{\mu\nu}$. Newton's second law of motion is the fundamental dynamical equation, and we re-write it in a form valid in accelerating frames. In the absence of non-gravitational forces the second law describes a body in free fall and we show that such a path is a *geodesic in spacetime*. A procedure for transforming to a frame in free fall is then described. Geodesics are *stationary* paths in spacetime with an integral equation that will be used later when calculating the trajectories of planets and light in curved spacetime. Finally the correspondence between metric connections and the inertial and gravitational forces of classical Newtonian mechanics is explored.

5.1 The Covariant Derivative

We consider the change in a local four-vector q^μ during some physical process. A concrete choice could be the four-momentum of a test body acted on by gravitational and non-gravitational forces. In Figure 5.1 the broken line indicates the path followed by a test body in curved spacetime, where s is the path length and $\mathbf{e}_\mu$ is a basis vector in the μ-direction. Changes in the coordinate frame along the path mean that the component q^μ would change, independent of any physical process. This change must be subtracted from the observed change to obtain the change that is the physical effect of any non-gravitational interaction experienced by the body. Over an element of path $\mathrm{d}s$ let the observed change be $\mathrm{d}q^\mu$. The change due solely to the curvature of spacetime will be linear in the distance, that is in the change in the components of $\mathrm{d}s$ along the axes, $\mathrm{d}x^\rho$. It will also be linear in the magnitude of the components of the vector q^σ, so the change takes the form

$$\delta q^\mu = -\Gamma^\mu_{\sigma\rho} q^\sigma \mathrm{d}x^\rho. \tag{5.1}$$

The new quantity $\Gamma^\mu_{\sigma\rho}$ introduced here is called the *metric connection*, and is clearly a function of the position in spacetime. All components have to be included to take account of the varieties of possible spacetime curvature: for example, the 1-component can change during a displacement in the 2-direction. Thus the change in q^μ due to the physical interaction is

$$\mathrm{d}q^\mu + \Gamma^\mu_{\sigma\rho} q^\sigma \mathrm{d}x^\rho. \tag{5.2}$$

The physical change in q^μ per unit path length at a point in spacetime is obtained by dividing this expression by $\mathrm{d}s$ and taking the limit as $\mathrm{d}s \to 0$. This gives the covariant derivative, a valid tensor (in this case a vector)

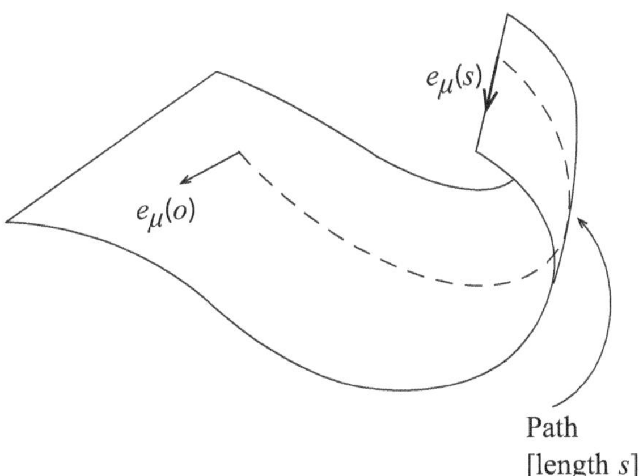

Figure 5.1. A path across a curved surface; $\mathbf{e}_\mu$ is a local basis vector drawn at two points along this path.

$$\frac{\mathrm{D}q^{\mu}}{\mathrm{D}s} = \frac{\mathrm{d}q^{\mu}}{\mathrm{d}s} + \Gamma^{\mu}{}_{\sigma\rho}q^{\sigma}\left(\frac{\mathrm{d}x^{\rho}}{\mathrm{d}s}\right). \tag{5.3}$$

$\mathrm{D}q^{\mu}/\mathrm{D}s$ is the μ-component of the covariant derivative of the vector with components q^{μ}. It is quite easy to see that the connections are not themselves tensors. In a frame in free fall the covariant derivative is the same as the standard derivative, so that the metric connection vanishes. In any other frame the connection is non-zero so it cannot be a tensor.

Expressions for covariant derivatives of tensors of any rank can be obtained, all of which are also tensors. For covector components (see problem 5.1)

$$\frac{\mathrm{D}q_{\mu}}{\mathrm{D}s} = \frac{\mathrm{d}q_{\mu}}{\mathrm{d}s} - \Gamma^{\nu}{}_{\mu\rho}\, q_{\nu}\frac{\mathrm{d}x^{\rho}}{\mathrm{d}s}.$$

The covariant derivative of a rank-2 tensor, $\mathrm{A}_{\mu\nu}$, is evaluated by considering the product with vectors $A_{\mu\nu}q^{\mu}q^{\nu}$. This product is invariant under parallel transport in free fall so that

$$\delta\left[A_{\mu\nu}q^{\mu}q^{\nu}\right] = 0.$$

Expanding this using Equation (5.1) gives

$$\frac{\mathrm{DA}_{\mu\nu}}{\mathrm{D}s} = \frac{\mathrm{dA}_{\mu\nu}}{\mathrm{d}s} - \Gamma^{\tau}{}_{\mu\rho}\mathrm{A}_{\tau\nu}\frac{\mathrm{d}x^{\rho}}{\mathrm{d}s} - \Gamma^{\tau}{}_{\nu\rho}\mathrm{A}_{\mu\tau}\frac{\mathrm{d}x^{\rho}}{\mathrm{d}s}$$

5.2 The Calculation of the Metric Connection

The full information about the structure of spacetime is embodied in the metric equation, and so it is reasonable to expect that the metric connections should be functions of the metric coefficients $g_{\mu\nu}$. Replacing $\mathrm{A}_{\mu\nu}$ in the previous equation by $g_{\mu\nu}$ and multiplying by $\partial s/\partial x^{\rho}$ gives

$$\frac{\mathrm{D}g_{\mu\nu}}{\mathrm{D}x^{\rho}} = \frac{\partial g_{\mu\nu}}{\partial x^{\rho}} - \Gamma^{\tau}{}_{\mu\rho}g_{\tau\nu} - \Gamma^{\tau}{}_{\nu\rho}g_{\mu\tau}. \tag{5.4}$$

Now let us specialize to the frame in free fall. There the connections vanish so that

$$\frac{\mathrm{D}g_{\mu\nu}}{\mathrm{D}x^{\rho}} = \frac{\partial g_{\mu\nu}}{\partial x^{\rho}}.$$

In addition the metric of the frame in free fall is locally that of SR. As discussed in Chapter 3 this condition is written

$$g_{\mu\nu} = \eta_{\mu\nu} \text{ and } \frac{\partial g_{\mu\nu}}{\partial x^{\rho}} = 0.$$

Then for a frame in free fall we have

$$\frac{\mathrm{D}g_{\mu\nu}}{\mathrm{D}x^{\rho}} = 0,$$

which, being a tensor equation, must hold in all frames. Then using Equation (5.4) we have in general:

$$\frac{\partial g_{\mu\nu}}{\partial x^\rho} = \Gamma^\tau{}_{\mu\rho} g_{\tau\nu} + \Gamma^\tau{}_{\nu\rho} g_{\mu\tau}.$$

From here onward we write $\Gamma^\tau{}_{\nu\rho}$ as $\Gamma^\tau_{\nu\rho}$ for compactness. Now define

$$\Gamma_{\mu\nu\rho} = g_{\mu\tau} \Gamma^\tau{}_{\nu\rho}, \tag{5.5}$$

and the previous equation can be written

$$\frac{\partial g_{\mu\nu}}{\partial x^\rho} = \Gamma_{\nu\mu\rho} + \Gamma_{\mu\nu\rho}.$$

For convenience we introduce the *subscript comma* notation to represent a derivative:

$$g_{\mu\nu,\rho} = \frac{\partial g_{\mu\nu}}{\partial x^\rho}.$$

Then the previous line becomes

$$g_{\mu\nu,\rho} = \Gamma_{\nu\mu\rho} + \Gamma_{\mu\nu\rho}. \tag{5.6}$$

Equation (5.6) can be inverted to give

$$2\Gamma_{\nu\mu\rho} = g_{\mu\nu,\rho} - g_{\rho\mu,\nu} + g_{\nu\rho,\mu}. \tag{5.7}$$

This expression is called the *fundamental theorem* of Riemannian geometry. It is clear that the connections are symmetric under interchange of the last two indices:

$$\Gamma_{\sigma\rho\mu} = \Gamma_{\sigma\mu\rho}. \tag{5.8}$$

The reader can check Equation (5.7) by substituting for $\Gamma_{\nu\mu\rho}$ and $\Gamma_{\mu\nu\rho}$ in Equation (5.6).

5.3 More on the Covariant Derivative

The covariant derivative can be used to describe the differential change of any local vector along a given path, because local vectors all transform in the same way. There are a number of useful forms equivalent to Equation (5.3). First s can be replaced by $c\tau$, where τ is the proper time measured on a clock traveling the same path. When the path is that of light some alternative parameter to τ can always be defined to specify the path consistently. Thus the covariant derivative can be rewritten in terms of a general path parameter λ, which is linearly related to τ when the path is non light-like:

$$\frac{Dq^\mu}{D\lambda} = \frac{dq^\mu}{d\lambda} + \Gamma^\mu_{\nu\rho}\frac{dx^\nu}{d\lambda}q^\rho. \tag{5.9}$$

For covariant components

$$\frac{Dq_\mu}{D\lambda} = \frac{dq_\mu}{d\lambda} - \Gamma^\nu_{\mu\rho}\frac{dx^\rho}{d\lambda}q_\nu.$$

Multiplying Equation (5.3) by $\partial s/\partial x^\nu$ gives another form:

$$\frac{Dq^\mu}{Dx^\nu} = \frac{\partial q^\mu}{\partial x^\nu} + \Gamma^\mu_{\nu\rho}q^\rho. \tag{5.10}$$

This can be written more compactly by using the subscript comma notation for derivatives, and introducing the *subscript semicolon* notation for covariant derivatives:

$$q^\mu_{;\nu} = \frac{Dq^\mu}{Dx^\nu}.$$

Then Equation (5.10) becomes

$$q^\mu_{;\nu} = q^\mu_{,\nu} + \Gamma^\mu_{\nu\rho}q^\rho.$$

When a vector is *parallel transported* along a path (as described in Chapter 3) it remains the same at each point on its path; thus

$$\frac{Dq^\mu}{Ds} = 0, \tag{5.11}$$

i.e.,

$$\frac{dq^\mu}{ds} = -\Gamma^\mu_{\nu\rho}q^\rho\frac{dx^\nu}{ds},$$

i.e.,

$$dq^\mu = -\Gamma^\mu_{\nu\rho}q^\rho\, dx^\nu \tag{5.12}$$

reproducing Equation (5.1).

5.4 The Principle of Generalized Covariance

This principle postulated by Einstein has two components. First, physical laws must be expressible as tensor equations so that they remain valid under transformations to any accelerated frame. Second, when specialized to a frame in free fall the physical laws should reproduce the established laws consistent with SR. Fundamental to any study of dynamics is Newton's second law. Expressed in a form consistent with SR we have

$$F^\mu = \frac{\mathrm{d}p^\mu}{\mathrm{d}\tau}$$

where F^μ is the four-vector force. Although the right-hand side is not a valid tensor, we have just learnt how to write a covariant derivative that is a valid tensor. Replacing the derivative on the right-hand side of this equation by a covariant derivative yields

$$F^\mu = \frac{\mathrm{D}p^\mu}{\mathrm{D}\tau}. \tag{5.13}$$

This is a valid tensor equation *and* reduces to SR form in a frame in free fall, so that it satisfies the principle of generalized covariance. Any other physical law consistent with SR in a frame in free fall can be converted to a physical law valid in accelerating frames by replacing spacetime derivatives by the equivalent covariant derivatives. Clearly we have here a very powerful tool.

Using the subscript comma and subscript semicolon notation a standard spacetime derivative is written $p^\mu_{;\nu}$, while a covariant derivative is written $p^\mu_{;\nu}$. Hence the procedure for converting an equation so that it becomes valid in accelerating frames can be expressed pithily as: "replace commas by semicolons."

5.5 The Geodesic Equation

The simplest dynamical question that can be posed for curved spacetime is to ask what the motion of a test body is in free fall, i.e., under gravitational forces alone. The gravitational forces are the manifestations of spacetime curvature due to the presence of matter. Therefore in Equation (5.13) F^μ must be set to zero:

$$\frac{\mathrm{D}p^\mu}{\mathrm{D}\tau} = 0. \tag{5.14}$$

Referring to Section 5.3 we see that p^μ is being parallel transported. In addition p^μ, being the test body's momentum, points along the path. p^μ is therefore being parallel transported along itself, which is the prescription given in Section 3.3 for generating a geodesic! Hence the geometric view of the path of a body in free fall is that it follows a geodesic in spacetime. Equation (5.14) is known as the *geodesic equation.* It can be converted into a useful equivalent form by making the replacement

$$p^\mu = m\frac{\mathrm{d}x^\mu}{\mathrm{d}\tau}$$

where m is the rest mass of the test body. Then the geodesic equation becomes

$$\frac{\mathrm{d}^2x^\mu}{\mathrm{d}\tau^2} + \Gamma^\mu_{\nu\rho}\frac{\mathrm{d}x^\nu}{\mathrm{d}\tau}\frac{\mathrm{d}x^\rho}{\mathrm{d}\tau} = 0. \tag{5.15}$$

If we wish to extend the equation to describe geodesics followed by light, we write

$$\frac{\mathrm{d}^2x^\mu}{\mathrm{d}\lambda^2} + \Gamma^\mu_{\nu\rho}\frac{\mathrm{d}x^\nu}{\mathrm{d}\lambda}\frac{\mathrm{d}x^\rho}{\mathrm{d}\lambda} = 0 \tag{5.16}$$

where λ is the path parameter discussed in Section 5.3. Notice that the geodesic Equation (5.15) does not depend on the mass of the test body. This shows that the weak equivalence principle is already built into the very structure of Riemann spacetime.

In Section 3.3 an alternative definition given for a geodesic was of a curve across space with no component of curvature in that space. This property of a geodesic is inherent in the geodesic equation, as we shall now show. The differential $\mathrm{d}x^\mu/\mathrm{d}s$ is a valid vector so we can write

$$\frac{\mathrm{D}x^\mu}{\mathrm{D}s} = \frac{\mathrm{d}x^\mu}{\mathrm{d}s}.$$

Then

$$\frac{\mathrm{D}^2x^\mu}{\mathrm{D}s^2} = \frac{\mathrm{D}}{\mathrm{D}s}\frac{\mathrm{d}x^\mu}{\mathrm{d}s} = \frac{\mathrm{d}^2x^\mu}{\mathrm{d}s^2} + \Gamma^\mu_{\nu\rho}\frac{\mathrm{d}x^\nu}{\mathrm{d}s}\frac{\mathrm{d}x^\rho}{\mathrm{d}s},$$

so that

$$\frac{\mathrm{D}^2x^\mu}{\mathrm{D}s^2} = 0 \tag{5.17}$$

is the compact form of the geodesic equation. In geometrical terms the $\mathrm{D}x^\mu/\mathrm{D}s$ are gradient components and the $\mathrm{D}^2x^\mu/\mathrm{D}s^2$ the curvature components of the path. Therefore the geodesic equation implies that a geodesic has no component of curvature in spacetime.

5.6 Geodesics as Stationary Paths

From the geometric point of view geodesics are the straightest paths between spacetime events; physically they are the paths of test bodies in free fall. A third equivalent property of geodesics is that they are *stationary* paths between spacetime events. This final property is useful because it simplifies the calculation of general relativistic orbits. In flat spacetime a straight line is the shortest path between two points: $S_{\mathrm{AC}} < S_{\mathrm{AB}} + S_{\mathrm{BC}}$ as in Figure 5.2(a). The corresponding case for Minkowski space is shown in Figure 5.2(b) for AC being time-like. A suitable Lorentz transformation makes the time axis lie along AC and then

$$S_{\mathrm{AC}} = ct_{\mathrm{AC}}, \quad S_{\mathrm{AB}} = \left(c^2t_{\mathrm{AB}}^2 - x^2\right)^{1/2}, \quad S_{\mathrm{BC}} = \left(c^2t_{\mathrm{BC}}^2 - x^2\right)^{1/2},$$

where x is the spatial coordinate of B. Then $S_{\mathrm{AC}} > S_{\mathrm{AB}} + S_{\mathrm{BC}}$. The straight line path is therefore the longest path in Minkowski space. A description that covers both cases is to say that geodesics are stationary paths; this means that any small deviation of path from the geodesic produces no change in length to first order in the deviation. Expressed formally this requirement is that

$$\delta S = 0$$

where the integrated path length is

$$S = \int \left(g_{\mu\nu}\,\mathrm{d}x^\mu\,\mathrm{d}x^\nu\right)^{1/2}.$$

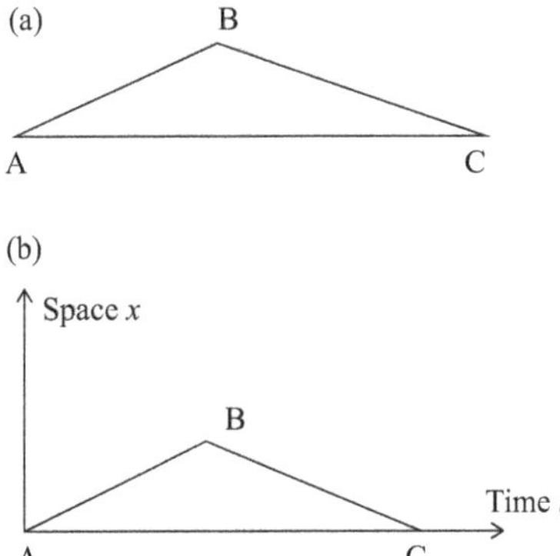

Figure 5.2. (a) A straight line path AC in space and another longer path ABC; (b) a straight line path in spacetime AC, and a displaced path ABC.

If the path length is parametrized by the proper path length ds then

$$S = \int \left(g_{\mu\nu} \frac{\mathrm{d}x^\mu}{\mathrm{d}s} \frac{\mathrm{d}x^\nu}{\mathrm{d}s} \right)^{1/2} \mathrm{d}s.$$

This integrand is unity all along the path so that if S is stationary then so too is a simpler integral

$$\delta I = 0 \tag{5.18}$$

where

$$I = \int \left(g_{\mu\nu} \frac{\mathrm{d}x^\mu}{\mathrm{d}s} \frac{\mathrm{d}x^\nu}{\mathrm{d}s} \right) \mathrm{d}s. \tag{5.19}$$

Equations (5.18) and (5.19) set a variational problem that is solved in the standard way in Appendix B. There it is shown that these equations are the integral form of the geodesic equation, entirely equivalent to Equation (5.10).

It is important to appreciate that a body in free fall from a fixed starting event will follow different geodesics if it is given different starting velocities. These geodesics will lie inside the forward light cone through the starting event; they are *time-like* with $\int \mathrm{d}s^2 > 0$. When the test body is a photon the path integral $\int \mathrm{d}s^2 = 0$, which defines a *null geodesic*. Finally the *space-like geodesics* that have $\int \mathrm{d}s^2 < 0$ would correspond to motion with velocity greater than c. Neither material particles nor light can follow space-like geodesics; however these geodesics are useful in setting up coordinate frames that span spacetime.

5.7 Familiar Quantities

In this section some familiar examples of gravitational and inertial forces will be used to bring out the correspondence between these forces and the metric connections. First let us establish how the geodesic equation simplifies when the motion can be described approximately by classical mechanics. Then the velocities must be small compared with c, the gravitational fields must be small and their rates of change must be slow. The geodesic equation is

$$\frac{d^2x^\mu}{ds^2} + \Gamma^\mu_{\alpha\beta}\frac{dx^\alpha}{ds}\frac{dx^\beta}{ds} = 0.$$

If the velocities are small, then

$$dx^0 = c\ dt \gg dx^i.$$

Therefore dx^0/ds is much larger than the derivatives of space components and in addition $dx^0/d\ s \approx 1$. Thus the dominant components of the geodesic equation are

$$\frac{d^2x^\mu}{dt^2} + c^2\Gamma^\mu_{00} = 0,$$

provided that not all the $\Gamma^\mu{}_{00}$ are negligible. The time component

$$\Gamma^0_{00} = g_{00}\frac{(dg_{00}/dt)}{2c}$$

is negligible in the classical limit of fields that vary slowly with time. Thus, provided that the $\Gamma^i{}_{00}$ are not negligible the dominant components of the geodesic equation in the classical limit are

$$\frac{d^2x^i}{dt^2} + c^2\Gamma^i_{00} = 0. \tag{5.20}$$

Now if there is a constant acceleration g in the i-direction we would write classically

$$\frac{d^2x^i}{dt^2} = g.$$

A comparison of the last two equations yields

$$\Gamma^i_{00} = -g/c^2$$

and shows the equivalence between an inertial force of classical mechanics and the metric connections. Both can be made to vanish by an appropriate coordinate transformation. If g is taken to be the local gravitational acceleration then we can see that the metric connections correspond to the components of the gravitational acceleration. This is a reasonable result when it is recalled that the metric components correspond to the classical gravitational potential (Section 3.7). More generally, if there is a gravitational potential φ, we have classically

$$\frac{d^2x^i}{dt^2} = -\frac{\partial\varphi}{\partial x^i}.$$

Comparison with Equation (5.20) shows that

$$\Gamma^i_{00} = (\partial\varphi/\partial x^i)/c^2.$$

It is emphasized here that no rotation is considered in a frame in free fall. Throughout what follows, either Cartesian or spherical polar coordinates are used when analyzing motion in free fall: the latter choice is made whenever the system has spherical symmetry.

5.8 Exercises

1. Show that for covector components the second term of Equation (5.10) becomes negative:

$$\frac{\mathrm{D}p_\mu}{\mathrm{D}x^\nu} = \frac{\partial p_\mu}{\partial x^\nu} - \Gamma^\rho_{\nu\mu}p_\rho$$

   (Take a scalar quantity $\phi = p_\alpha q^\alpha$ and calculate its derivative).
2. Show that a geodesic that is time-/space-/light-like at a given point remains time-/space-/light-like on its journey through spacetime. Hint: consider parallel-transporting a vector A_μ and show that $A_\mu A^\mu$ is constant.
3. Take a space-like geodesic in Minkowski space and show that it is neither maximal nor minimal.
4. The metric equation for free fall is

$$\mathrm{d}s^2 = c^2\mathrm{d}t^2 - \mathrm{d}r^2 - r^2\mathrm{d}\theta^2.$$

   Deduce the radial and angular components of the geodesic equation:

$$\frac{\mathrm{d}^2r}{\mathrm{d}\tau^2} - r\left(\frac{\mathrm{d}\theta}{\mathrm{d}\tau}\right)^2 = 0$$

$$\frac{\mathrm{d}^2\theta}{\mathrm{d}\tau^2} + \frac{2}{r}\frac{\mathrm{d}\theta}{\mathrm{d}\tau}\frac{\mathrm{d}r}{\mathrm{d}\tau} = 0.$$

Further Reading

Schutz B 2008 *A First Course in General Relativity* (2nd ed.; Cambridge: Cambridge Univ. Press). This is a longer text by an expert and is useful in the areas covered by this and the following chapter.

Carlip S 2019 *General Relativity: A Concise Introduction*. This book is pitched at more advanced students, despite the title. It brings out interesting points and is modern in its approach.

Misner C W, Thorne K S and Wheeler J A 1971 *Gravitation* (W H Freeman and company, San Francisco). A 1000 page long groundbreaking book that established the notation widely used thereafter. It is worth dipping in for enlightenment on material in this or the following chapter. One author, Kip Thorne, shared the Nobel Prize in Physics in 2017 for the discovery of gravitational waves.

AAS | IOP Astronomy

Introduction to General Relativity and Cosmology (Second Edition)

Ian R Kenyon

Chapter 6

Einstein's Theory II

The general technique for converting an equation valid for SR into one valid in all frames cannot be applied to gravitation because we lack the starting point—a law of gravitation consistent with SR. Newton's law of gravitation implies that gravitational effects are transmitted instantaneously to all parts of the universe, faster than the speed of light. Einstein perceived that there must be a direct link between the distribution of mass/energy and the curvature of spacetime, expressible in tensor form. Einstein also recognized that the stress–energy tensor provided the appropriate tensor description for the distribution and flow of energy in spacetime. Later he identified a curvature tensor (the Einstein tensor) having formal properties that match those of the stress–energy tensor. Einstein simply equated these two tensors, making "curvature" at a given point in spacetime proportional to "energy density" at the same point.

In Section 6.1 the Riemann curvature tensor is introduced; this is a tensor that provides a complete description of curvature in multi-dimensional spaces. In Section 6.2 attention is focused on the stress–energy tensor $T_{\mu\nu}$ and its properties. The conjecture that the stress–energy tensor is proportional to some curvature tensor leads to the selection for this role of a unique contraction of the Riemann tensor (Section 6.3) called the Einstein tensor $G_{\mu\nu}$.

6.1 The Riemann Curvature Tensor

The description of curvature in terms of Gaussian curvatures becomes unwieldy beyond two dimensions. Riemann developed more elegant techniques to analyze curvature in higher-dimensional spaces. A full description of curvature at a given point in such a space is provided by a rank-4 tensor called the Riemann curvature tensor. It is not difficult to infer where the information on curvature must come from. First we note that by going to a frame in free fall at an event (point in spacetime) x the spacetime is made flat locally, and we have

$$g_{\mu\nu} = \eta_{\mu\nu}, \quad g_{\mu\nu,\rho} = 0 \text{ both at } x. \tag{6.1}$$

doi:10.1088/2514-3433/acc3ffch6

Then the metric tensor at $x + \Delta x$ and its first derivative are given by expansions around $g_{\mu\nu}(x)$ and $g_{\mu\nu,\rho}(x)$:

$$g_{\mu\nu}(x + \Delta x) = \eta_{\mu\nu} + \frac{1}{2} g_{\mu\nu,\rho\sigma} \Delta x^\rho \Delta x^\sigma$$

and

$$g_{\mu\nu,\rho}\left(x + \Delta x\right) = g_{\mu\nu,\rho\sigma} \Delta x^\sigma .$$

The change in $g_{\mu\nu}$ depends only on the *second* derivatives $g_{\mu\nu,\rho\sigma}$ at x, and so these derivatives must embody the curvature information.

> In Section 5.7 it was seen that the metric connection and hence the first derivatives of the metric components correspond to the gravitational acceleration. A tidal acceleration is the difference between gravitational acceleration at different locations and so it involves the *second* derivatives of the metric connections. There is therefore a fundamental connection between the curvature of spacetime and classical tidal forces.

The curvature tensor can be identified by parallel transporting a vector around an infinitesimal closed path and calculating the change this operation produces in it. Figure 6.1 shows the route. AB is a vector path element da and BC is a vector path element db; the return path consists of the vector elements $-$da(CD) and $-$db(DA). Parallel transporting a vector q^α around the loop ABCDA induces changes that are calculated using Equation (5.12). The changes induced over each of the four segments of path are, respectively,

$$-\Gamma^\alpha_{\beta\nu}(x) q^\nu(x) \mathrm{d}a^\beta,$$

$$-\Gamma^\alpha_{\beta\nu}\left(x + \mathrm{d}a\right) q^\nu\left(x + \mathrm{d}a\right) \mathrm{d}b^\beta,$$

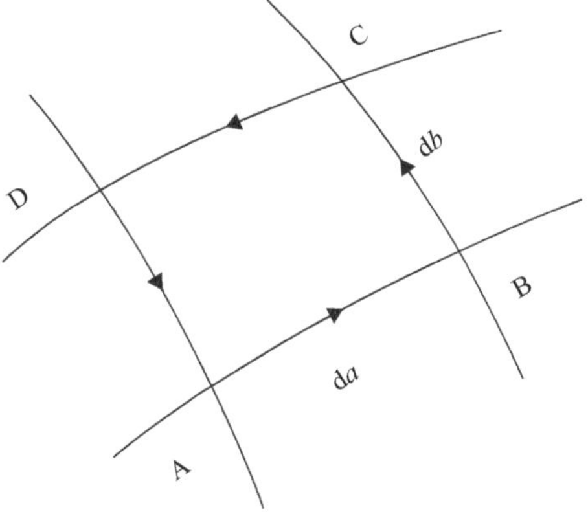

Figure 6.1. A closed path in curved spacetime around which a vector is parallel transported.

$$+\Gamma^{\alpha}_{\beta\nu}(x+\mathrm{d}b)q^{\nu}(x+\mathrm{d}b)\mathrm{d}a^{\beta},$$

$$+\Gamma^{\alpha}_{\beta\nu}(x)q^{\nu}(x)\mathrm{d}b^{\beta}.$$

$\Gamma^{\alpha}_{\beta\nu}$ and q^{β} are evaluated at the location indicated within parentheses. Collecting these terms yields the overall change suffered by q^{α}.

$$\mathrm{d}q^{\alpha}=\frac{\partial\left(\Gamma^{\alpha}_{\beta\nu}q^{\nu}\right)}{\partial x^{\gamma}}\mathrm{d}b^{\gamma}\mathrm{d}a^{\beta}-\frac{\partial\left(\Gamma^{\alpha}_{\beta\nu}q^{\nu}\right)}{\partial x^{\delta}}\mathrm{d}a^{\delta}\mathrm{d}b^{\beta}.$$

Using Equation (5.12) again we obtain

$$\begin{aligned}\mathrm{d}q^{\alpha}&=\mathrm{d}a^{\beta}\ \mathrm{d}b^{\gamma}\left(\Gamma^{\alpha}_{\beta\nu,\,\gamma}q^{\nu}-\Gamma^{\alpha}_{\beta\nu}\Gamma^{\nu}_{\sigma\gamma}q^{\sigma}\right)-\mathrm{d}a^{\delta}\ \mathrm{d}b^{\beta}\left(\Gamma^{\alpha}_{\beta\nu,\,\delta}q^{\nu}-\Gamma^{\alpha}_{\beta\nu}\Gamma^{\nu}_{\sigma\delta}q^{\sigma}\right)\\&=\mathrm{d}a^{\delta}\ \mathrm{d}b^{\gamma}\ q^{\beta}\left(\Gamma^{\alpha}_{\beta\delta,\,\gamma}-\Gamma^{\alpha}_{\beta\gamma,\,\delta}+\Gamma^{\alpha}_{\sigma\gamma}\Gamma^{\sigma}_{\beta\delta}-\Gamma^{\alpha}_{\sigma\delta}\Gamma^{\sigma}_{\beta\gamma}\right).\end{aligned}$$

In writing the last line use has been made of identities such as

$$\mathrm{d}a^{\beta}\Gamma^{\alpha}_{\beta\nu,\,\gamma}=\mathrm{d}a^{\delta}\Gamma^{\alpha}_{\delta\nu,\,\gamma}$$

which result from changing a repeated suffix, in this case $\beta\to\delta$. The expression for $\mathrm{d}q^{\alpha}$ can be written compactly as

$$\mathrm{d}q^{\alpha}=\mathrm{d}a^{\delta}\ \mathrm{d}b^{\gamma}\ q^{\beta}R^{\ \alpha}_{\ \beta\gamma\delta} \tag{6.2}$$

where $R^{\ \alpha}_{\ \beta\gamma\delta}$, or more compactly $R^{\alpha}_{\beta\gamma\delta}$, is defined by

$$R^{\alpha}_{\beta\gamma\delta}=\Gamma^{\alpha}_{\beta\delta,\,\gamma}-\Gamma^{\alpha}_{\beta\gamma,\,\delta}+\Gamma^{\alpha}_{\sigma\gamma}\Gamma^{\sigma}_{\beta\delta}-\Gamma^{\alpha}_{\sigma\delta}\Gamma^{\sigma}_{\beta\gamma}. \tag{6.3}$$

Contracting $R^{\ \alpha}_{\ \beta\gamma\delta}$ with the arbitrary tensors in Equation (6.2) has produced a tensor; which means, using the quotient theorem, that it too is a tensor. This, the *Riemann curvature tensor*, quantifies spacetime curvature. Specializing to a frame in free fall causes all the connections to vanish but not their derivatives, and Equation (6.3) collapses to

$$R^{\alpha}_{\beta\gamma\delta}=\Gamma^{\alpha}_{\beta\delta,\,\gamma}-\Gamma^{\alpha}_{\beta\gamma,\,\delta}. \tag{6.4}$$

Other associated forms of the Riemann tensor are also of interest, e.g.,

$$R_{\alpha\beta\gamma\delta}=g_{\alpha\mu}R^{\mu}_{\beta\gamma\delta}.$$

One useful result that can be proved is that Equation (6.4) can be written

$$2R_{\alpha\beta\gamma\delta}=g_{\alpha\delta,\beta\gamma}-g_{\beta\delta,\alpha\gamma}+g_{\beta\gamma,\alpha\delta}-g_{\alpha\gamma,\beta\delta} \tag{6.5}$$

in which the anticipated dependence of curvature on the second derivatives of the metric coefficients is made explicit. If spacetime is flat everywhere these derivatives and the Riemann tensor will all vanish. Other properties of the Riemann tensor $R_{\alpha\beta\gamma\delta}$ are that it reverses under interchange of α with β, or δ with γ:

$$R_{\beta\alpha\gamma\delta}=R_{\alpha\beta\delta\gamma}=-R_{\alpha\beta\gamma\delta}; \tag{6.6}$$

and exchanging $\alpha\beta$ for $\gamma\delta$ leaves it unchanged

$$R_{\gamma\delta\alpha\beta} = R_{\alpha\beta\gamma\delta}. \tag{6.7}$$

The behavior of the Riemann tensor is contrasted with that of the metric connections in Figure 6.2. A suitable choice of frame can always be made such that the connections and the accelerations vanish locally. This is not the case for the Riemann tensor: it vanishes wherever spacetime is flat, whatever generalized coordinates are chosen. The Riemann tensor for spacetime has 4^4 components; thanks to the symmetry of this tensor it can be shown that only 20 components are linearly independent.

Another property of curved spaces is that two nearby and initially parallel geodesics do not continue parallel indefinitely, but converge or diverge depending on the local curvature. For example, a pair of lines of longitude are parallel at the equator but converge toward the poles. Suppose that a vector ξ is drawn in curved spacetime linking points on two nearby geodesics at x and $x + \xi$. The two geodesics have equations

$$0 = \frac{d^2x^\mu}{d\tau^2} + \Gamma^\mu_{\nu\lambda}(x)\frac{dx^\nu}{d\tau}\frac{dx^\lambda}{d\tau}$$

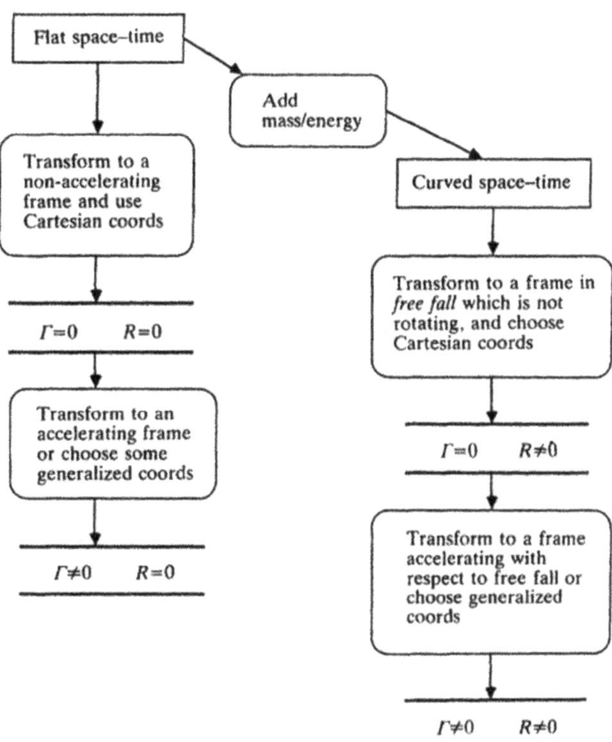

Figure 6.2. Summary of the effects of both spacetime curvature and coordinate transformations on the metric connection and the Riemann curvature tensor.

and a similar expression with x replaced by $x + \xi$. Their deviation is given by the *equation of geodesic deviation*:

$$0 = \frac{D^2\xi^\mu}{D\tau^2} + R^\mu_{\ \nu\rho\lambda}\xi^\rho \frac{dx^\nu}{d\tau}\frac{dx^\lambda}{d\tau} \tag{6.8}$$

which, being expressed entirely in terms of tensors, will hold true under general transformations of coordinates. This Equation (6.8), is called the *equation of geodesic deviation.*

The next stage along our road toward Einstein's general theory will be to discover how to present the density of mass/energy in a way compatible with SR, that is as a tensor. Once this result is in place it will be possible to discuss how to relate spacetime curvature mathematically to mass–energy.

6.2 The Stress–Energy Tensor

In SR the energy and the linear momentum are two aspects of a single entity, the four-momentum. They are connected to the rest mass through the relation given in Section 1.14

$$E^2 - c^2p^2 = m^2c^4.$$

This intimate connection between mass, energy and momentum suggests that a general law of gravitation will contain not just mass but energy and momentum as well. In order to proceed we need to use Newton's law of gravitation in its differential form:

$$\nabla^2\varphi = 4\pi G\rho, \tag{6.9}$$

where φ is the local gravitational potential energy per unit rest mass and ρ is the local rest mass density. In order to see how ρ can be generalized in SR we consider a simple example of matter, namely a dust cloud. A cloud of dust in its rest frame S has energy density

$$\rho_0c^2 = m_0n_0c^2,$$

where m_0 is the rest mass of the average dust grain and n_0 is the number of dust grains per unit volume. Viewed in a frame S' moving with velocity $v = \beta c$ with respect to the cloud, each grain becomes more massive, and the volume containing a fixed number of grains is Lorentz contracted along the direction of motion. Explicitly

$$m_0 \Rightarrow m = m_0\gamma \ \text{ and } \ n_0 \Rightarrow n = n_0\gamma,$$

where $\gamma = 1/(1 - \beta^2)^{1/2}$. Thus

$$\rho_0 \Rightarrow \rho = \rho_0\gamma^2.$$

Obviously ρc^2 is neither a scalar nor the component of a four-vector: if it were the first it would remain constant under a Lorentz transformation to S'; if it were the second the change would be linear in γ. The behavior of ρ is, however, exactly that of the time-time component of a second-rank tensor $T^{\mu\nu}$:

$$T^{\mu\nu} = \rho_0 v^\mu v^\nu, \tag{6.10}$$

where v^μ is the four-vector velocity of the cloud. In the frame S only the time-time component of this tensor is non-zero; it is $(T^{00})_S = \rho_0 c^2$. Under the transformation from S to S'

$$(T^{00})_S \Rightarrow T^{00} = \gamma^2 (T^{00})_0.$$

The tensor $T^{\mu\nu}$ is called the *stress–energy tensor*. A definition of its components that applies equally to dust clouds or more complex systems is the following: $T^{\mu\nu}$ is the flow of the μ component of the four-momentum along the ν direction.

Some examples should help to make this clearer.

- T^{00} is the energy density.
- cT^{0i} is the energy flow per unit area parallel to the i direction. In the case of the dust cloud this would constitute heat flow.
- T^{ii} is the flow of momentum component i per unit area in the i direction, i.e., the pressure across the i plane.
- T^{ij} is the flow of the i component of momentum per unit area in the j direction. This is another way of describing a component of the viscous drag across the j plane.
- $T^{i0}c$ is the density of the i component of momentum.

It is apparent that in SR the stress–energy tensor embodies a compact description of energy and momentum density. In contrast, the underlying connection between energy, momentum, pressure, and heat flow is only latent in the classical (Newtonian) view. Referring to the dust cloud it can be seen that if a Lorentz transformation is made to another frame, the spatial components of $T^{\mu\nu}$ generally receive contributions from $(T^{00})_S$: thus the spatial components must also affect the curvature of spacetime. Consequently *all* the terms appearing in the stress–energy tensor must help to warp the fabric of spacetime!

Herein lies a paradox for the classical view. It is well known that the pressure inside a star resists gravitational collapse, and yet it emerges that pressure can, by virtue of contributing to a component of the stress–energy tensor, hasten gravitational collapse. This is indeed the case, and pressure contributes to the contraction of sufficiently massive stars to black holes.

Conservation laws for energy and momentum take particularly simple forms when expressed in terms of the stress–energy tensor. The conservation of energy offers a good example of the way this happens. Figure 6.3 shows a cube of edge length l with its edges parallel to the x-, y- and z-axes in a medium whose stress–energy tensor is $T^{\mu\nu}$. The rate of change in the energy content of the box is

$$l^3 \frac{\partial T^{00}}{\partial t}.$$

This change is produced by the net energy inflow through the six faces of the cube. The energy flows through the faces $x = x$ and $x = x + l$ are, respectively,

$$\begin{array}{ll} l^2 c T^{01}(x) & \text{inward per unit time} \\ l^2 c T^{01}(x + l) & \text{outward per unit time.} \end{array} \tag{6.11}$$

The net flow inward from these two faces is

$$l^2 c\left[T^{01}(x) - T^{01}(x + l)\right] = -l^3 c \frac{\partial T^{01}}{\partial x}.$$

There are similar contributions from the other pairs of faces:

$$-l^3 c\left(\frac{\partial T^{02}}{\partial y}\right) \text{ and } -l^3 c\left(\frac{\partial T^{03}}{\partial z}\right).$$

Summing the three contributions gives the total inflow, and so we have

$$l^3 \frac{\partial T^{00}}{\partial t} = -l^3 c\left(\frac{\partial T^{01}}{\partial x} + \frac{\partial T^{02}}{\partial y} + \frac{\partial T^{03}}{\partial z}\right),$$

which can be rearranged to become

$$\frac{\partial T^{00}}{\partial x^0} + \frac{\partial T^{0i}}{\partial x^i} = 0,$$

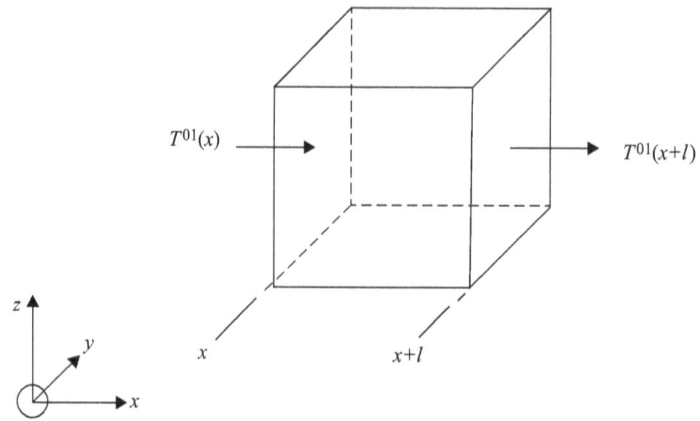

Figure 6.3. A cube of side l located in a material whose stress–energy tensor is $T^{\alpha\beta}$.

that is,

$$\frac{\partial T^{0\alpha}}{\partial x^\alpha} = 0.$$

This last result can be expressed in a more compact form using the subscript comma notation for spacetime derivatives:

$$T^{0\alpha}_{,\alpha} = 0. \tag{6.12}$$

This type of derivative is known as the *divergence* of $T^{0\alpha}$, the derivative is taken with respect to the same spacetime component as appears in the tensor (here α), and the sum over the expression on the left-hand side is made for all values α. A parallel procedure applied for the conservation of linear momentum yields the three equations

$$T^{i\alpha}_{,\alpha} = 0 \text{ for } i = 1,\ 2, \text{ or } 3. \tag{6.13}$$

Finally the conservation laws of Equations (6.12) and (6.13) can be combined into a single equation:

$$T^{\beta\alpha}_{,\alpha} = 0 \text{ for } \beta = 0, 1, 2, \text{ or } 3. \tag{6.14}$$

This equation summarizes the conservation laws for four-momentum in SR. Put into words they require that the divergences of the stress–energy tensor vanish everywhere. This result from SR can be converted to a form of general validity in curved spacetime if the simple derivatives are replaced by covariant derivatives. Thus the laws of conservation of energy and momentum in curved spacetime take the form

$$T^{\beta\alpha}_{\ \ ;\alpha} = 0, \tag{6.15}$$

where the subscript semicolon indicates a covariant derivative. This can be written in more detail as

$$\frac{\partial T^{\beta\alpha}}{\partial x^\alpha} + \Gamma^{\alpha}_{\ \mu\alpha} T^{\beta\mu} + \Gamma^{\beta}_{\ \mu\alpha} T^{\mu\alpha} = 0.$$

The properties of the energy-momentum tensor are now summarized:

- it vanishes in the absence of matter;
- it is of second rank;
- its divergences all vanish;
- it is symmetric, i.e., $T^{\mu\nu} = T^{\nu\mu}$.

6.3 Einstein's equation

Einstein identified the stress–energy tensor as the *source* of spacetime curvature and suggested the simplest possible relationship between it and the curvature:

$$KT_{\mu\nu} = G_{\mu\nu},$$

where $G_{\mu\nu}$ is the tensor describing spacetime curvature and K is some scalar constant whose magnitude determines how effective the energy density is in distorting spacetime.

An immediate consequence of this Ansatz is that $G_{\mu\nu}$ should be a symmetric, divergenceless second rank tensor to match the stress–energy tensor. The Riemann curvature tensor of rank 4 quantifies spacetime curvature, and it is reasonable to expect $G_{\mu\nu}$ should be a contraction of the Riemann curvature tensor. A second rank tensor can easily be constructed by contraction:

$$R_{\beta\delta} = R^{\alpha}{}_{\beta\alpha\delta} = g^{\alpha\sigma} R_{\sigma\beta\alpha\delta}, \tag{6.16}$$

which is called the Ricci tensor. Using Equation (6.6) the contraction $R^{\alpha}{}_{\alpha\gamma\delta}$ vanishes, while $R^{\alpha}{}_{\delta\gamma\alpha}$ gives an equivalent result: hence the Ricci tensor is the *unique* contraction of the Riemann tensor. The Ricci tensor has a non-zero divergence which can be removed by a simple subtraction. The result is the divergenceless second rank Einstein tensor

$$G_{\beta\delta} = R_{\beta\delta} - g_{\beta\delta} R/2 \tag{6.17}$$

where R is called the Ricci scalar

$$R = g^{\beta\delta} R_{\beta\delta}. \tag{6.18}$$

$R_{\beta\delta}$ and $g_{\beta\delta}$ are both symmetric so that the Einstein tensor is also symmetric. Like its progenitors, the Riemann and Ricci tensors, the Einstein tensor vanishes in the absence of any material to warp spacetime. Referring to the summary of the properties of the stress–energy tensor given at the head of the previous section, we see that the Einstein tensor matches these precisely. With this tensor we can now rewrite Einstein's Ansatz as

$$G_{\alpha\beta} = \frac{8\pi G}{c^4} T_{\alpha\beta}. \tag{6.19}$$

The value of the constant $8\pi G/c^4$ is fixed by the requirement that, in the limit of weak slowly varying gravitational fields, the Einstein equation should reduce to Newton's law of gravitation. This will be shown explicitly in Section 6.4. Note that the G appearing on the right-hand side of Equation (6.19) is the usual gravitational constant and not some contraction of a tensor. As with all such tensor equations the equality will remain true if we simultaneously raise one or more indices on both sides of the equation.

When Einstein's equation was applied to calculate the behavior of the universe on the large scale it was apparent that the universe could contract or expand indefinitely depending on the starting condition and the amount of matter present. However at that time it was known that some stars moved toward the Earth and some away, with no sign of universal contraction or expansion. Einstein presumed that the universe was static and was therefore in a dilemma.

Einstein found that by adding a constant term to his equation he could obtain a static universe

$$G_{\alpha\beta} - \Lambda g_{\alpha\beta} = \frac{8\pi G}{c^4} T_{\alpha\beta} \tag{6.20}$$

where Λ is a universal constant called the *cosmological constant*. As we can see, it would lead to curvature of spacetime in the absence of any matter and radiation ($T_{\alpha\beta} = 0$). It is therefore possible by choosing Λ appropriately to obtain a static universe. By 1931 Hubble and Humason had convincingly demonstrated that the universe is currently expanding and Einstein was then happy to disown the cosmological constant. Ironically, modern cosmological observations show that something very like a cosmological constant is an essential feature of the universe.

In addition the quantum field theory of fundamental particles has taught us that the vacuum is not a featureless void, but is in a continuous ferment involving particle–antiparticle production and absorption, making a non-zero cosmological constant altogether reasonable. One point worth consideration is that the cosmological constant is like a constant of integration, it is there and the only question has to be: how large is it? Finally the contribution of the cosmological constant $\Lambda g_{\alpha\beta}$ defines its equation of state. Going to the frame in free fall it only has diagonal terms with the time or energy component and the spatial or momentum component being equal and of opposite sign. Thus the energy density due to the cosmological constant is positive, but totally unlike matter or radiation the pressure is negative. The argument is refined in Appendix F. This unexpected behavior has fundamental implications for the evolution of the universe.

6.4 The Newtonian Limit

It is instructive to check that in the limit of weak slowly varying gravitational fields Einstein's equation reduces to Newton's law of gravitation. This will show how the new variables of Einstein's general theory are related to the more familiar classical non-relativistic variables. The starting point is Einstein's Equation (6.19). In the classical limit when the gravitational field is weak and slowly varying

$$g_{\mu\nu} = \eta_{\mu\nu} + h_{\mu\nu}$$

where all the components of the tensor h are much less than unity. If we choose Cartesian coordinates $\eta_{00} = +1, \eta_{11} = \eta_{22} = \eta_{33} = -1$. In the same limit velocities are small ($\ll c$) and the spatial components of momentum are much less than energies. The dominant term in the stress–energy tensor is therefore the energy density T_{00}. Thus the important part of Einstein's equation, Equation (6.19), in the classical non-relativistic limit is

$$R_{00} = \frac{8\pi G(T_{00} - T^0_{\ 0}\, g_{00}/2)}{c^4} - \Lambda g_{00}. \tag{6.21}$$

In evaluating this expression we first note that the metric connections are linear in h; hence to a first approximation in h the form Equation (6.4) or (6.5) can be used for the Riemann curvature tensor rather than Equation (6.3). To a first order approximation in h, in the classical slow-moving limit where the time derivative is negligible compared to the spatial derivatives the Ricci tensor reduces to

$$R_{00} = \frac{h_{00,ii}}{2},$$

summing over only the spatial components i. The result of the gravitational redshift experiment determines h_{00}: from Equation 2.12 for instance,

$$h_{00} = -\frac{2GM}{rc^2} = \frac{2\varphi}{c^2},$$

where φ is the gravitational potential. If this assignment is carried through here, then

$$R_{00} = \frac{\varphi_{,ii}}{c^2} = \frac{\nabla^2\varphi}{c^2}.$$

Suppose that the rest density of matter is ρ; then

$$T_{00} = \rho c^2.$$

It is a sufficient approximation to take $g_{00} = 1$ on the right-hand side of Equation (6.21), so that

$$T_{00} - \frac{T^0_0 g_{00}}{2} = \frac{\rho c^2}{2}.$$

Finally, substituting on both sides in Equation (6.21) gives

$$\nabla^2\varphi = 4\pi G\rho - \Lambda c^2 = 4\pi G\left[\rho - \rho_\Lambda\right],$$

where $\rho_\Lambda = \Lambda c^2/[4\pi G]$. Ignoring Λ for the moment, this reproduces Equation (6.9), the differential form of Newton's law of gravitation. The comparison verifies that the constant appearing in Einstein's equation has magnitude $8\pi G/c^4$; any other choice would spoil the agreement with Newton's law in the classical non-relativistic limit. Λ is hard to interpret in Newtonian terms: it remains constant as space expands, while the density of matter and radiation fall; its force is repulsive, but because the source is everywhere the net force on any mass is zero; finally we shall find that the cosmological constant expands space itself, not something that has a Newtonian equivalent. If you could confine the cosmological constant to a sphere of radius r centered on a mass M, then the force on a unit mass at the boundary would be

$$F = \frac{-GM}{r^2} + \frac{c^2\Lambda r}{3}. \tag{6.22}$$

Observed from our local perspective the impact of Λ is not felt in any direct physical way. However about 5 Gyrs ago the effect of the repulsion of Λ had already grown with the universe to the point of balancing the attraction of matter. From that moment onward, the expansion of the universe accelerated, so that eventually all cosmic structures will be pulled apart.

At this point the equivalences between classical non-relativistic and general relativistic quantities are collected and reviewed. Making use of the result of the gravitational redshift experiment we have shown that

$$g_{00} = 1 + h_{00} = 1 + 2\varphi/c^2$$

so that the metric coefficients replace the classical gravitational potential. However $g_{\mu\nu}$ (and $h_{\mu\nu}$) have six independent components compared with a single classical potential. In the Newtonian limit, $\mathrm{d}x^i/\mathrm{d}\tau \ll \mathrm{d}x^0/\mathrm{d}\tau$ and $\tau \approx t$, so that the geodesic Equation (5.15) reduces to Equation (5.20)

$$\frac{\mathrm{d}^2x^i}{\mathrm{d}t^2} = -c^2\Gamma^i_{00},$$

where the Γ^i_{00} are the important metric connections in the Newtonian limit. To first order in h,

$$\Gamma^i_{00} = -\frac{1}{2}\eta^{ii}\frac{\partial h_{00}}{\partial x^i} = \frac{\partial\varphi/\partial x_i}{c^2}$$

(with no summation over i). We can reiterate the conclusion drawn in Section 5.7 that the metric connections replace both the inertial and the gravitational forces of classical non-relativistic mechanics. Finally, Equation (6.4) permits an interpretation of the Riemann curvature tensor in classical terms. In the Newtonian limit all the time derivatives are small compared with spatial derivatives, and so Equation (6.4) reduces to

$$R_{0k0j} = -\frac{1}{2}g_{00,kj} = -\frac{(\partial^2\varphi/\partial x^k\partial x^j)}{c^2},$$

which is a component of the *tidal* force. Table 6.1 summarizes the correspondence between general relativistic and Newtonian kinematic quantities.

Table 6.1. General Relativistic Quantities and Their Newtonian Analogs

GR Quantities	Newtonian Analogs
$g_{\mu\nu} = 1 + h_{\mu\nu}$	$1 + 2\varphi/c^2$
Metric tensor	Gravitational potential φ/c^2
$g_{\mu\nu,\alpha}$, $\Gamma^{\mu}{}_{\nu\alpha}$	$(\partial\varphi/\partial x^\alpha)/c^2$
	Gravitational force/c^2
$g_{\mu\nu,\alpha\beta}$, $R^{\mu}_{\nu\alpha\beta}$	$(\partial^2\varphi/\partial x^\alpha\partial x^\beta)/c^2$
	Tidal force/c^2

6.5 Exercises

1. Obtain the expressions given for Γ^1_{00}, Γ^0_{10}, Γ^0_{10} and Γ^1_{11} for the Schwarzschild metric. Then calculate $\Gamma^1_{00,\,1}$, $\Gamma^1_{01,\,1}$, $\Gamma^1_{01,\,0}$. Finally, calculate R^1_{010}.
2. Starting from the force given by Equation (6.22) show that the corresponding gravitational potential satisfies

$$\nabla^2\varphi = 4\pi G\rho - \Lambda c^2.$$

3. Show that the Riemann curvature tensor vanishes for the metric equation

$$\mathrm{d}s^2 = c^2\,\mathrm{d}t^2 - \mathrm{d}r^2 - r^2\,\mathrm{d}\theta^2.$$

Ian R Kenyon

Chapter 7

Tests of General Relativity

Einstein's theory passes the zeroth test because, as we saw in the last chapter, it reproduces Newton's laws in the limit of low velocity and weak gravitational fields. Developments in technology opened the way to very precise tests of GR: making use of space probes, satellites, atomic clocks, huge telescopes, and laser ranging to the Moon and radar ranging to other planets. This hardware is supported by onboard computers on spacecraft, and by high-rate data links to terabyte data storage and processor farms on Earth.

In this chapter the measurements made within the solar system to test GR are described. The earlier tests used the advance of the perihelion of Mercury, the deflection of radiation by the Sun, and the time delay of radar signals passing by the Sun. Together with the gravitational redshift measurements described in Chapter 2, these are the successful classical tests of GR. More recent tests measured the frame-dragging and geodetic precession that massive bodies, in this case the Earth, produce according to Einstein's theory. The observation of gravitational lensing by galaxies, another effect predicted by GR, is also discussed here. All these effects involve velocities much less than c and modest gravitational potentials. The physical effects in more extreme conditions predicted by GR, involving black holes and gravitational radiation, are covered in the two following chapters. Gravitational lensing is now used to infer the mass distribution of galaxies and galaxy clusters: the method is described in Chapter 17.

7.1 The Perihelion Advance of Mercury

Mercury, the innermost planet of the solar system, follows an elliptical orbit at a mean distance of 58 million km from the Sun. Other planets attract Mercury and perturb its orbit so that the long axis of the ellipse slowly rotates in its plane with respect to the frame of the distant galaxies. The point at which the planet is nearest to the Sun is called its perihelion, and so this motion is known as the precession of the perihelion. Calculations using Newtonian mechanics predict a precession of

doi:10.1088/2514-3433/acc3ffch7

532 arcsec per century. Observation shows, however, that the precession is 43.11 arcsec larger with an error of only 0.45 arcsec, a discrepancy that was recognized as long ago as 1859 by Leverrier.

After three centuries of telescopic measurements, the more precise radar echo detection is now used to directly determine the distance to Mercury and other planets. Figure 7.1 illustrates techniques using one of the 64 m antennae of the Deep Space Network (DSN). A radar carrier pulse of 1 ms duration (300 km long) is phase modulated with a random 255 bit code. When the pulse echo $E(t + t_0)$ returns it is cross-correlated with the pattern transmitted $T(t)$ by forming the sum of the products $S(t_0) = \sum_t [T(t)E(t + t_0)]$. The delay t_0 is only known approximately, and so the cross-correlation is repeated for 3 μs steps in t_0 (every 1 km). If the echo and the pulse pattern match, then the correlator output is large; otherwise, even for a 3 μs offset, the output is small. In practice the output rises sharply when the echo returns from the nearest point on the planet but dies away slowly as echoes arrive from surrounding regions of the planet's surface. The delay until the initial sharp rise in output from the correlator gives the distance to the planet. Corrugations of terrain blur the precision to ± 1 km.

The analysis of Mercury's motion begins with the statement that Mercury follows a geodesic in Schwarzschild spacetime around the Sun. Perturbations of the predicted GR-induced precession due to the other planets are negligible. The opportunity is taken to include some relevant comments in what follows, beyond the bare calculation. We can assume that Mercury's orbital plane is the Sun's equatorial plane ($\theta = \pi/2$) so that the metric Equation (3.48) becomes

$$c^2\, \mathrm{d}\tau^2 = c^2 Z\, \mathrm{d}t^2 - \mathrm{d}r^2/Z - r^2\, \mathrm{d}\varphi^2, \tag{7.1}$$

where $Z = (1 - 2GM/rc^2)$, M is the Sun's mass, and r is the distance of Mercury from the Sun. Then $Z \approx 1 - 5 \times 10^{-7}$. Multiplying this equation by the planet's mass squared, m^2, and dividing by $\mathrm{d}\tau^2$ gives

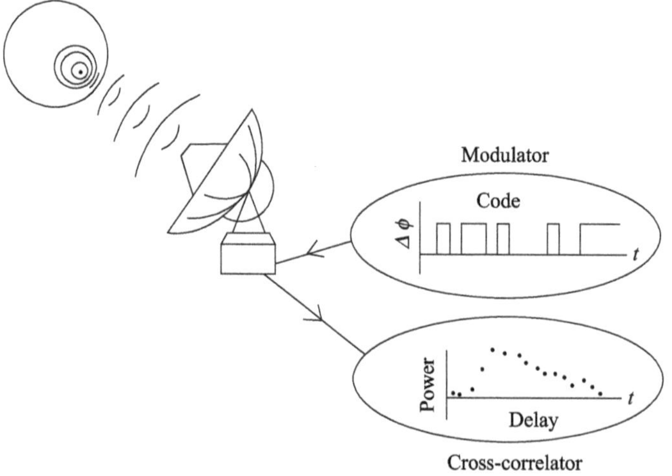

Figure 7.1. The use of radar ranging to measure planetary distances (adapted from Hellings (1984)).

$$m^2c^2 = m^2c^2Z\left(\frac{\mathrm{d}t}{\mathrm{d}\tau}\right)^2 - \frac{m^2(\mathrm{d}r/\mathrm{d}\tau)^2}{Z} - m^2r^2\left(\frac{\mathrm{d}\varphi}{\mathrm{d}\tau}\right)^2. \tag{7.2}$$

In flat spacetime this reduces to

$$m^2c^2 = m^2c^2\left(\frac{\mathrm{d}t}{\mathrm{d}\tau}\right)^2 - m^2\left(\frac{\mathrm{d}r}{\mathrm{d}\tau}\right)^2 - m^2r^2\left(\frac{\mathrm{d}\varphi}{\mathrm{d}\tau}\right)^2,$$

that is to

$$m^2c^2 = m^2c^2\gamma^2 - m^2v_r{}^2\gamma^2 - m^2v_\varphi{}^2\gamma^2,$$

where v_r and v_φ are the radial and tangential components of the velocity v and $\gamma = 1/(1 - v^2/c^2)^{1/2}$. This last equation is the standard SR formula relating rest mass to four-momentum:

$$m^2c^2 = E^2/c^2 - p^2,$$

and Equation (7.2) is its equivalent in Schwarzschild spacetime. The geodesic equation for Mercury in its integral form is given by Equations (5.18) and (5.19):

$$\delta\int \mathrm{d}\tau\left[Zc^2\left(\frac{\mathrm{d}t}{\mathrm{d}\tau}\right)^2 - \frac{(\mathrm{d}r/\mathrm{d}\tau)^2}{Z} - r^2\left(\frac{\mathrm{d}\varphi}{\mathrm{d}\tau}\right)^2\right] = 0.$$

General solutions for such equations are discussed in Appendix A. In particular if L, the quantity within the square brackets, is independent of a coordinate x^μ, then $\partial L/\partial(\mathrm{d}x^\mu/\mathrm{d}\tau)$ is conserved. Here L is independent of t giving the conservation law

$$\frac{\partial L}{\partial(c\ \mathrm{d}t/\mathrm{d}\tau)} = \text{constant},$$

so that

$$2Zc\frac{\mathrm{d}t}{\mathrm{d}\tau} = \text{constant}. \tag{7.3}$$

Re-expressing this result in terms of the momentum component p_0, we have

$$cp_0 = Zcp^0 = Zmc^2\frac{\mathrm{d}t}{\mathrm{d}\tau} = E,$$

where E is a constant. In words, E is a constant of motion for a body in free fall, which in the absence of a gravitational force reduces to $mc^2\gamma$, the usual SR energy. In SR E would be constant in the absence of any forces. L is also independent of φ, so that

$$\frac{\partial L}{\partial(\mathrm{d}\varphi/\mathrm{d}\tau)} = \text{constant},$$

equivalently

$$r^2\frac{d\varphi}{d\tau} = J, \text{ a constant.} \tag{7.4}$$

Equation (7.4) is the equivalent of the Newtonian law of conservation of angular momentum for Schwarzschild spacetime. Replacing $dt/d\tau$ in Equation (7.2) by E/Zmc^2 gives

$$\frac{E^2}{Zc^2} - \frac{m^2(dr/d\tau)^2}{Z} - m^2r^2\left(\frac{d\varphi}{d\tau}\right)^2 = m^2c^2. \tag{7.5}$$

Multiplying this by Z and dropping a factor m throughout we obtain

$$\frac{E^2}{mc^2} - m\left(\frac{dr}{d\tau}\right)^2 - Zr^2m\left(\frac{d\varphi}{d\tau}\right)^2 = mc^2 - \frac{2GMm}{r},$$

which can be rearranged as

$$\frac{m(dr/d\tau)^2}{2} + \frac{mr^2(d\varphi/d\tau)^2Z}{2} - \frac{GMm}{r} = \frac{(E^2/mc^2 - mc^2)}{2} = T, \tag{7.6}$$

where T is also a conserved quantity. Equation (7.6) is the equivalent in Schwarzschild spacetime of the Newtonian conservation law for energy. There is a radial kinetic energy term $m(dr/\tau)^2/2$, a transverse kinetic energy term $mr^2(d\varphi/d\tau)^2Z/2$, and a gravitational energy term $-GMm/r$. Together, Equations (7.3), (7.4), and (7.6) fully describe the motion of Mercury, or for that matter any test mass in free fall in Schwarzschild spacetime. The quantities $E = cp_0$ and $J = r^2(d\varphi/d\tau)$ are invariants of motion, like their SR counterparts. We next go on to solve these equations of motion. Using Equation (7.4) gives

$$\frac{dr}{d\tau} = \frac{dr}{d\varphi}\frac{d\varphi}{d\tau} = \frac{J}{r^2}\frac{dr}{d\varphi}$$

and putting $u = 1/r$ this becomes

$$\frac{dr}{d\tau} = -J\frac{du}{d\varphi}.$$

Substituting this expression for $dr/d\tau$ into Equation (7.6) gives

$$\frac{J^2(du/d\varphi)^2}{2} + \frac{u^2J^2Z}{2} - GMu = \frac{T}{m}.$$

Differentiating this with respect to φ and canceling a factor $du/d\varphi$, we obtain

$$J^2\left(\frac{d^2u}{d\varphi^2}\right) + J^2u - \frac{3GMu^2J^2}{c^2} - GM = 0.$$

Rearrangement gives

$$\frac{d^2u}{d\varphi^2} + u - \frac{GM}{J^2} = \frac{3GMu^2}{c^2}. \tag{7.7}$$

This result can be compared to the Newtonian equation for orbits in the gravitational potential of a mass M:

$$\frac{\mathrm{d}^2u}{\mathrm{d}\varphi^2} + u - \frac{GM}{J^2} = 0. \tag{7.8}$$

The solution of Equation (7.8) is well known:

$$u = \frac{1 + e\cos\varphi}{l},$$

where $l = a(1 - e^2)$. Figure 7.2 shows a bound orbit: an ellipse with eccentricity $0 \leqslant e < 1$. At aphelion $\varphi = \pi, r = a(1 + e)$, at perihelion $\varphi = 0, r = a(1 - e)$. Hence the long (major) axis is $2a$ in length. In addition, although less easy to prove,

$$l = J^2/GM.$$

A solution of Equation (7.8) is clearly a very good approximate solution of Equation (7.7) because Mercury's orbit is nearly Newtonian. Consequently we can rewrite the small term on the right-hand side of Equation (7.7) as

$$3GM(1 + e\cos\varphi)^2/l^2c^2$$

and make an entirely negligible error. With this substitution Equation (7.7) becomes

$$\frac{\mathrm{d}^2u}{\mathrm{d}\varphi^2} + u - \frac{GM}{J^2} = \frac{3GM}{l^2c^2}(1 + 2e\cos\varphi + e^2\cos^2\varphi). \tag{7.9}$$

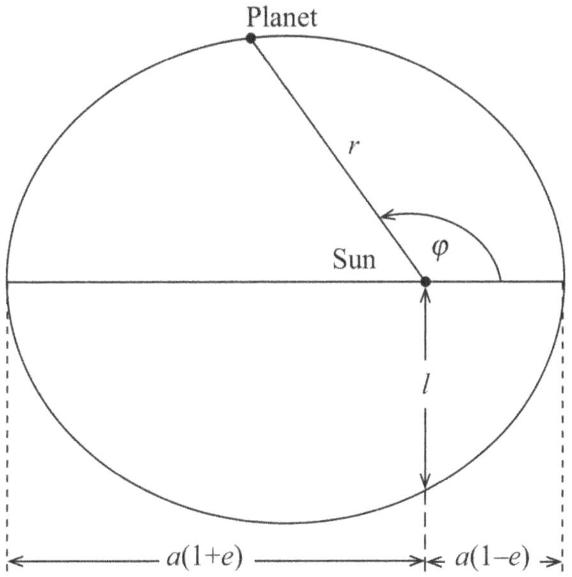

Figure 7.2. The bound elliptical orbit of a planet.

The solution to Equation (7.9) is similar to that for Equation (7.8) with extra particular integral terms arising from the three terms on the right-hand side. Explicitly

$$u = \frac{(1 + e\cos\varphi)}{l} + \frac{3GM}{l^2c^2}\left[\left(1 + \frac{e^2}{2}\right) - \left(\frac{e^2}{6}\right)\cos 2\varphi + e\varphi\sin\varphi\right].$$

Of the additional terms, the first is a constant and the second oscillates through two cycles on each orbit; both these terms are immeasurably small. However, the last term increases steadily in amplitude with φ, and hence with time, while oscillating once per orbit, and so this term is responsible for the precession. Dropping the unimportant terms we have

$$u = \frac{1 + e\cos\varphi + e\alpha\varphi\sin\varphi}{l},$$

where the factor $\alpha = 3GM/lc^2$ is extremely small. Thus

$$u = \frac{1 + e\cos[(1 - \alpha)\varphi]}{l} \tag{7.10}$$

is the general relativistic solution. At perihelion we have

$$(1 - \alpha)\varphi = 2n\pi,$$

that is

$$\varphi = 2n\pi + 6n\pi\frac{GM}{lc^2},$$

where n is an integer. This shows that the perihelion advances by $\Delta\varphi = 6\pi GM/lc^2$ per rotation, in time τ, and so the rate of precession is

$$\frac{\Delta\varphi}{\tau} = \frac{6\pi GM}{a(1 - e^2)\tau c^2}. \tag{7.11}$$

Table 7.1 compares the observed and predicted planetary precessions measured in arcsec per century. Among the planets, Mercury, with the smallest radius and greatest eccentricity (which also makes it easier to locate the perihelion precisely), has the largest precession. More recently the much larger precession of a pulsar

Table 7.1. Comparison of Observed and Predicted Planetary Precessions

Planet	Observed Precession (arcsec/century)	Predicted Precession (arcsec/century)
Mercury	43.11 ± 0.45	43.03
Venus	8.4 ± 4.8	8.6
Earth	5.0 ± 1.0	3.8

PSR1913+16 orbiting a companion star was measured by Hulse and Taylor (1975). The pulsar and its companion star both have mass 1.4 $M_{\odot}$, the orbital period is 7.75 hours and the orbital eccentricity is large, 0.617. The resulting precession of the periastron, that is the point of closest approach of the binary pair, is 4.23 degrees per year. Measurements of the precession of several such pairs agree precisely with the GR predictions. More details about PSR1913+16 will be given in the context of gravitational wave detection.

7.2 The Deflection of Light by the Sun

Measurements of this deflection were discussed in Chapter 2. The calculation of the orbit of a photon in Schwarzschild spacetime around the Sun follows the same steps as for matter in the previous section. For a photon the left-hand side of Equation (7.1) is zero and the equivalent of Equation (7.7) is

$$\frac{\mathrm{d}^2u}{\mathrm{d}\varphi^2} + u = \frac{3GMu^2}{c^2} \tag{7.12}$$

for light rays traveling in the equatorial plane. The approximate solution of this equation, when the small term on the right-hand side is neglected, is

$$u = [\cos\varphi]/b$$

where b is a constant of integration. This is the equation of a straight line with b being the distance of closest approach to the Sun, also called the impact parameter, as shown in Figure 7.3. Then proceeding as in the previous section we substitute this value for u in Equation (7.12) to give

$$\frac{\mathrm{d}^2u}{\mathrm{d}\varphi^2} + u = \frac{3GM\cos^2\varphi}{b^2c^2}. \tag{7.13}$$

This equation has the particular integral

$$u = \frac{GM(2 - \cos^2\varphi)}{b^2c^2},$$

and its full solution is

$$u = \frac{\cos\varphi}{b} + \frac{GM(2 - \cos^2\varphi)}{b^2c^2}. \tag{7.14}$$

At distant points $u \to 0$ and Equation (7.14) becomes

$$\frac{\cos\varphi}{b} + \frac{2GM}{b^2c^2} - \frac{GM\cos^2\varphi}{b^2c^2} = 0.$$

Also $\cos\varphi \to 0$ at distant points; therefore the $\cos^2\varphi$ term can be omitted. Setting $\varphi = \alpha + \pi/2$ for the path beyond the Sun, gives asymptotically

$$\alpha = \sin\alpha = 2GM/bc^2.$$

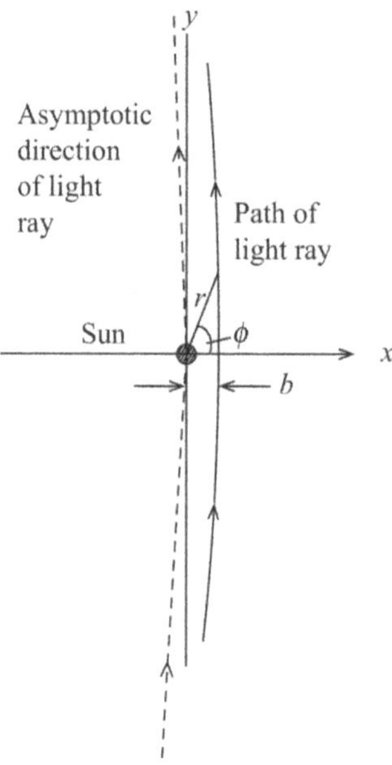

Figure 7.3. The deflection of a light ray passing near the Sun.

Now the path before and after the Sun are reflections in the x-axis. This makes the total deflection

$$\Delta\varphi = 4GM/bc^2 \tag{7.15}$$

which for light just grazing the Sun's limb ($b = b_0 = 6.96 \times 10^8$ m) is 1.750 arcsec. At any larger impact parameter b, the deflection $\Delta\varphi$ is scaled down by b_0/b. This prediction of GR is confirmed by both the optical and radio measurements described in Chapter 2. If the Newtonian calculation is made taking into account only the time distortion implied by the equivalence principle, the resultant deviation is only half as large. The GR prediction takes into account both frame and time distortion.

In 1971 Shapiro and colleagues (Shapiro et al. 2004) measured the deflection of 541 compact radio sources by the gravitational field of the Sun, using them as targets in place of Mercury. The authors made use of the generalization of the deflection formula to cover sources far from the Sun's direction due to Ward (Ward 1970). Equation (7.15) becomes

$$\Delta\varphi = \frac{(1+\gamma)GM}{bc^2}[1 + \cos\alpha], \tag{7.16}$$

where α is the angle between the Sun and the source as seen from Earth. γ would be zero (unity) in Newtonian mechanics (GR). The 2500 measurement sessions each extended over 24 hours to cover a range of α values, thus allowing both b and α to be determined. These sessions were recorded using very large baseline interferometer (VLBI) arrays between 1979 and 1999. A global fit made to this data set gave $\gamma = 0.9998 \pm 0.0004$, again in strong support of Einstein's theory.

7.3 Radar Echo Delays

Light and radar beams are not only deflected but also delayed when passing massive objects like the Sun. The delay is almost entirely due to spacetime curvature; the slight increase in path length due to the deflection contributes very little to the delay. Here we calculate this time delay for radar echoes from Venus at superior conjunction of Venus and the Earth, as shown in Figure 7.4.

We take the path to be a straight line in space in the x-direction, with impact parameter b from the Sun. This null geodesic followed by the radar beam is given by the Schwarzschild metric (3.48). The path is near radial to the Sun and we make the approximation

$$\mathrm{d}s^2 = 0 \approx [1 - 2GM/(rc^2)]c^2\mathrm{d}t^2 - \mathrm{d}x^2/[1 - 2GM/(rc^2)]. \tag{7.17}$$

Then to first order in $GM/(rc^2)$ we have

$$c\mathrm{d}t \approx \mathrm{d}x/[1 - 2GM/(rc^2)]. \tag{7.18}$$

Integrating this over the path from Earth to the point of nearest approach to the Sun gives

$$\begin{aligned} ct_\mathrm{E} &\approx \int_0^{x_\mathrm{E}} \left[1 + \frac{2GM/c^2}{\sqrt{b^2 + x^2}}\right]\mathrm{d}x \\ &\approx x_\mathrm{E} + \frac{2GM}{c^2}\ln\frac{x_\mathrm{E} + \sqrt{x_\mathrm{E}^2 + b^2}}{b} \\ &\approx x_\mathrm{E} + \frac{2GM}{c^2}\ln\frac{2x_\mathrm{E}}{b}, \end{aligned} \tag{7.19}$$

where the second term is the delay predicted by GR. A similar expression holds for the Sun–Venus section of the radar's path, of length x_V, making a total predicted delay of

$$\Delta t \approx \frac{2GM}{c^3}\ln\left[\frac{4x_\mathrm{E}x_\mathrm{V}}{b^2}\right]. \tag{7.20}$$

Making a more precise calculation gives

$$\Delta t \approx \frac{2GM}{c^3}\left(\ln\left[\frac{4x_\mathrm{E}x_\mathrm{V}}{b^2}\right] + 1\right), \tag{7.21}$$

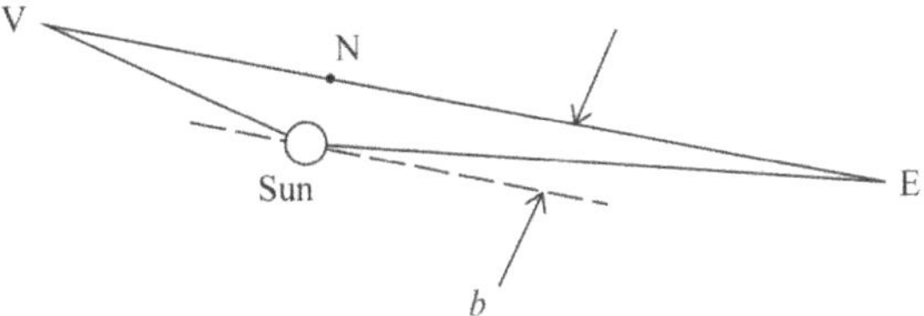

Figure 7.4. A radar beam from Earth to Venus near superior conjunction.

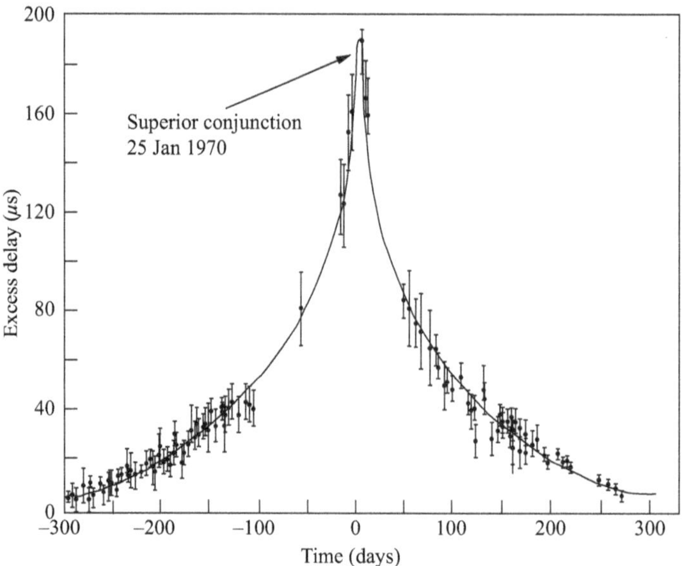

Figure 7.5. A sample of post-fit residuals for Earth–Venus time-delay measurements. The solid line is the prediction using GR. Adapted from Figure 1 in Shapiro et al. (1971). Courtesy of the American Physical Society.

which increases the time delay by 10%. The round trip takes 1300s and the predicted delay is merely 220 μs. Figure 7.5 shows over 600 days' measurements of time delays of reflections from Venus made by Shapiro and colleagues in 1971 (Shapiro et al. 1971), using two radiotelescopes: the Haystack Observatory, Massachusetts at 7.84 GHz and another at Arecibo, Puerto Rico at 430 MHz. The figure shows how the delay changes from a year before to a year after superior conjunction. The solid line giving the GR prediction over that time span is in excellent agreement with the data. One experimental difficulty is that the solar corona has a refractive index different from unity (varying as 1/(frequency)^2), and the effect of the corona is to increase the delay. A correction for this effect was applied before producing Figure 7.5. A second difficulty lies in the uncertainty of our knowledge of the topography of Venus, at a precision of 1500 m or 10 μs in timing. In 1979 Reasenberg and colleagues (Reasenberg et al. 1979) overcame these difficulties when measuring the time delay for reflections from Mars over a period of 14 months. The uncertainty in topography was avoided by receiving and retransmitting the radar signal from the Viking Lander on Mars. The effect of the corona was studied at two frequencies (2.3 and 8.4 GHz) using a transponder on a Viking Orbiter in Mars orbit. Knowing the frequency variation of the refractive index, a comparison of the delays could be used to correct for the second effect. Timing uncertainty was reduced to about 0.1 μs. The ratio of the observed delay to that predicted by GR was 1.000 ± 0.002, another clear success of Einstein's theory.

7.4 Geodetic and Frame Dragging Effects

The precession of the planets' orbits are examples on the large scale of the effect of a gravitational field (in this case of the Sun) on angular momentum. Equally the angular momentum (spin for short) of a gyroscope in free fall near the Earth will precess. The full analysis is very involved and here only the basic points are established.

Let the gyro's spin four-vector in its translational rest frame be $S^\alpha = (0, \mathbf{S})$. If there are no forces acting d $S^\alpha/\mathrm{d}\tau = 0$ with τ being the proper time. This equality remains valid in any other inertial frame. In order to generalize this equation to be valid in GR, that is in gravitational fields, the derivative must be replaced by the covariant derivative as in Equation (5.3):

$$\frac{\mathrm{d}S^\mu}{\mathrm{d}\tau} - \Gamma^\mu_{\sigma\rho} q^\rho S^\sigma = 0, \tag{7.22}$$

where q^ρ is the four-vector velocity of the gyro. The Earth's gravitational field causes spacetime to curve and consequently the connections $\Gamma^\mu_{\sigma\rho}$ are non-zero. Hence the gyro spin precesses, which is the *geodetic effect* deduced from GR by de Sitter in 1916, but inaccessible with the technology of the time. A second much weaker effect comes from the Earth itself rotating, which drags the spacetime frame seen by the gyro. This effect was deduced by Lense and Thirring in 1918.

Both the geodetic and frame-dragging effects were measured using the Gravity probe B (GP-B) satellite (Everitt et al. 2011), pictured in Figure 7.6. Conveniently these two effects are orthogonal, being gravitational analogs of electrical and magnetic effects. This experiment, like others described here, illustrates the many advantages and the sophistication of space borne experiments. There were four

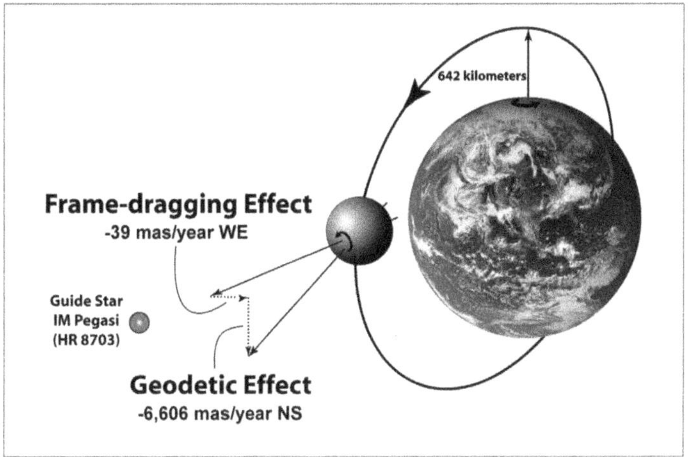

Figure 7.6. The satellite Gravity Probe B. Figure from Everitt et al. (2011). Courtesy of the American Physical Society.

gyros that floated freely inside the GP-B. Each was a 3.8 cm diameter homogeneous fused silica near-perfect sphere coated with a niobium film. These were contained in an enclosure cooled by liquid helium at 1.8 K rendering the niobium superconducting. The spheres were spun up by blasts of gas to 4300rpm. As predicted by Fritz London, the magnetic moment of a superconductor aligns precisely with the spin axis. A first step was to point this London moment at a guide star IM Pegasi using an onboard telescope. IM Pegasi is particularly stable with proper motion of under 0.15 mas/yr. The axes of the gyros were continuously monitored, over the 17 months that the coolant lasted, to determine the spin axis orientation and so measure its precession. Any external magnetic field was excluded by shielding. Analysis of the data transmitted to Earth gave values for the two precessions of 6601.8 ± 18.3 mas/yr and 37.2 ± 7.2 mas/yr in excellent agreement with the GR predictions of 6606.1 and 39.2 mas/yr. This remarkable experiment, 45 years in design and development before launch, achieved the astonishing angular precision of one one-hundred-thousandth of a degree per year.

A thought-provoking comment made by one of the experimenters: if space were nothing you couldn't twist it.

7.5 Gravitational Lensing

Einstein himself pointed out that if a galaxy lay between Earth and a distant source of light, the light would be focused by the gravitational field of the galaxy. Two or more images would then be seen, and if the alignment were exact a ring image would encircle that of the focusing galaxy. Many examples have been seen including the near-perfect Einstein ring shown in Figure 1.1. The deflection angle is simply that given by Equation (7.16). Note that the mass M is the sum of all the matter in the galaxy: later we shall see how lensing has been used to give an independent way to determine the total mass of visible plus the invisible dark matter in galaxies. Surdej et al. (1987) provided the first unequivocal evidence that two apparently separate sources were in fact images of a single source. These are the quasar pair UM673A and UM673B at a redshift of 2.72 separated by 2.2 arcsec. Galaxies are generally understood to be each centered on a black hole with mass equal to 10^6–10^{10} $M_\odot$. In their juvenile active state, matter is drawn into the hole and to conserve angular momentum this matter rotates in a disk surrounding the hole. The energy released when matter plunges into the black hole is funneled into jets of radiation perpendicular to the disk. Quasars are examples of such active galactic nuclei (AGN) where the Earth lies close to the jet axis. The luminosity is then many thousand times that of a quiescent galaxy and hence quasars are visible as compact sources at high redshifts (large distances) like the pair in question. Figure 7.7 shows how the spectra of UM673A and UM673B match precisely, with a tiny offset in redshift, corresponding to a velocity difference of only 24 ± 109 km s^{-1} in 281,600 km s^{-1}. It is inescapable that they are images of one and the same quasar. The ubiquity of examples of Einstein imaging, seen for example by the Hubble Space Telescope, is clear evidence that GR applies over the whole universe, not just locally.

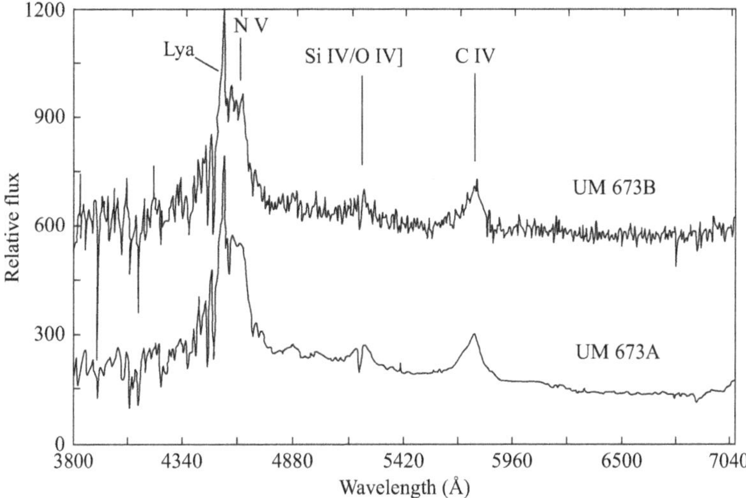

Figure 7.7. The low dispersion spectra of UM673A and UM673B recorded in December 1986. The resolution is about 1.3 nm. Figure 2 from Surdej et al. (1987). Courtesy Springer.

7.6 Exercises

1. What is the radar echo time delay for reflection from Mars at superior conjunction? Assume that the orbit of Mars is a circle of radius 228 million kilometers.
2. An observer is at rest at a distance r from a star of mass M. Show that his four-velocity in this frame (S) is $u^{\alpha} = (c/\sqrt{Z}, 0, 0, 0)$, where $Z = 1 - 2GM/rc^2$. Then consider a frame in free fall (F) which coincides momentarily with the observer's rest frame. Show that his four-velocity in frame F is $v^{\alpha} = (c, 0, 0, 0)$. Suppose a body with four-momentum p^{α} in frame S passes close the observer. Show that E^*, the energy of the body in frame F, is $p^{\alpha}u_{\alpha}$.

   This question illustrates that the energy measured by an observer is the invariant $p^{\alpha}u_{\alpha}$ where p^{α} is the body's four-momentum and u_{α} is the observer's four-velocity, both referred to the same frame.
3. Show that the angular radius of an Einstein ring is $\sqrt{\frac{4GM}{c^2}\frac{D_{LS}}{D_L D_S}}$, where M is the mass of the lens, D_L its distance from the Earth, D_S the distance of the source from the Earth, and D_{LS} the source–lens separation.
4. Calculate the precession of Mars' orbit given that the semimajor axis is 228 10^9 m, the orbital period is 687 days, and the orbit's eccentricity is 0.0934.
5. In the Schwarzschild metric the coordinate distances in the radial direction differ from those in the angular directions. Use the transformation $r = w[1 + k]^2$, where $k = GM/[2c^2w]$, to produce *isotropic coordinates* in which there is no difference between the radial and angular directions.

Further Reading

Will C M 1986 *Was Einstein Right?* (New York: Basic Books). A readable account of tests of Einstein's theory, backed by expert knowledge of the subject.

Will C M 2018 *Theory and Experiment in Gravitational Physics* (2nd ed.; Cambridge: Cambridge Univ. Press). This gives a thorough presentation by an expert gives a deep, exhaustive account of experimental tests.

References

Everitt, C. W. F., DeBra, D. B., Parkinson, B. W., et al. 2011, PhRvL, 106, 221101

Hellings, R. W. 1984, in General Relativity and Gravitation, ed. B. Bertotti, F. Felice, & A. Pascolini (Dordrecht: Reidel), 365

Hulse, R. A., & Taylor, J. H. 1975, ApJ, 195, L51

Reasenberg, R. D., Shapiro, I. I., MacNeil, P. E., et al. 1979, ApJ, 234, L219

Shapiro, I. I., Ash, M. E., Ingalls, R. P., et al. 1971, PhRvL, 26, 1132

Shapiro, S. S., Davis, J. L., Lebach, D. E., & Gregory, J. S. 2004, PhRvL, 92, 121101

Surdej, J., Magain, P., Swings, J.-P., et al. 1987, Natur, 329, 695

Ward, W. R. 1970, ApJ, 162, 345

Introduction to General Relativity and Cosmology
(Second Edition)

Ian R Kenyon

Chapter 8

Black Holes

The general theory of relativity provides a precise, well-tested description of spacetime within the solar system, where spacetime is nearly flat—with the parameter quantifying spacetime curvature, $2GM/rc^2$, reaching only 10^{-5} at the Sun's surface. In regions where this parameter becomes of order unity the strong curvature isolates these regions from the rest of the universe. GR provides the means to go some way to determine the properties of these isolated regions of spacetime.

In 1784 the Rev. J. Michell became the first person to draw attention to the implications of the gravitational potential GM/r becoming large (Michell 1784). For a body of mass m to be able to escape to infinity from a star of mass M and radius r its kinetic energy must exceed the gravitational potential. It requires an initial velocity v, such that

$$mv^2/2 \geqslant GMm/r, \quad \text{i. e.,} \quad v \geqslant (2GM/r)^{1/2},$$

so that escape is only possible for velocities greater than $\sqrt{2GM/r}$. With a dense enough star the escape velocity reaches the velocity of light: the star's radius is

$$r_0 \equiv 2GM/c^2. \tag{8.1}$$

A more compact star would be invisible. The curvature of spacetime is so severe that we can only hope to give a consistent account of conditions using GR. For light traveling radially in a region described by the Schwarzschild metric, Equation (3.48), with $ds^2 = d\Omega^2 = 0$, becomes

$$0 = c^2\, dt^2(1 - r_0/r) - \frac{dr^2}{(1 - r_0/r)}.$$

Thus

$$c\, dt = \frac{dr}{1 - r_0/r}. \tag{8.2}$$

If a star shrinks to a radius less than $r_0 \equiv 2GM/c^2$ the time taken for light to emerge from the spherical surface at r_0 becomes infinite. An observer will never receive any light emitted from within that radius. The spherical surface at r_0 is called an *event horizon*, and r_0 is known as the *Schwarzschild radius* of the star. Any star shrinking within its Schwarzschild radius becomes invisible, a *black hole*. Black holes are in one sense simple: lack of contact with the universe outside limits the properties they can exhibit to their mass, angular momentum, and charge.

The Schwarzschild radius of the Sun is 2.96 km and that of the Earth is 8.9 mm.

For simplicity the analysis here will be carried through for non-rotating electrically neutral (Schwarzschild) black holes. It is likely that most black holes are in fact rotating, and given the name *Kerr black holes*. Where necessary, the effects of this rotation on the analysis will be pointed out. Space is populated by matter so that any charge on a black hole would soon be neutralized. In 1931 Chandrasekhar (Chandrasekhar 1931) deduced that the gravitational self-attraction of a sufficiently massive star leads inevitably to its collapse to a point.[1] The mass limit depends on the equation of state assumed for the star: Chandrasekhar gave a limit of 1.4 $M_\odot$, and it is now considered to be around 2.1 $M_\odot$. A star that is 1.4 times more massive than our Sun has a Schwarzschild radius of only 2 km, and such a star would then attain a mean density of 10^{20} kg m^{-3}, far beyond the density of nuclei. It seems difficult to imagine how this could come about. Surely the mounting pressure would eventually halt the contraction? In the last chapter we saw that the curvature tensor is proportional to the stress–energy tensor, so that *pressure also* contributes to the gravitational self-attraction of a star. Another paradoxical property of black holes, treated below, is that they can radiate through a quantum field effect discovered by Hawking (1974).

The first topic treated here is the unusual and sometimes counter-intuitive properties of spacetime near a black hole and the orbits followed there by matter and radiation. The interplay of gravitational and quantum physics is the second topic: this includes an account of Hawking radiation from black holes, an outline of the thermodynamic properties of black holes, and a brief section to expose the puzzle over whether information is lost when matter enters a black hole. The third topic, last but not least, recounts the now convincing experimental evidence for the existence of both stellar and galactic black holes.

8.1 The Spacetime Structure

The event horizon at the radius $r_0 = 2GM/c^2$ is a fundamental feature of a Schwarzschild black hole. Electromagnetic radiation originating within the horizon can never escape, so that spacetime inside the horizon is effectively isolated from the rest of the universe.[2] In order to obtain an idea of how spacetime behaves across the

[1] Chandrasekhar won the Nobel Prize in Physics for this work, 52 years later—a record.
[2] Hawking radiation escapes at the black hole's surface.

horizon, consider a probe falling radially toward a star that has collapsed inside its Schwarzschild radius. Equation (7.6) is the appropriate orbital equation. If the probe moves along a radius with $\theta = \pi/2$, $\phi = 0$, then

$$\frac{(\mathrm{d}r/\mathrm{d}\tau)^2}{2} - \frac{GM}{r} = \frac{T}{m}$$

For simplicity the probe is taken to have started at rest at an infinite distance from the hole, making its kinetic energy T vanish. The equation then reduces to

$$c\ \mathrm{d}\tau = \pm\left(\frac{r}{r_0}\right)^{1/2} \mathrm{d}r \tag{8.3}$$

whereas defined, r_0 is $2GM/rc^2$. Taking the negative sign for travel inward and integrating gives

$$\tau = \tau_0 - \frac{r_0}{c}\left(\frac{2}{3}\right)\left(\frac{r}{r_0}\right)^{3/2}, \tag{8.4}$$

where τ_0 is the proper time when the probe reaches the center ($r = 0$). This path is plotted in Figure 8.1 as the solid curve. A key feature to note is that the *proper time* τ, recorded by an onboard clock, changes smoothly on crossing the horizon. Once across the horizon the probe soon reaches the center of the hole: in the case of a hole of mass 10 $M_\odot$ this interval $\tau_0 - \tau$ is 10^{-4} s. In practice instruments could not survive, being torn apart by the gravitational field gradients (tides). Next consider

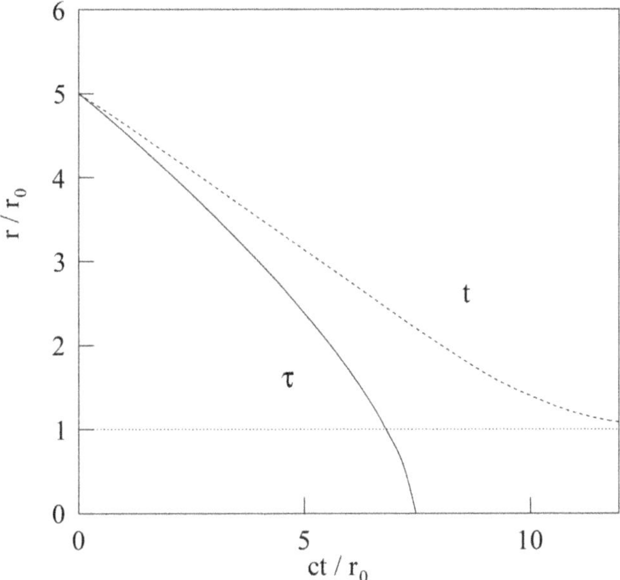

Figure 8.1. The proper time τ and the coordinate time t plotted as a function of the radial coordinate r for a probe falling radially into a black hole. τ and t are set equal at $r = 5r_0$.

how the journey appears to a distant observer. The *coordinate time t* measured by this remote observer is related to the proper time through Equation (7.3):

$$\left(1 - \frac{r_0}{r}\right)\frac{\mathrm{d}t}{\mathrm{d}\tau} = K \quad \text{a constant.}$$

Imposing the initial conditions, with the distant probe at rest, gives $K = 1$. Thus

$$\left(1 - \frac{r_0}{r}\right)\mathrm{d}t = \mathrm{d}\tau. \tag{8.5}$$

Equation (8.4) can be used to replace $\mathrm{d}\tau$ in Equation (8.5) and rearranging we obtain

$$c\mathrm{d}t = -\frac{\mathrm{d}r\ r^{3/2}}{(r - r_0)r_0^{1/2}}.$$

Integration over the inward journey yields

$$t = t_0 + \frac{r_0}{c}\left[-\frac{2}{3}\left(\frac{r}{r_0}\right)^{3/2} - 2\left(\frac{r}{r_0}\right)^{1/2} + \ln\left|\frac{(r/r_0)^{1/2} + 1}{(r/r_0)^{1/2} - 1}\right|\right]. \tag{8.6}$$

Remote from the black hole this reduces to

$$t \approx t_0 - \frac{r_0}{c}\left(\frac{2}{3}\right)\left(\frac{r}{r_0}\right)^{3/2} - 2\left(\frac{r_0}{c}\right)\left(\frac{r}{r_0}\right)^{1/2}.$$

By choosing t_0 suitably it is possible to arrange that t and τ are equal at some large distance R. This choice is

$$t_0 \approx \tau_0 + 2\,\frac{r_0}{c}\left(\frac{R}{r_0}\right)^{1/2}.$$

From Equation (8.6) it is clear that as r tends toward r_0 then t tends to infinity. The world line described by Equation (8.6) is shown by the broken line in Figure 8.1. The very different behavior of coordinate and proper time as the probe approaches and crosses the horizon illustrates vividly how the curvature of spacetime makes it impossible to cover all spacetime with one set of Cartesian coordinates.

We can show in a few steps that signals received from a probe crossing the horizon fade away quickly. Suppose the probe emits signals at constant frequency (from an onboard quartz crystal oscillator): as it approaches the horizon, the photons traveling to the distant observer arrive less frequently and increasingly redshifted. Both effects diminish the energy received so that eventually the signals are undetectable. The photons in question follow radial paths for which the time between emission and detection is given by Equation (8.2)

$$\begin{aligned} t' &= \int \frac{\mathrm{d}r}{c(1 - r_0/r)} \\ &= \frac{r - r_0}{c} + \frac{r_0}{c}\ln\left(\frac{R - r_0}{r - r_0}\right), \end{aligned} \tag{8.7}$$

where the probe is at radius r and the observer is at radius R. As far as the observer is concerned the arrival time T is measured relative to some fixed event, which can be the departure of the probe. This time interval is the sum of the inward travel time of the probe given by Equation (8.6) and the return time of the photons from probe to observer is given by Equation (8.7). The expressions simplify a good deal once it is noted that the most significant contributions come from terms containing $\ln[\sqrt{r} - \sqrt{r_0}]$, which dominate when the probe approaches the horizon and $r \to r_0$: in comparison terms like $r - r_0$, $\ln[\sqrt{r} + \sqrt{r_0}]$ and $\ln[\sqrt{r_0}]$ can be ignored. First from Equation (8.6)

$$t \approx -\left(\frac{r_0}{c}\right)\ln\left[\left(\frac{r}{r_0}\right)^{1/2} - 1\right] \approx -\frac{r_0}{c}\ln(r^{1/2} - r_0^{1/2})$$
$$\approx -\left(\frac{r_0}{c}\right)\ln(r - r_0).$$

Next, from Equation (8.7)

$$t' \approx -\left(\frac{r_0}{c}\right)\ln(r - r_0).$$

Thus the total elapsed time is

$$T = t + t' \approx -2\left(\frac{r_0}{c}\right)\ln(r - r_0). \tag{8.8}$$

Now the energy $L(T)$ received per unit time contains a factor $\sqrt{1 - r_0/r}$ due to the increase in time interval between photons and another factor $\sqrt{1 - r_0/r}$ due to their redshift. Thus

$$L(T) \propto \left(1 - \frac{r_0}{r}\right) = \frac{r - r_0}{r}.$$

Using Equation (8.8) this becomes

$$L(T) \propto \exp\left(-\frac{cT}{2r_0}\right).$$

This analysis applies equally to a star in the act of collapsing through its horizon: it reddens and fades on a timescale $2r_0/c$. In the case of a star of mass 5 $M_\odot$ it would blink out in a microsecond.

The critical difference between spacetime inside and outside the horizon lies in the sign reversal of the metric coefficients $g_{00} = (1 - r_0/r)$ and $g_{11} = -1/(1 - r_0/r)$ at the surface $r = r_0$. Table 8.1 illustrates this. Therefore if a small change in t is made at constant radius inside the horizon ($r < r_0$),

$$\frac{ds^2}{c^2} = d\tau^2 = g_{00}\, dt^2 < 0.$$

Table 8.1. Sign of the metric coefficients outside and inside the horizon (r_0)

	$r > r_0$	$r < r_0$
g_{00}	+	−
g_{11}	−	+

This is opposite in sign to the effect of a similar small change in t outside the horizon ($r > r_0$), namely

$$\frac{\mathrm{d}s^2}{c^2} = \mathrm{d}\tau^2 = g_{00}\,\mathrm{d}t^2 > 0.$$

Inside the horizon a separation in coordinate time has become *space-like* ($\mathrm{d}\,s^2 < 0$). Similar considerations show that inside the horizon a separation in radial coordinate only has become *time-like* ($\mathrm{d}\,s^2 > 0$). The curvature is so intense that if we insist on using coordinates appropriate to distant flat spacetime, then we find that space and time inside the horizon interchange the properties normally associated with them.

This, unsurprisingly, has significant consequences. One is that if the coordinate r is held fixed for any length of time the interval $\mathrm{d}s^2$ is negative and hence *space-like*. This would require a body taking such a path to follow a space-like geodesic; a logical impossibility. Hence static equilibrium inside the horizon must be impossible physically. Viewed from inside the horizon the external universe would appear equally strange. It is worth remembering here that spacetime as seen by someone in free fall would locally still be a Minkowski space. However, the region over which the Minkowski frame matches actual spacetime would be small because of the intense curvature, and would become smaller as the center of the black hole is approached. Crossing the horizon of a small black hole would be marked by a tidal force, tearing extended objects apart radially. This tidal force at the horizon diminishes as the black hole mass increases. When the black hole is one as massive as our Galaxy then going across the horizon would not be especially noticeable, until you tried to leave.

The absence of any indication in the world line of the onboard clock that it is crossing the horizon was illustrated in Figure 8.1 and shows that the horizon is not a *physical* singularity; rather it is a *mathematical* singularity.

What then is the import of the mathematical singularity with g_{11} diverging to infinity as r goes to r_0? The mathematical singularity arises because the set of coordinates imposed everywhere is best suited to regions of small curvature. A set of coordinates more appropriate to the locale of the black hole was invented by Eddington in 1924 (Eddington 1924) and rediscovered by Finkelstein in 1958 (Finkelstein 1958); the time coordinate t is replaced by $\tilde{t}$ such that

$$\tilde{t} = t + \frac{r_0}{c} \ln \left| \frac{r}{r_0} - 1 \right|$$

and

$$\mathrm{d}\tilde{t} = \mathrm{d}t - \frac{\mathrm{d}r}{c(1 - r/r_0)}.$$

In terms of this new time coordinate the Schwarzschild metric Equation (3.48)

$$\mathrm{d}s^2 = c^2\mathrm{d}t^2(1 - r_0/r) - \mathrm{d}r^2/(1 - r_0/r) - r^2\mathrm{d}\Omega^2$$

becomes

$$\mathrm{d}s^2 = \left(1 - \frac{r_0}{r}\right)c^2\,\mathrm{d}\tilde{t}^2 - 2c\,\mathrm{d}r\,\mathrm{d}\tilde{t}\left(\frac{r_0}{r}\right) - \mathrm{d}r^2\left(1 + \frac{r_0}{r}\right) - r^2\,\mathrm{d}\Omega^2. \tag{8.9}$$

The metric coefficients of Equation (8.9) have no mathematical singularity at the horizon. We shall show next that Eddington's coordinates (also called Eddington–Finkelstein coordinates) provide the basis for a clearer understanding of spacetime structure close to the black hole, although such coordinates would be a strange choice in a region remote from the black hole.

The radial path of a light ray in these new coordinates is

$$0 = \left(1 - \frac{r_0}{r}\right)c^2\,\mathrm{d}\tilde{t}^2 - 2c\,\frac{r_0}{r}\,\mathrm{d}r\,\mathrm{d}\tilde{t} - \left(1 + \frac{r_0}{r}\right)\mathrm{d}r^2,$$

which has two solutions

$$\frac{\mathrm{d}\tilde{t}}{\mathrm{d}r} = -\frac{1}{c} \quad \text{and} \quad \frac{\mathrm{d}\tilde{t}}{\mathrm{d}r} = \frac{1}{c}\,\frac{1 + r_0/r}{1 - r_0/r}. \tag{8.10}$$

These solutions describe the paths of ingoing and outgoing rays respectively, and in the absence of a black hole would become

$$\frac{\mathrm{d}\tilde{t}}{\mathrm{d}r} = -\frac{1}{c} \quad \text{and} \quad \frac{\mathrm{d}\tilde{t}}{\mathrm{d}r} = +\frac{1}{c}.$$

Light cones constructed according to Equation (8.10) are drawn in Figure 8.2 at various points on the trajectory of a source falling into a black hole. The maximum inward component of the velocity of light is c throughout: the left-hand edges of the light cones drawn in the figure have the same length and make the same angle with the time axis. On the other hand the maximum outward component varies with the radial distance. When the source is far from the horizon $\mathrm{d}\tilde{t}/\,\mathrm{d}r$ is $+1/c$ for light directed outwards. Then as the source approaches the horizon $\mathrm{d}\tilde{t}/\,\mathrm{d}r$ increases until, at the horizon it points along the time axis; the maximum outward component of the

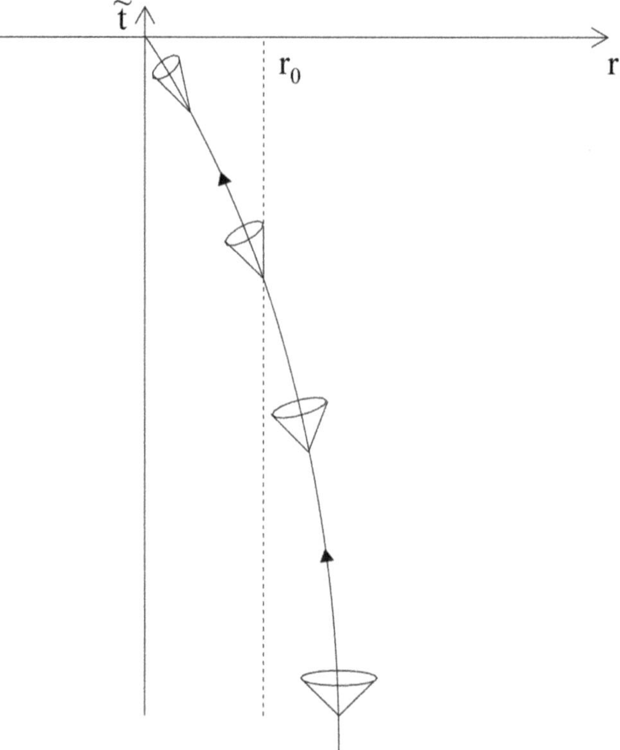

Figure 8.2. The light cones of a probe falling into a black hole. $\tilde{t}$ is the Eddington time coordinate.

velocity of light therefore falls to zero at the horizon. When the source goes inside the horizon $d\tilde{t}/dr$ becomes negative for "outgoing" as well as ingoing rays so that the whole light cone is tilted inward. This also means that once the horizon is crossed the source will inevitably head toward the center of the black hole. However powerful the rocket engine may be, the probe's velocity vector must lie *inside* the light cone, and this seals the fate of the probe.

There exists an equally valid alternative choice for Eddington coordinates with

$$\tilde{t} = t - \frac{r_0}{c} \ln \left| \frac{r}{r_0} - 1 \right|.$$

Now the light cone of a source within the horizon tilts so that it points *outward*. The physical situation is that of a *white* hole ejecting material from the singularity at $r = 0$ into spacetime. Gravitational collapse may create a black hole, but we don't know of any physical process to generate its converse, the white hole: mathematically they are equally valid. We have learnt too that the choice of coordinate scheme can bring with it an apparent singularity that in fact does not exist.

However the singularity associated with the center of a black hole cannot be removed by the choice of coordinates. In 1965 Penrose showed that physical singularities are a feature of a general relativity (Penrose 1965). Such a singularity marks the abrupt end of an entire region of spacetime.

In non-rigorous language Penrose's theorem states that if a continuous surface (Cauchy surface) can be drawn across spacetime (the universe) marking a moment in (cosmic) time, which is crossed once only by all causal paths, *and* a closed future trapped surface (for example a Schwarzschild horizon) exists, then there are future light paths that do not continue for infinite time, but instead stop (thus forming a singularity). There is no need for spherical or other symmetry to complete this proof.

Hawking & Penrose (1970) refined the proof that physical singularities are broadly expected in general relativity. Penrose was awarded the Nobel Prize in physics in 2020 for his work showing that general relativity leads to singularities. There is no accepted theory for describing the conditions at the center of the black hole ($r = 0$), a singularity where the curvature of spacetime becomes infinite. Penrose conjectured that there are no *naked singularities*, that is to say singularities that we can directly observe, all would be hidden behind some horizon: this is called *cosmic censorship*. Of course these proofs and speculation are purely valid *classically*. What happens in practice depends on what the quantum theory of gravitation might be, which is so far not understood.

8.2 Orbits Around Black Holes

The analysis of stable orbits developed in Chapter 7 can also be applied when the parent body is a Schwarzschild black hole. Combining Equations (7.4) and (7.5) to eliminate $\mathrm{d}\phi/\,\mathrm{d}\tau$ gives

$$\frac{E^2}{Zc^2} - \frac{m^2(\mathrm{d}r/\mathrm{d}\tau)^2}{Z} - \frac{m^2J^2}{r^2} = m^2c^2,$$

where E is a constant of motion equivalent to the classical energy, and J the angular momentum per unit mass is another constant of motion. Rearranging the above equation gives

$$\frac{E^2}{m^2c^4} = \frac{(\mathrm{d}r/\mathrm{d}\tau)^2}{c^2} + Z\left(1 + \frac{J^2}{c^2r^2}\right). \tag{8.11}$$

We can make the substitutions

$$\epsilon = \frac{E^2}{m^2c^4} \quad \text{and} \quad V(r) = Z\left(1 + \frac{J^2}{c^2r^2}\right). \tag{8.12}$$

and the equation becomes:

$$\epsilon = (\mathrm{d}r/\mathrm{d}\tau)^2/c^2 + V(r), \tag{8.13}$$

which we recognize is formally an equation for one-dimensional motion in a potential $V(r)$ with total energy ϵ. The equivalent potential energy $V(r)$ is plotted in Figure 8.3 against r/r_0 for several values of $[J/r_0c]^2$. In those cases where $[J/r_0c]^2 > 3.0$ the potential has a trough in which stable orbits can exist. Referring to the potential with $[J/r_0c]^2 = 4.0$: at the energy indicated by the broken line the elliptical orbit has a semimajor axis length of r_B and a semiminor axis length r_A. When the energy is exactly at the peak, or at the minimum of the trough, the orbits are circular, but only the latter orbit is stable. At these circular orbits

$$dV/dr = 0,$$

that is

$$r = r_J\left[1 \pm \left(1 - \frac{3r_0}{r_J}\right)^{1/2}\right], \tag{8.14}$$

where $r_J = J^2/(c^2r_0)$. The radius of the *smallest* stable circular orbit is found by taking the factor under the square root sign to be zero:

$$1 - 3r_0/r_J = 0 \quad \text{and} \quad J^2 = 3r_0^2c^2. \tag{8.15}$$

Then

$$r = r_J = 3r_0. \tag{8.16}$$

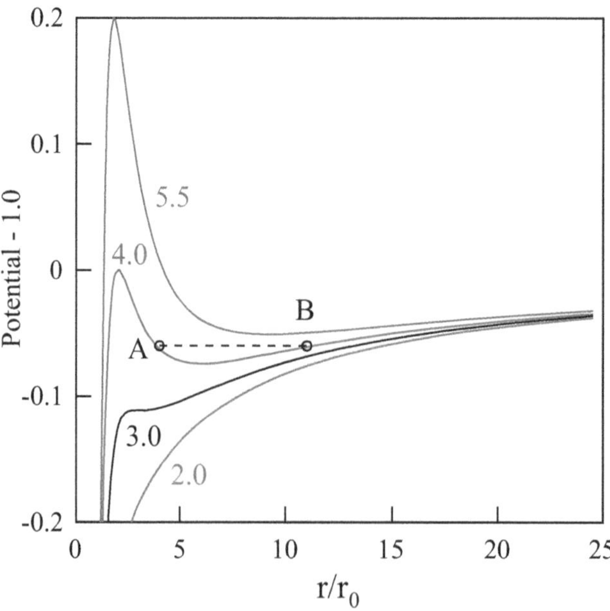

Figure 8.3. Equivalent radial potential seen by a massive body in orbit around a Schwarzschild black hole. The labels on the potentials refer to the values of $J^2/r_0^2c^2$.

The energy of this smallest stable circular orbit is given by Equation (8.11) as

$$E^2(\text{min}) = \frac{8m^2c^4}{9} \quad \text{and} \quad E(\text{min}) = \left(\frac{8}{9}\right)^{1/2} mc^2, \tag{8.17}$$

and lies on the potential curve labeled 3 in the figure. Therefore the binding energy amounts to a fraction $1 - \sqrt{8/9}$ or 5.72% of the rest mass energy. This quantity is of cosmic importance: it is the energy released when material accreted around a black hole spirals into the stable orbit with lowest energy. By comparison the maximum energy release in thermonuclear fusion when hydrogen burns to ^{56}Fe is only 0.84% of the rest mass energy. Gravitational energy release is therefore the most potent energy source contributing to stellar processes. When a body orbits a rotating black hole the gravitational energy release can go much higher, reaching 42% of the rest mass energy in favorable cases. In most galaxies there is a massive black hole, in our Galaxy of mass $\sim 4 \times 10^6\ M_\odot$. In many galaxies, but not ours, matter is accreting in a disk around the black hole under the intense gravitational attraction, and then falls into the black hole with a huge energy release in radiation. Such galactic cores are known as active galactic nuclei (AGNs). Those AGNs whose intense radiation is observed without obscuration by the accreting disk are known as *quasars*.

In the image of the black hole M87* shown in Figure 1.9 the bright ring matches the circular *unstable* photon orbits around the black hole. Matter in the accretion disk lying along these obits emits photons that after orbiting then escape and have been detected by the *Event Horizon Telescope* (EHT) Collaboration (Beckenstein 1972). We show now that the radius of this unstable orbit is $3r_0/2$.

The Schwarzschild metric equation, Equation (3.48), in the case of a photon traveling in the equatorial plane around a static black hole is

$$Zc^2(\mathrm{d}t/\mathrm{d}\lambda)^2 - (\mathrm{d}r/\mathrm{d}\lambda)^2/Z - r^2(\mathrm{d}\phi/\mathrm{d}\lambda)^2 = 0, \tag{8.18}$$

where λ is the path length parameter. In this orbit the components of the Schwarzschild metric are independent of both the azimuthal angle ϕ and time t. Hence taking results from analyzing the precession of Mercury there are two invariants given by Equations (7.3) and (7.4):

$$e = Zc\mathrm{d}t/\mathrm{d}\lambda \quad \text{and} \quad \ell = r^2\mathrm{d}\phi/\mathrm{d}\lambda.$$

These are equivalent to the conservation, respectively, of energy and angular momentum. Then the equation for the photon orbit round the black hole reduces to

$$e^2/Z - (\mathrm{d}r/\mathrm{d}\lambda)^2/Z - \ell^2/r^2 = 0. \tag{8.19}$$

Multiplying by Z/ℓ^2 and putting $b = |\ell/e|$ gives

$$b^{-2} = \ell^{-2}(\mathrm{d}r/\mathrm{d}\lambda)^2 + Z/r^2. \tag{8.20}$$

We can interpret the parameter b physically using the top-right inset in Figure 8.4. This inset shows a section of a photon's path from the neighborhood of the black hole out to a remote observer. We write R to distinguish the large values of r remote

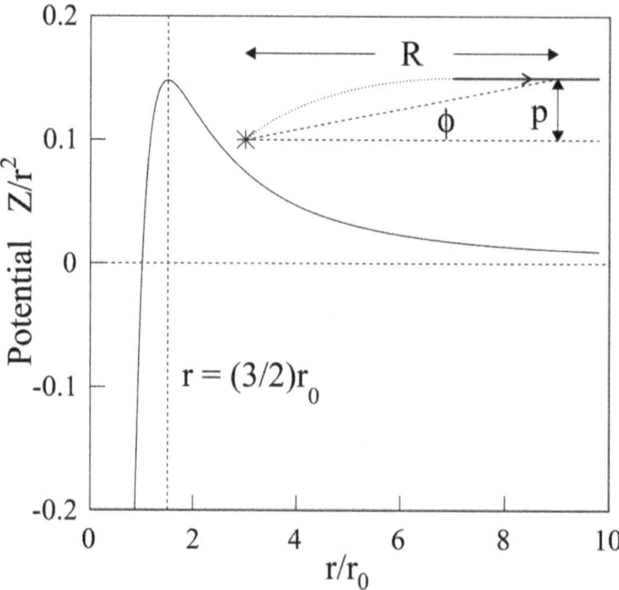

Figure 8.4. Equivalent potential for photon near a Schwarzschild black hole. The inset shows the path from the photon ring to the observer. The symbols are used in the text.

from the black hole. Then at remote points: $Z = 1$ and $\phi = p/R$, where p is called the *impact parameter*. Equally

$$b = |\ell/e| = \frac{R^2(\mathrm{d}\phi/\mathrm{d}\lambda)}{Z(c\mathrm{d}t/\mathrm{d}\lambda)} = R^2\frac{\mathrm{d}\phi}{c\mathrm{d}t} = R^2\frac{\mathrm{d}(p/R)}{c\mathrm{d}t} = p\frac{\mathrm{d}R}{c\mathrm{d}t} = p. \tag{8.21}$$

The main body of Figure 8.4 shows the potential from Equation (8.20). It peaks where $\mathrm{d}(Z/r^2)/\mathrm{d}r = 0$, that is where

$$r = 3GM/c^2 = 3r_0/2 = r_\gamma. \tag{8.22}$$

This is the radius of an *unstable* circular photon orbit; any perturbation would propel the photon either inward or outward. We can obtain the impact parameter for this unstable orbit by substituting r_γ for r in Equation (8.20): the orbit is circular so that $\mathrm{d}r/\ \mathrm{d}\lambda$ is zero and we get

$$b_\gamma = p_\gamma = \sqrt{r_\gamma^2/(1 - r_0/r_\gamma)} = 3\sqrt{3}\,r_0/2. \tag{8.23}$$

Then, seen from the Earth, a distance R away, the bright ring around the black hole's shadow has an angular radius b_γ/R. If, as seems probable, the black hole is rotating, the Kerr metric replaces the Schwarzschild metric to describe spacetime around the black hole. The angular size of the image is reduced, but only by 5% even if the rotation is maximal. Figure 8.5 shows photon paths from around the black hole in a plane containing the center of the black hole. That in red is the unstable orbit with the path to the Earth. Material anywhere along the

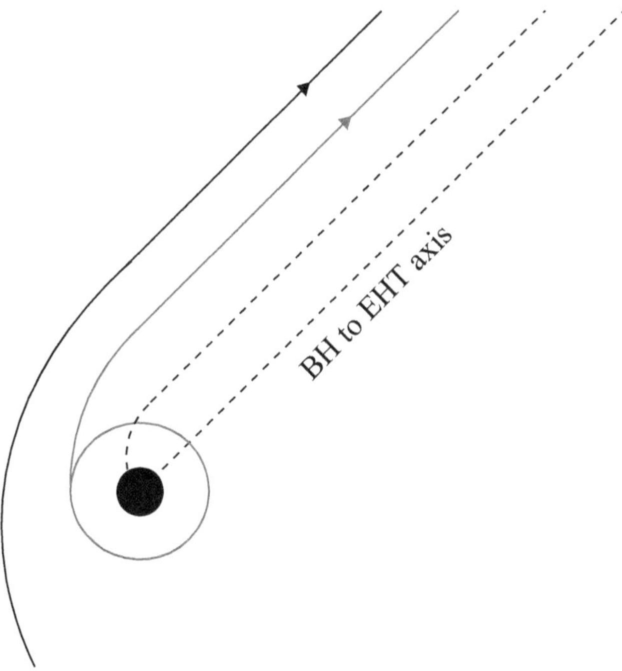

Figure 8.5. Photon paths to the Earth from near a black hole. The unstable circular path and its link to the telescope are indicated in red. The broken lines are ingoing paths only. The components are not drawn to scale.

unstable orbit can emit photons into this orbit: thus radiation detected with impact parameter b_γ should be intense. The broken lines ending on the horizon are ingoing paths only. Observations made by the EHT of black holes at the centers of the nearby M87 galaxy and our own Galaxy are described in Section 8.10.

8.3 Rotating Black Holes

Anyone outside a black hole cannot follow what happens to material once it crosses the horizon. This raises the question of whether it is possible to make any distinction between black holes, beyond the difference in mass. There is a conjecture that black holes have only three observable properties: mass, spin J (their angular momentum) and charge. Kerr's solution of Einstein's equation for spinning black holes shows that the event horizon is smaller than for Schwarzschild black holes: the radius is

$$r_- = r_0/2 + \sqrt{(r_0/2)^2 - a^2} \tag{8.24}$$

where $a = J/Mc$. The larger the angular momentum the smaller the radius of the horizon becomes. In the limit that J/Mc exceeds $r_0/2$ it would become complex and the singularity at the center of the black hole might become accessible. What would then happen is unclear; Penrose conjectured that some physical principle (cosmic censorship) would guarantee that singularities remain hidden behind horizons. A second important surface defines a region outside the event horizon from which escape is possible but within which static equilibrium cannot be maintained. Any matter in the intermediate volume, the *ergosphere*, will rotate with the black hole. The outer surface of the ergosphere is a spheroid of revolution with a radial coordinate r_+, which depends on the polar angle θ it makes with the spin axis

$$r_+(\theta) = r_0/2 + \sqrt{(r_0/2)^2 - a^2 \cos^2 \theta}\,. \tag{8.25}$$

On this surface it is just possible to remain at rest. Penrose proposed a technique to extract energy from a rotating black hole. In one form this involves scattering radiation from the ergosphere; some energy from the rotation of the black hole is then transferred to the scattered radiation.

The quanta of electromagnetic fields are photons that cannot escape from within a black hole horizon. However black holes can carry charge and exert an electrostatic field, which appears to contradict the previous statement. The answer to this puzzle takes us into the realm of quantum field theory. Usually we picture light and hence photons as transversely polarized. However photons can equally have polarization in the longitudinal or time-like directions. When studying the propagation of electromagnetic waves the contributions of these polarization components are seen to cancel. They do however contribute constructively to the *electrostatic* field and this remains when charge enters the horizon. Hence a charged black hole can exert an external electrostatic field. The same argument has implications if the gravitational field due to the black hole is quantized and is carried by massless field particles, analogous to the massless photons.

8.4 The Planck Scale

Figure 8.6 displays the realms of general relativistic and quantum effects on a plot of an object's mass versus its linear size. General relativistic effects dominate when a star collapses within its Schwarzschild radius $2GM/c^2$. The upper sloping line $r = 2GM/c^2$ marks this transition; anything to the left is a black hole. The black * symbol marks the location of Sgr A*, the black hole at the center of the Galaxy. The lower sloping line has the equation $R = \hbar/Mc$, defining the Compton wavelength of a mass M. According to the uncertainty principle a mass M cannot be located more precisely than its Compton wavelength. Therefore quantum fluctuations dominate to the left of this line. At sufficiently small distances and large enough masses gravitational and quantum effects both become important. That condition is reached where the two lines intersect on Figure 8.6. Then at this point, known as the *Planck scale*, where the Schwarzschild radius equals the Compton wavelength:

$$GM_P/c^2 = \hbar/M_P c,$$

whereas customary a factor of 2 is omitted. This gives the Planck mass

$$M_P = (\hbar c/G)^{1/2} = 2.18 \times 10^{-8}\ \text{kg},$$

the Planck momentum

$$P_P = M_P c = 6.54\ \text{kg m s}^{-1}, \tag{8.26}$$

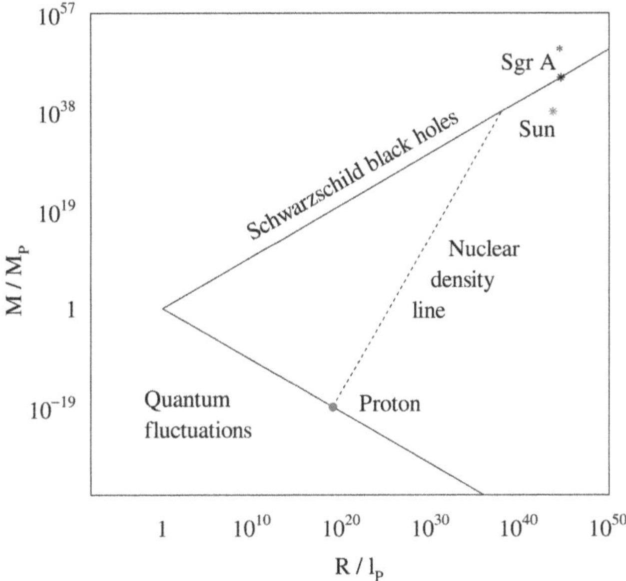

Figure 8.6. Mass-radius diagram showing the regions where general relativistic effects and quantum effects are important. M_P is the Planck mass and ℓ_P the Planck length.

and the Planck distance

$$\ell_P = (\hbar G/c^3)^{1/2} = 1.62 \times 10^{-35} \text{ m}. \tag{8.27}$$

Planck was the first to notice that this combination of the fundamental constants provided a natural scale for length. The corresponding Planck energy measured in GeV (10^9 eV) is

$$E_P = M_P c^2 = 1.22 \times 10^{19} \text{ GeV},$$

and quantum fluctuations of this size can persist for a time, also named after Planck:

$$t_P = \hbar / M_P c^2 = 5.39 \times 10^{-44} \text{ s}.$$

On the Planck scale a quantum fluctuation can generate a black hole, so that consistency between physical theories seems to require that GR should be quantized. Quantities on the Planck scale are remote from our experience; nonetheless if the history of the universe is followed back toward the Big Bang, an epoch will be reached in which the energy per particle exceeded the Planck energy. This *Planck era* endured for a time of order t_P. It is expected that a complete description of the evolution of the very early universe and of black holes will be based on a quantum theory of gravity: only tantalizing indications of the form of such a theory exist at present.

8.5 Hawking Radiation

In 1974 Hawking (Hawking 1974) made the astonishing prediction that black holes, previously regarded simply as absorbers of radiation, can in fact radiate, through a purely quantum effect. Quantum mechanics has revealed that the vacuum, rather than being inert, is in a state of constant activity due to the continuous creation and annihilation of particle–antiparticle pairs. Thus a pair of photons can be created close to, but outside the black hole horizon with four-momenta $(pc, -\mathbf{p})$ and $(-pc, \mathbf{p})$. The net four-momentum is zero, but the negative energy of one photon violates the requirement that real photons have positive energy. According to Heisenberg's uncertainty principle this *virtual* photon can only exist for a time

$$\Delta t \sim \hbar/pc.$$

For some directions of emission the negative-energy photon will cross the horizon. Once across the horizon and inside the black hole the space-like components of four-vectors become time-like, and vice-versa. Thus the photon's negative energy becomes an acceptable spatial momentum and its momentum converts into an acceptable *positive* energy. Its lifetime is no longer restricted and so it travels quite freely within the black hole. Its positive-energy partner travels freely outward, and such photons make up the Hawking radiation emerging from near the horizon. An estimate of the temperature of this radiation can be obtained as follows using the uncertainty principle. The position of a photon emitted from the surface of the black hole is uncertain to $\sim r_0$. Accordingly the uncertainty in the photon momentum is $\Delta p \sim \hbar/r_0$. This momentum can be expressed in terms of a thermal energy $k_B T$ at a temperature T where k_B is the Boltzmann constant: $\Delta p \approx k_B T/c$. Combining these two relations for momentum gives

$$\frac{k_B T}{c} \approx \frac{\hbar}{r_0} = \frac{\hbar}{2GM/c^2}.$$

Thus $T \approx \hbar c^3/(2k_B GM)$. The exact expression obtained by Hawking only differs by a factor 4π:

$$T = \frac{\hbar c^3}{8\pi k_B GM}. \tag{8.28}$$

Putting in numbers we have:

$$T = 6 \times 10^{-8}(M_\odot/M)\ \mathrm{K}. \tag{8.29}$$

The rate at which a black hole loses energy through Hawking radiation is

$$\frac{\mathrm{d}(Mc^2)}{\mathrm{d}t} = \sigma T^4(\text{surface area}) \propto M^{-2},$$

where σ is the Stefan–Boltzmann constant. The lifetime is consequently proportional to M^{-3}. Small black holes therefore have higher temperatures and radiate their energy more rapidly than larger black holes. The lifetime of a black hole is given approximately by

$$\tau \approx \left(\frac{M}{10^{11}\ \mathrm{kg}}\right)^3 \times 10^{10}\ \text{years}. \tag{8.30}$$

Thus any black hole of one solar mass would not have had time to evaporate since the origin of the universe; however, very small black holes could have formed early in the life of the universe and subsequently evaporated.

Another important result proved by Hawking will be used the next section. His theorem maintains that in any process the surface area of a black hole cannot shrink.

8.6 Black Hole Thermodynamics

The thermal nature of Hawking radiation suggests that a black hole could possess other thermodynamic properties. Beckenstein (1972) was the first to appreciate that a black hole should have entropy. Noting that neither classical entropy nor the area of a black hole may grow smaller, he conjectured that the entropy of a black hole was proportional to the area of its horizon. This idea was incorporated by Bardeen, Carter and Hawking (Bardeen et al. 1973) in the four laws summarizing black hole thermodynamics.

These laws are presented here and compared with the corresponding laws of classical thermodynamics.[3]

- The first classical law is that the temperature of a body in thermal equilibrium should be the same throughout the body. In the case of a black hole the parallel law is that the interface with the rest of nature, the horizon, should be at a uniform temperature. Locally the gravitational force at the surface of a black hole is

$$\frac{GM}{r_0^2} = \frac{c^4}{4GM} = \frac{2\pi c}{\hbar} k_{\mathrm{B}} T,$$

  demonstrating that the Hawking temperature must also be uniform over the horizon's surface.
- The second classical law states that energy E is conserved. In the case of a volume of gas V: $\mathrm{d}E = T\mathrm{d}S - P\mathrm{d}V$, where S is the entropy, P the pressure. The equivalent relationship for a black hole can be obtained by considering the energy change when a mass $\mathrm{d}M$ falls into the black hole,

$$\mathrm{d}Mc^2 = T\mathrm{d}S_{\mathrm{BH}} = \frac{\hbar c^3}{8\pi k_{\mathrm{B}} GM} \mathrm{d}S_{\mathrm{BH}}. \tag{8.31}$$

  Rearranging and integrating gives

$$M^2 = \frac{\hbar c}{4\pi k_{\mathrm{B}} G} S_{\mathrm{BH}}. \tag{8.32}$$

[3] These laws apply equally to rotating and charged black holes; proofs are then more complicated. I am grateful to Jeff Forshaw for pointing this out.

Now introducing the surface area of the horizon

$$A = 4\pi r_0^2 = 16\pi G^2M^2/c^4. \tag{8.33}$$

Using this equality the previous equation can be re-written in terms of A

$$S_{\rm BH} = \frac{4\pi k_{\rm B}GM^2}{\hbar c} = \frac{k_{\rm B}c^3}{4\hbar G}A. \tag{8.34}$$

This result is open to further interpretation. If we take the quantum of area to be the square of the Planck length $\ell_{\rm P}^2$, then

$$S_{\rm BH} = k_{\rm B}\frac{A}{4\ell_{\rm P}^2} \tag{8.35}$$

making the black hole entropy proportional to the number of area quanta that make up the area of the horizon.

- The third classical law maintains that entropy never reduces, and in the case of black holes the area of the horizon always grows when black holes merge.
- The final, fourth classical law states that it is impossible to reach the absolute zero of temperature. Correspondingly, it can be proved that it would be impossible to reduce the surface gravity of a black hole to zero in a finite number of steps.

Black hole entropy being proportional to surface area seems at best puzzling, because for anything else entropy is proportional to volume. However it can be argued that this comes about because there is no shielding from gravity. Thus adding matter to normal objects affects their internal structure only weakly, but has a pervasive effect on black holes. 't Hooft and Susskind proposed that entropy being proportional to surface area, the *holographic principle*, is a general property of gravitating systems.

8.7 The Information Paradox

Hawking radiation is thermal radiation, and as a result the information carried by matter falling into a black hole is lost. To highlight this, imagine that a shell of matter in a pure quantum state collapses to form a black hole, which subsequently evaporates. The pure state has a wavefunction ψ and according to quantum theory it should evolve into a final pure quantum state with a different wavefunction: instead the evaporated black hole leaves thermal radiation, an incoherent mix of states. Evidently the predictions, as they stand, of general relativity and quantum mechanics are incompatible. The paradox revealed may, in the future, be resolved by a quantum theory of gravity. In the absence of such a theory clues are being sought on how to proceed. One proposal is that a nugget remains after evaporation containing the information, another that the information may leak out somehow in the Hawking radiation. Alternatively it is proposed that the horizon has a complex nature that accounts for, or contains, the lost information: the possibilities include a firewall caused by the violent entropy change when information is lost as matter enters the black hole (Almheiri et al. 2013) or that the surface is a fuzzball of string microstates encoding the information (Chowdhury & Mathur 2008). At any event the resolution of the paradox of information loss is theoretical work in progress.

8.8 Stellar Black Holes

From the theory of black holes we now move on to describe their formation and detection in the universe: first stellar mass black holes, and then the black holes with masses in the range $\sim 10^5$–10^{10} $M_\odot$ found at the centers of galaxies. Stellar black holes start life as stars with masses greater than about 10 $M_\odot$. Stars are initially formed from gas, mostly hydrogen, and contract under their gravitational self-attraction. The temperature and pressure rise and eventually conditions are reached at which thermonuclear fusion commences. In this phase hydrogen is converted to helium and the resulting kinetic motion opposes further collapse. After the hydrogen supply is depleted the gravitational collapse resumes and fusion from helium to carbon and oxygen occurs and so on until after several million years the core is converted to iron. With iron the binding energy per nucleon reaches a peak at iron and burning ceases. The several solar mass core has a density 10^{11} kg m^{-3} and a temperature of 10^9 K. Further collapse is resisted by electron degeneracy pressure, the result of the Pauli exclusion principle that no two electrons may share the same quantum eigenstate.

However in 1931 Chandrasekhar discovered that the degeneracy pressure is inadequate to resist continued contraction if the core mass exceeds 1.4 $M_\odot$. The argument proceeds as follows. Suppose that a stellar core of mass M and radius R contains N electrons, then their average spacing is

$$\Delta x \approx R/N^{1/3}.$$

The uncertainty in the individual electron momentum must be at least

$$p \approx \hbar/\Delta x \approx \hbar N^{1/3}/R.$$

Therefore if nearby electrons are to remain in distinct eigenstates their momenta must differ by at least p. This momentum is a measure of the average electron momentum if the gas of electrons is in its lowest energy state, i.e., it is degenerate. With a large enough core the pressure and temperature rise so that the electrons are moving relativistically. Their energies are then

$$E_{\mathrm{R}} \approx pc \approx \frac{\hbar N^{1/3} c}{R}, \tag{8.36}$$

The gravitational energy of the core is due almost entirely to nucleons of mass m_{n}. Therefore, assuming one nucleon per electron, the gravitational energy per electron is

$$E_{\mathrm{G}} \approx -\frac{GNm_{\mathrm{n}}^2}{R}. \tag{8.37}$$

The first law of thermodynamics gives for adiabatic contraction

$$P = -\frac{\mathrm{d}(NE)}{\mathrm{d}V} = -\frac{N(\mathrm{d}[E_{\mathrm{R}} + E_{\mathrm{G}}]/\mathrm{d}R)}{4\pi R^2} = [\hbar c N^{1/3} - G m_{\mathrm{n}}^2 N]\frac{N}{4\pi R^4}.$$

The first term in the square bracket gives the electron degeneracy pressure, the second term gives the gravitational pressure. Contraction under gravity has the stronger dependence on the number of particles and hence on the mass. The degeneracy pressure can only resist the gravitational pressure if

$$\hbar c N^{1/3} \geqslant G m_{\mathrm{n}}^2 N, \quad \text{i. e.,} \quad \hbar c \geqslant G m_{\mathrm{n}}^2 N^{2/3},$$

otherwise there is continuous collapse under gravity. Equilibrium rather than collapse is thus only possible if the number of electrons (and hence nucleons) is less than

$$N(\max) = \left(\frac{\hbar c}{G m_{\mathrm{n}}^2}\right)^{3/2} = 2 \times 10^{57}.$$

The mass of a core at this limit of stability is

$$M(\text{Chandrasekhar}) = N(\max) m_{\mathrm{n}} = 1.4\ M_{\odot}.$$

Heavier cores continue to collapse with consequent rises in pressure and temperature. When the mean electron energy reaches an MeV, electrons initiate *neutronization* reactions such as

$$\mathrm{e}^{-} + {}^{56}\mathrm{Fe} \rightarrow {}^{56}\mathrm{Mn} + \nu,$$

which requires a threshold energy of 3.7 MeV. Neutronization is swift, taking about 1 s, and is accompanied by further core collapse. When the density reaches about 10^{17} kg m^{-3}, the nucleons have mostly become free neutrons. The closely packed neutrons form a degenerate gas and, being fermions, they can also exert a degeneracy pressure. A direct repetition of the analysis shows that stellar cores of mass less than the Chandrasekhar limit are stable against further collapse and form neutron stars. More massive cores inexorably collapse to form black holes. Hydrodynamic calculations, taking general relativity into account, starting with those of Tolman, Oppenheimer, and Volkov, give an upper limit for the neutron star mass of 2.1 $M_{\odot}$ and diameter 10 km. A crucial feature underlying this analysis is that pressure as well as density contributes to the gravitational attraction. Pressure has a dual role in GR. It opposes collapse, but, as a component of the stress–energy tensor, it also contributes to hastening the collapse.

Thus far the history of the core has been followed. The most spectacular external manifestation of the process occurs when the collapse is checked by the stiffening of the core due to nuclear repulsion. Then the imploding outer layers of the star strike the core and rebound under the shock. These layers are ejected as a type-II supernova explosion, of which SN1987a was one example. Supernovae have an important cosmological impact: it is during these uniquely energetic events that heavier elements beyond iron are formed and ejected into the interstellar medium.

In the explosion the accumulation of elements remaining from the earlier fusion sequence up to iron are similarly ejected. Any stars, like the Sun, containing such elements are therefore in at least the second generation of stars.

8.9 Cygnus X-1

The stellar mass black holes that have been identified are all members of a binary pair thanks to two helpful characteristics. First, there is intense X-ray emission accompanying the conversion of gravitational to kinetic energy as matter is dragged from the partner into the black hole. Second, the partner's orbital motion is measurable. Detectors on orbiting satellites have been necessary because the Earth's atmosphere absorbs X-rays strongly. The first example, Cygnus X-1, was discovered during a rocket flight in 1964, and it is still the best example.

Located in our Galaxy, Cygnus X-1 is a particularly strong emitter of X-rays and radiowaves. A visible star HDE 226868 is located within the one arcsec error box defined by the radio frequency measurements giving the position of Cygnus X-1. This star has a spectrum identifying it as a blue supergiant, making it too cold to be the source of the X-rays. Equally important, the wavelengths of the lines in the spectrum of HDE 226868 change with time following a regular cycle that repeats every 5.6 days. The distance of Cygnus X-1 from the Earth obtained by parallax measurements with an array of radio telescopes is 1.86 kpc.

An economical explanation of the features just presented is that HDE 226868 and Cygnus X-1 form a binary pair with an orbital period of 5.6 days; the time varying spectral shift of the optical partner is then simply the Doppler shift produced by its orbital motion. Material around the blue supergiant fills its gravitational well (Roche lobe) and spills over into that around the black hole. The transfer shown in Figure 8.7 feeds the flat rapidly rotating accretion disk around the black hole. Matter from the accretion disk falls into the black hole with a large energy release (see Section 8.1). This process easily heats the accretion disk to a temperature at which it can emit X-rays.

When the observed Doppler shift of the spectral lines emitted by HDE 226868 is converted to a relative velocity with respect to its partner, this velocity varies sinusoidally with time, showing that the orbits are almost *circular*. Suppose the

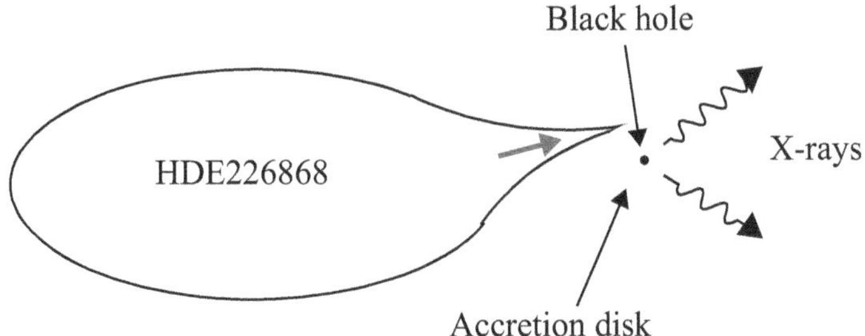

Figure 8.7. A section taken in the orbital plane of the X-ray binary Cygnus X-1. The red arrow indicates the flow of material from the blue supergiant HDE 226868 onto the accretion disk surrounding its black hole partner.

following: that the orbital radii are r_X and r_{bg} for Cygnus X-1 and the blue supergiant, respectively, that the normal to the orbital plane is tilted at an angle i with respect to our line of sight, that τ is the orbital period and that the maximum velocity of HDE 226868 toward us is v_{bg}. Then simple application of Kepler's laws to the orbital motion determines the *mass function*:

$$\begin{aligned}\frac{(m_X \sin i)^3}{(m_X + m_{bg})^2} &= \frac{(r_{bg} \sin i)^3 (2\pi/\tau)^3}{G} \\ &= \frac{v_{bg}^3 \tau}{2\pi G}.\end{aligned} \tag{8.38}$$

A lower limit on the mass of Cygnus X-1 can be obtained from this equation by setting m_{bg} to 20 $M_{\odot}$, an approximate lower limit for blue supergiant masses, and i to 90°. Then we insert the measured quantities in the equation: the period of 5.6 days and the maximum velocity, from Doppler shift data, of 95 km s^{-1}. The emerging lower limit for m_X is 10 $M_{\odot}$, well above the upper limit (2.1 $M_{\odot}$) predicted for the mass of a neutron star. Orosz et al. (2011) made a fit to all measurements: they find that the orbits of the two stars are only 0.2 AU across; that the mass of HDE 226868 is 19.2 $M_{\odot}$; and that the black hole has mass 14.8 $M_{\odot}$, and that its Schwarzschild radius is 57 km, while i is 27°. Finally the X-ray emission from Cygnus X-1 flickers on a timescale of milliseconds: in order to show this degree of coherence the source must be less than milli-lightseconds across, that is less than 300 km across. This neatly matches the X-ray source to the expected region of energy release and excitation, namely the accretion disk. Estimates of the total number of stellar black holes in our Galaxy lies between 10^6 and 10^9.

The black holes found in X-ray binaries all have masses of order 10 $M_{\odot}$. In the following chapter the growing number of mergers of stellar black holes detected by the gravitational wave detectors LIGO and Virgo are discussed. These have been events with the merging black hole masses ranging from 10 $M_{\odot}$ to 85 $M_{\odot}$. In the event labeled GW190321 black holes of masses 85 $M_{\odot}$ and 65 $M_{\odot}$ merged giving a black hole of mass 142 $M_{\odot}$ releasing an energy 8 $M_{\odot}c^2$ in gravitational waves. Apart from that instance there is only tentative evidence at present for black holes with masses in the range 100–10^5 $M_{\odot}$. This gap impacts on the question of how supermassive black holes were formed at the centers of galaxies.

Processes discussed that would inhibit stars ending as black holes with intermediate masses roughly of order 100 $M_{\odot}$ (apart from mergers) take account of the production during nuclear burning of a large flux of energetic photons. Once the temperature exceeds ~10^9 K (0.1 MeV) the fraction of photons capable of electron–positron pair production (requiring 1.022 MeV) becomes significant. Pair production reduces the internal photon pressure, allowing further collapse; this leads to increased pressure and temperature with the cycle repeating and ending with a final collapse to a black hole. Alternatively if the star is massive enough it would explode at the end of the first cycle as a supernova, blowing itself apart without a remnant.

8.10 Supermassive Black Holes

There is an enormous jump in mass scale to the second class of identified black holes, which lie at the centers of galaxies. These galactic black holes have masses in the range $\sim 10^5$–10^{10} $M_\odot$, with lighter galaxies harboring the lighter black holes. In the case of our Galaxy researchers have deduced from the trajectories of stars close to the strong, compact radio source Sgr A* that this is the location of a massive black hole. The two teams independently recorded the orbits of a dozen stars passing close to Sgr A* for more than 20 years: they used infrared imaging to penetrate the dust and gas clouds shrouding the galactic center. Both groups also used adaptive optics: a corrective mirror in the light path after the primary mirror is warped to keep a guide star in the field of view in sharp focus. The system was able to compensate for the distortion due to atmospheric turbulence by having a response fast enough to respond to the typical 100 Hz variation of this turbulence. Kinematic calculation using the observed orbits of these stars around Sgr A* fixes the mass of the black hole to be about 4.4×10^6 $M_\odot$. The left-hand panel of Figure 8.8 shows the orbit of the star S2 tracked from 1992 to 2018, approaching Sgr A* to within a distance comparable to the size of the solar system. The determination of the stellar orbits, especially S2, requires that the central mass density is $\sim 10^{16} M_\odot$ pc^{-3}. If there were a dark astrophysical cluster of this density rather than a black hole, then at such a density the cluster would have dispersed through collisions in $\sim 10^6$ yrs, far shorter than the

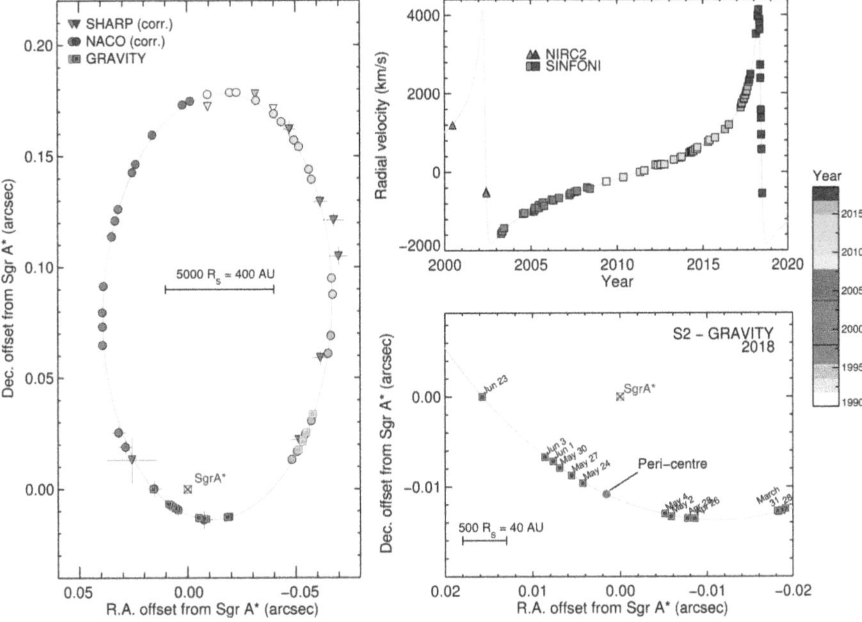

Figure 8.8. Observations of the orbit of S2 made from 1992 to 2018 covering a full orbit around Sgr A*. The left-hand panel shows the orbit projected on the sky. The upper right-hand panel shows the radial velocity as a function of time. The panel below is an expanded view of the portion of the orbit nearest to Sgr A*. The cyan curve is the best fit including special and general relativistic effects. Figure from Abuter et al. (2018). Reproduced with permission © ESO.

lifetime of the solar system (Maoz 1998). The leaders of these two research teams, Andrea Ghez and Reinhard Genzel, shared the 2020 Nobel Prize in physics (with Roger Penrose) for their research contributions demonstrating the existence of a supermassive black hole at the center of our Galaxy.

Returning to Figure 8.9 the top right-hand panel shows the radial velocity of S2, and below this an expanded view of the orbit when nearest Sgr A*. There are deviations of the measured stellar velocities from an overall Newtonian fit, in the case of S2 amounting to 200 km s^{-1} at the closest approach to Sgr A*. These deviations disappear when the calculations are repeated with GR, which is illustrated in Figure 8.9 for S2. This comparison extends the successful tests of GR to conditions beyond the reach of solar system tests, to the *strong coupling* regime, where GR effects are large.

In parallel the motion of Sgr A* has been measured using long baseline radio interferometry. Its velocity in the galactic plane is about 7 km s^{-1}, and about 0.4 km s^{-1} perpendicular to the plane: a thousand times slower than the motion of those nearby stars used to infer the presence of a black hole. These velocities are consistent with the motion expected for a 4.4×10^6 $M_\odot$ black hole under the fluctuating gravitational effect of the stars in the Galaxy. Energy released from material falling into the black hole is responsible for the intense radio source Sgr A*. Applying Equation (3.52) to the galactic black hole gives a Schwarzschild radius of 1.30×10^{10} m or 0.087 AU. Using Equation (8.29) gives a Hawking temperature of 1.4×10^{-14} K.

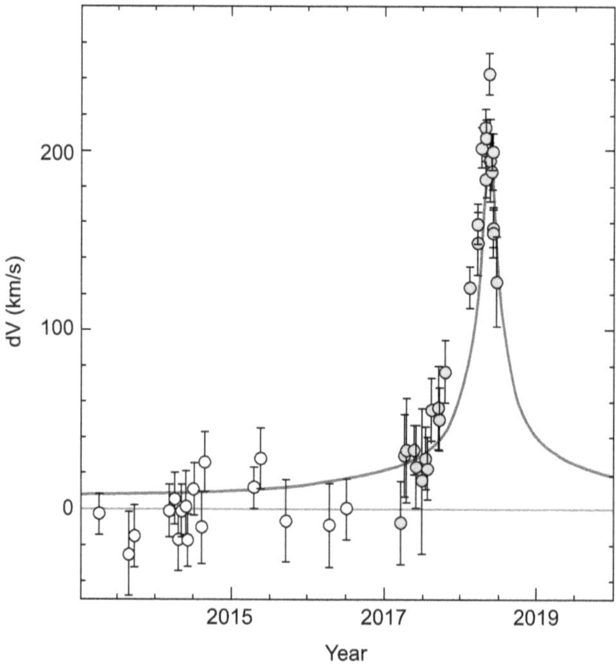

Figure 8.9. The departure of the radial velocity of S2 around Sgr A* from the flat gray line showing the Newtonian prediction. The cusped GR prediction is seen to agree with the data. Figure from Abuter et al. (2018). Reproduced with permission © ESO.

The next nearest black hole of interest, M87*, is located at the center of the spiral galaxy M87, itself at the center of the ~1500 member Virgo cluster of galaxies. The M87 galaxy is proportionately larger, comprising $\sim 2 \times 10^{12}$ stars within a radius of 80 kpc. M87* lies 16.4 Mpc from the Earth, but still close enough for the orbital dynamics of stars and gas close to the gravitational center of the galaxy to be measured with the HST and Gemini telescopes.[4] Analysis of the dynamics reveals that M87* is a monster compared to Sgr A*, weighing in at around $6.5 \times 10^9\ M_\odot$. M87* has a correspondingly larger Schwarzschild radius ($2GM/c^2$) of 1.9×10^{13} m or 130 AU, comparable to the size of the solar system. There is strong emission across the γ-ray and X-ray spectrum from M87*, the former fluctuating over times as short as days. In order that radiation from the whole source may show this degree of temporal correlation the source must be at most light-days across, consistent with the inferred Schwarzschild radius. We found in Section 8.2 that a black hole is encircled by possible photon orbits terminating in the closest, an unstable circular orbit. This means that though we receive copious radiation from material on its way into the black hole, none will come from within this orbit. Therefore the image of a black hole should be a central shadow whose edge is defined by the innermost, unstable photon orbit. It was shown that this orbit has a radius $3\sqrt{3}\,r_0/2 = \sqrt{27}\,GM_{\text{Black Hole}}/c^2$. The corresponding light rings for Sgr A* and for M87* subtend similar opening angles on Earth, ~40 μas: the larger mass of M87* is compensated in the case of Sgr A* by the shorter distance. The Event Horizon Telescope Collaboration (The Event Horizon Telescope Collaboration et al. 2019) set out to observe the light ring of M87* using radio telescopes to achieve the required microarc-second angular resolution. They used the shortest wavelength, 1.3 mm, and an array of eight telescopes whose sites enclose an area near to one hemisphere of the Earth. The angular resolution is thus ~1.3 mm/10,000 km, that is 10^{-10} rad or 20μas: just enough! Digitized images were recorded on memory at each site and time stamped with GPS. Later the interference pattern was produced at a central site by synchronizing, playing through and superposing the contents of the hard drives carrying the individual signals.[5] This produced the world scale interferometer's image of M87*. Figure 8.10 shows the results. On the left the image of M87* reconstructed from data; at the center the predicted image with perfect resolution showing the sharp photon ring; on the right the outcome of convolving this predicted image with the expected resolution. This is the first image of a black hole and as such adds the stamp of direct observation to the wealth of indirect evidence for the existence of black holes. In 2022 the EHT collaboration imaged the black hole at the center of our Galaxy.

8.11 Active Galactic Nuclei

Some galactic black holes can excite spectacular activity in the core (nucleus) of the parent galaxy, far removed from the relative passivity of the cores around Sgr A* and M87*. The sites are called active galactic nuclei, (AGNs). The first type recognized,

[4] These are twin mountain-based telescopes at Maunakea in Hawaii and Cerro Pachon in Chile with detection from the infrared to the ultraviolet. Both have 8 m diameter aperture primary mirrors.

[5] Recall that radio telescopes respond to the electromagnetic field amplitude, not intensity.

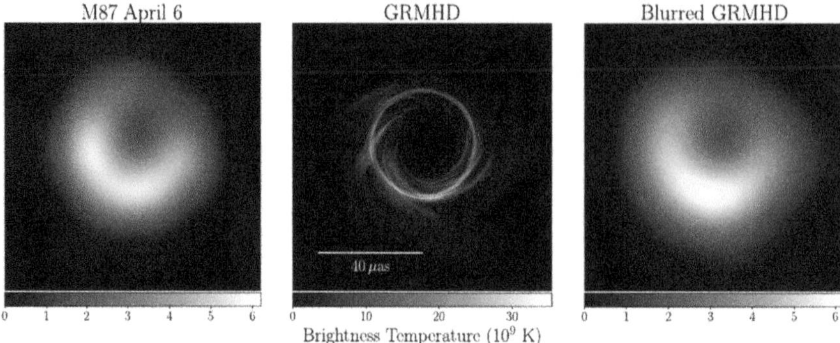

Figure 8.10. Left: the image of the black hole M87* recorded by the EHT; Center: the predicted image with perfect resolution; Right: the prediction taking account of the interferometer's intrinsic resolution. Figure from The Event Horizon Telescope Collaboration et al. (2019). Reproduced with permission under CC-BY-SA-3.0.

the *quasars*, were detected using radio telescopes as unresolved bright point-like sources. Their intensity varied rapidly, which showed that they were surprisingly compact, only as large as the solar system. In 1963 Maarten Schmidt discovered a matching optical source for the brightest quasar 3C273, and this optical source was used to determine the redshift. It turned out to have a high redshift (for that era) of $z = 0.16$ and hence 3C273 lies at a distance from the Earth $cz/H_0 = 690$ Mpc. Using this distance as input the luminosity calculated for 3C273 had the astonishingly large value 2.2×10^{40} W. For comparison, our own typical galaxy of around 2×10^{11} stars radiates just 5×10^{36} W. Later on, after the introduction of X-ray telescopes on rockets and satellites, space-based observations revealed the existence of narrow jets of energetic material, mainly relativistic electrons, emerging from such quasars. Often there are two back-to-back jets from individual quasars. Figure 8.11 is a composite image of the jet from 3C273 (moving rightward) starting 12 arcsec and ending 22 arcsec from 3C273. This figure was constructed from images taken with detectors sensitive to parts of the spectrum: from X-rays at the left to infrared radiation at the right. The further the jet travels from its source, the more it is decelerated by intergalactic gas, and hence the cooler the radiation it emits. This radiation is plane polarized as expected for synchrotron radiation emitted by relativistic electrons. 3C273 is at an angular distance[6] of 590 Mpc; its jet subtends 22 arcsec so the jet length is 60 kpc (196 lyr). The average velocity of matter in the jet is well below c, so we can infer that 3C273 has been ejecting material for well over 200 kyr. Jets from some other AGNs extend to several 100 kpc.

From its luminosity of 2.2×10^{40} W it follows that over a million years 2C273 would radiate around 0.7×10^{54} J or $3.9 \times 10^6\ \mathrm{M}_\odot c^2$. The mass of the source has been measured by the GRAVITY collaboration (Sturm et al. 2018) to be $3 \times 10^8\ M_\odot$. Converting hydrogen to helium releases 0.7% of the mass as free energy, which in the case of 2C273 would give an energy release of $2.1 \times 10^6\ \mathrm{M}_\odot c^2$ if all the mass were

[6] In Chapter 11 we will learn that the angular distance is smaller than the proper distance 690 Mpc by a factor $(1 + z)$.

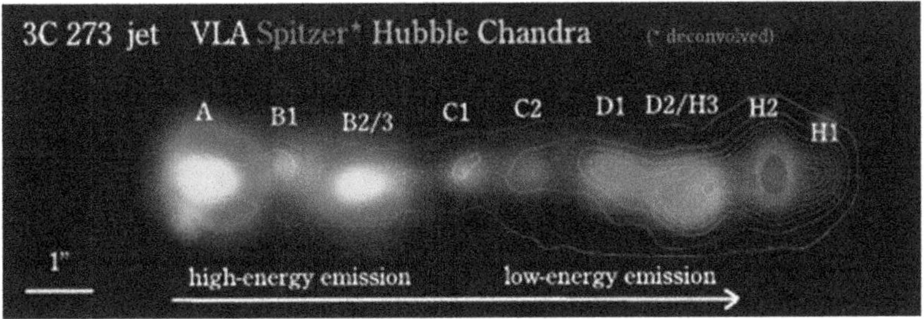

Figure 8.11. The image of one jet from 3C273. Blue identifies the 0.4–6 keV emission detected with the *Chandra* X-ray space telescope; green identifies UV emission detected by the HST; red identifies the 3.6 μm radiation detected by the Spitzer space telescope; the contours are drawn for 2 cm radio emission observed with the VLA. The latter is truncated on H2, the strongest source. Figure 7 from Uchiyama et al. (2006). Courtesy Professor C. Megan Urry for the copyright holders.

converted. This upper estimate is still well short of the energy radiated by the quasar. On the other hand, we have seen that when matter falls into a black hole between 5% and 42% of the initial mass is released as energy, depending on the angular momentum of the black hole. Evidently quasars are powered by black holes. On its way inbound matter ultimately entering the black hole will carry angular momentum with it. This has two effects: it provides the spin that gives the jets; and it makes it certain that the black holes in the universe are predominantly Kerr black holes. Angular momentum, lost by matter falling into the black hole, and intense magnetic fields both play some part in spinning up the AGN and launching the jet; though the details are yet to be worked out.

Figure 8.12 sketches the principal features of an AGN. The gravitational attraction of the black hole and the requirement to conserve angular momentum build a rotating accretion disk feeding the black hole. In turn this disk is fed from a torus of gas and dust. Around the disk are clouds of matter in violent motion and radiating Doppler broadened atomic lines. Depending on whether we are viewing the AGN head-on along the spin axis, or sideways-on through the accretion disk, or otherwise, AGNs exhibit very different features. For this reason it took decades to recognize their common origin. Seen from the side the active core is hidden by the torus which re-radiates in the radio spectrum. As the viewing angle steepens toward the jet the energetic core comes into sight, and such objects are classed as Seyfert galaxies. Quasars are AGNs seen head-on either along or near a jet. Typically a quasar's total energy flux exceeds that from the parent galaxy by a factor as large as 10^4. Thanks to their huge luminosity quasars are sources visible to us now from the earliest times in the life of the universe. Currently the most distant quasar, J1342 +0928, with redshift 7.54, is 4.2 Gpc away. The radiation we are receiving now from this quasar was therefore emitted within a billion years after the Big Bang. Once an AGN has swept up all the accessible local gas, dust, and stars, it becomes quiescent like M87* and Sgr A*. Figure 8.13 shows the plot of the quasar comoving density as a function of redshift. At redshift ~6 they are sparse, only of order 1 Gpc^{-3}.

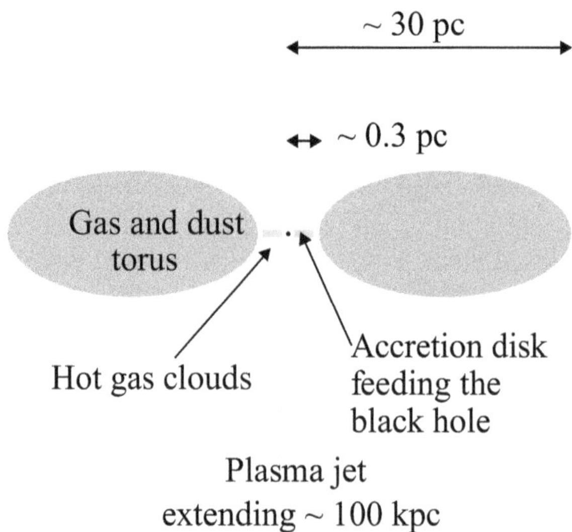

Figure 8.12. A sketch, not to scale, of a section through an AGN; the black dot marks the location of the black hole.

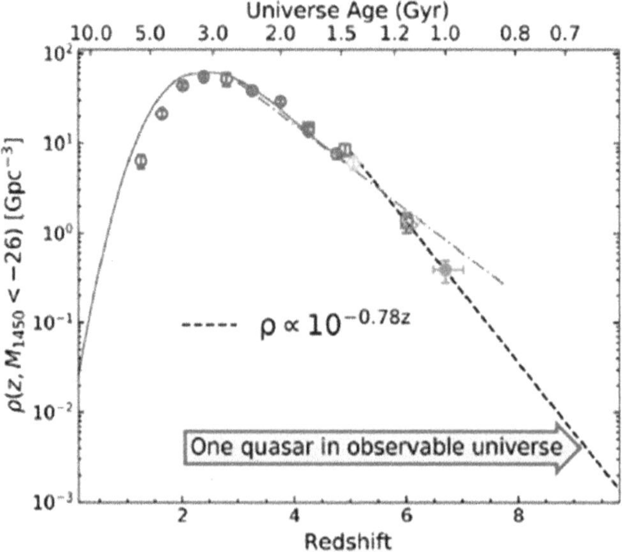

Figure 8.13. The density of luminous quasars in the early universe. The reference volume is the comoving volume. Making the extrapolation indicated by a broken line, there would only be one luminous quasar powered by a black hole of a billion solar masses or more in the observable universe at redshift of 9. Figure 1 (a) reproduced from Fan et al. (2019) with permission under CC-BY 4.0.

8.12 Exercises

1. Calculate the radius of the horizon of a neutral non-rotating black hole whose mass is 10^8 $M_\odot$. Show that the velocity of a body moving in the smallest stable orbit around this Schwarzschild black hole is $c/2$. Hence calculate the period measured locally (proper time) and that measured on a clock remote from the black hole (coordinate time).
2. If someone is in free fall radially toward a black hole, which has a mass equal to 10 $M_\odot$, he would feel a lateral tidal force squeezing him. Calculate the radial distance from the center of the black hole at which the tidal acceleration has grown to 400 m s^{-2}. At this point our traveler will be crushed. We assume that the traveler starts from rest at a place remote from the black hole.
3. What is the Schwarzschild radius of a galactic black hole of 10^8 $M_\odot$? How much mass would need to be consumed by the black hole per annum in order to generate 10^{39} W?
4. Show that when two black holes merge there is an increase in Beckenstein entropy. According to Beckenstein how many bits does a Schwarzschild black hole of mass 20 $M_\odot$ contain? What is the temperature of the Hawking radiation from it?
5. An astronaut falls under gravity only toward a black hole of mass 10^4 $M_\odot$ starting from rest at effectively infinite distance. How long does it take according to his onboard clock to travel from being at twice the Schwarzschild radius from the black hole center until he crosses the horizon?
6. Unruh (1976) used quantum field theory to argue that an observer moving at constant acceleration a with respect to an inertial frame would perceive black body radiation at a temperature

$$T = \frac{\hbar a}{2\pi k_B c}. \tag{8.39}$$

Calculate the Unruh temperature expected for an observer remaining stationary at the horizon of a black hole. Is this consistent with the Hawking temperature?

Further Reading

Susskind L and Lindesay J 2005 *The Holographic Universe: An Introduction to Black Holes, Information and the String Revolution* (Singapore: World Scientific) gives a less technical account of this modern approach to cosmology.

Haardt F, Gorini V, Moschella U, Treves A and Colpi M (editors) 2016 *Astrophysical Black Holes: Lecture Notes in Physics* (Berlin: Springer). This text is more wide ranging.

Blundell K 2015 *Black Holes: A Very Short introduction*. A short lively account.

References

Abuter, R., Amorim, A., Anugu, N., et al. 2018, A&A, 615, L15
Almheiri, A., Marolf, D., Polchinski, J., & Sully, J. 2013, JHEP, 2013, 18
Bardeen, J. M., Carter, B., & Hawking, S. W. 1973, CMaPh, 31, 161
Beckenstein, A. 1972, NCimL, 4, 99
Chandrasekhar, S. 1931, ApJ, 74, 81
Chowdhury, B. D., & Mathur, S. D. 2008, CQGra, 25, 225021
Eddington, A. S. 1924, Natur, 113, 192
Fan, X., Barth, A., Banados, E., et al. 2019, BAAS, 51, 121
Finkelstein, D. 1958, PhRv, 110, 965
Hawking, S. W. 1974, Natur, 248, 30
Hawking, S. W., & Penrose, R. 1970, RSPSA, 314, 529
Maoz, E. 1998, ApJ, 494, L181
Michell, J. 1784, RSPT, 74, 35
Orosz, J. A., McClintock, J. E., Aufdenberg, J. P., et al. 2011, ApJ, 742, 84
Penrose, R. 1965, PhRvL, 14, 57
Sturm, E., Dexter, J., Pfuhl, O., et al. 2018, Natur, 563, 657
The Event Horizon Telescope CollaborationAkiyama, K., Alberdi, A., et al. 2019, ApJ, 875, L5
Uchiyama, Y., Urry, C. M., Cheung, C., et al. 2006, ApJ, 648, 910
Unruh, W. G. 1976, PhRvD, 14, 870

AAS | IOP Astronomy

Introduction to General Relativity and Cosmology (Second Edition)

Ian R Kenyon

Chapter 9

The Discovery and Study of Gravitational Waves

Einstein showed that gravitational radiation is a natural consequence of the general theory of relativity and that the waves travel with the speed of light. He considered waves making small disturbances from flat spacetime, which is appropriate for the gravitational waves so far detected on the Earth. In this limit the Einstein equation reduces to a linear wave equation. This has plane wave solutions that are transverse waves traveling with velocity c, properties that electromagnetic waves also possess. During the passage of gravitational waves it is the structure of spacetime itself which oscillates. Putting this more precisely, the proper time taken by light to pass to and fro between two fixed points in space oscillates. There is no effect at a single point, only a change of the *separation* between points in space. The dominant mode of gravitational radiation from a source compact compared to the wavelength is quadrupole; the law of conservation of momentum requires the dipole component to vanish.

The constant $c^4/8\pi G$ appearing in Einstein's Equation (6.19),

$$G_{\alpha\beta} = \frac{8\pi G}{c^4} T_{\alpha\beta},$$

is the force per unit area required to give spacetime unit curvature: it requires $0.48\ 10^{43}$ Pa (N m^{-2}) to give a curvature of 1 m^{-2}. Spacetime is therefore an extremely stiff medium, and by the same token small amplitude waves carry large energies. Between 2015 and 2020 the LIGO and Virgo collaborations detected the gravitational waves from some 90 mergers of binary pairs, either black holes or neutron stars. Their detectors are evacuated Michelson interferometers with multi-kilometer long arms. The inferred masses of the merging black holes lie in the range 5–85 $M_{\odot}$. These mergers occurred at distances of typically 0.3 Gpc from Earth and hence one billion years ago. The strains in spacetime detected are $\sim 10^{-21}$ or one nuclear diameter in 1000 kilometers! This amazing sensitivity was only achieved after decades of development work, and requires that the interferometers work at the limit for measurement precision imposed by quantum mechanics. The LIGO team leaders, Barry Barish, Kip Thorne, and Rainer Weiss, were awarded the Nobel Prize in physics in 2017. In the following sections gravitational wave motion will be discussed and quantified, and then their detection.

9.1 Properties of Gravitational Radiation

The basic equation to describe empty spacetime is Einstein's equation, Equation (6.19) with the right-hand side set to zero:

$$G_{\mu\nu} = R_{\mu\nu} = 0. \tag{9.1}$$

A linear approximation can be made to the metric where gravitational disturbances from an overall flat spacetime are small

$$g_{\mu\nu} = \eta_{\mu\nu} + h_{\mu\nu} \tag{9.2}$$

where $\eta_{\mu\nu}$ is the Minkowski metric and all the components $h_{\mu\nu}$ are very much less than unity. $g_{\mu\nu}$ and $\eta_{\mu\nu}$ are symmetric tensors, and hence $h_{\mu\nu}$ is also a symmetric tensor. Expanding Equation (9.1) to first order in $h_{\mu\nu}$ gives

$$h^{\alpha}_{\ \nu,\,\mu\alpha} - h_{\nu\mu,\ \alpha}^{\ \ \ \ \alpha} + h_{\alpha\mu,\ \nu}^{\ \ \ \ \alpha} - h^{\alpha}_{\ \alpha,\,\mu\nu} = 0 \tag{9.3}$$

A symmetric tensor like $h_{\mu\nu}$ has in general ten independent components, but gravitational waves are quadrupole, with only two independent polarization states. This is similar to the situation with electromagnetism.[1] The same approach is used in both cases: gauge conditions are applied that restrict the solutions of the wave equation, here Equation (9.3), to those of physical importance. The first, the traceless condition, is to set the fourth term to zero; the next condition is to require the divergence to vanish, which eliminates the first and third terms. Applying these conditions together reduces Equation (9.3) to a wave equation

$$h_{\mu\nu,\ \alpha}^{\ \ \ \ \alpha} = \frac{\partial^2 h_{\mu\nu}}{\partial x_\alpha \partial x^\alpha} = 0, \tag{9.4}$$

A further and final requirement can be made on the coordinate choice

$$h_{\alpha 0} = 0. \tag{9.5}$$

The eight constraints that have been applied have left just two independent components of $h_{\mu\nu}$. These solutions appear below as the two independent polarization states presented in Equations (9.8) and (9.9). Writing Equation (9.4) in detail gives

[1] The underlying reason is that the quanta of electromagnetism and gravitational fields are massless. For massive particles all the degrees of freedom are necessary to describe the particle state.

$$\frac{\partial^2 h_{\mu\nu}}{(\partial ct)^2} = \frac{\partial^2 h_{\mu\nu}}{\partial x^2} + \frac{\partial^2 h_{\mu\nu}}{\partial y^2} + \frac{\partial^2 h_{\mu\nu}}{\partial z^2} = 0,$$

which is the familiar wave equation for waves with velocity c. A plane wave solution for a wave traveling along the z direction in space is

$$\begin{aligned} h_{\mu\nu} &= A_{\mu\nu} \cos\left[k(ct - z)\right] \\ &= A_{\mu\nu} \cos(\omega t - kz). \end{aligned} \tag{9.6}$$

Here $\lambda = 2\pi/k$ is the wavelength and $\omega = kc$ is the angular frequency of the wave. These solutions must satisfy the stated constraints: $A_{\mu\nu}$ must be symmetric, its trace must vanish and both $A_{0\beta}$ and $A_{\beta 0}$ should vanish for all values of β. Applying these constraints we have

$$A_{\mu\nu} = \begin{bmatrix} 0 & 0 & 0 & 0 \\ 0 & A_{11} & A_{12} & 0 \\ 0 & A_{12} & -A_{11} & 0 \\ 0 & 0 & 0 & 0 \end{bmatrix}. \tag{9.7}$$

The choice of gauge has made the wave amplitudes $A_{\mu\nu}$ both *transverse* and *traceless*; therefore this is called the transverse-traceless gauge. Quantities defined in this gauge carry a superscript TT: this will only be inserted from time to time in what follows as a reminder. The analysis for weak gravitational waves, and slow-moving weak (near Newtonian) sources can be carried through using transverse-traceless gravitational wave solutions of Einstein's equation. The general solution of the form of Equation (9.6) is made up of a linear combination of the two orthogonal states

$$h_{\mu\nu} = h_+(e_+)_{\mu\nu} \cos(\omega t - kz) \tag{9.8}$$

and

$$h_{\mu\nu} = h_\times(e_\times)_{\mu\nu} \cos(\omega t - kz), \tag{9.9}$$

where

$$(e_+)_{\mu\nu} = \begin{bmatrix} 0 & 0 & 0 & 0 \\ 0 & 1 & 0 & 0 \\ 0 & 0 & -1 & 0 \\ 0 & 0 & 0 & 0 \end{bmatrix} \tag{9.10}$$

$$(e_\times)_{\mu\nu} = \begin{bmatrix} 0 & 0 & 0 & 0 \\ 0 & 0 & 1 & 0 \\ 0 & 1 & 0 & 0 \\ 0 & 0 & 0 & 0 \end{bmatrix} \tag{9.11}$$

and where $h_+ = A_{11}$ and $h_\times = A_{12}$. e_+ and $e_\times$ are two independent polarization states of gravitational radiation. Their tensor form implies a more complicated polarization than the linear polarization met with in the case of light.

9.2 The Effects of Gravitational Waves

We can build a picture of the effect of gravitational wave states by following the motion of two test masses, nearby one another, when a wave passes. Suppose the masses are located at A($x = \xi$, $y = z = 0$) and B($x = y = z = 0$) at time $t = 0$ before any disturbance. When a wave arrives traveling in the z-direction their separation in the x-direction changes at time t to

$$\xi' = [\,|g_{11}(t, 0)|\,]^{1/2}\xi \approx \left[1 + \frac{h_{11}(t, 0)}{2}\right]\xi. \tag{9.12}$$

The separation AB undergoes strain of amplitude

$$\frac{\xi' - \xi}{\xi} = \frac{h_{11}(0, 0)}{2} = h_+/2, \tag{9.13}$$

where the e_+ polarization state of Equation (9.8) is chosen to get the second equality. This strain is the differential (tidal) effect along the x-direction. Along the y-direction $-h_+$ replaces h_+. If the incident e_+ polarized wave is a plane sinusoidal wave the separation of test masses will change from $(x_0, y_0, 0)$ in the absence of the wave to

$$x = x_0\left(1 + \frac{h_+}{2}\cos\omega t\right),$$

$$y = y_0\left(1 - \frac{h_+}{2}\cos\omega t\right).$$

This motion is shown in the left-hand column of Figure 9.1 for a set of test masses originally at rest in a circle in the xy plane: their displacements are followed over a complete cycle. There are two orthogonal symmetry axes and so the motion is quadrupole. The movement obtained with the orthogonal $e_\times$ polarization state of Equation (9.9) is

$$x = x_0 + y_0 h_\times \cos\omega t,$$

$$y = y_0 + x_0 h_\times \cos\omega t,$$

which can be manipulated to give

$$x + y = \left(1 + \frac{h_\times}{2}\cos\omega t\right)(x_0 + y_0)$$

and

$$x - y = \left(1 - \frac{h_\times}{2}\cos\omega t\right)(x_0 - y_0)$$

The complete cycle is drawn in the right-hand column of Figure 9.1. It is not possible to construct the e_+ pattern from the $e_\times$ pattern or vice versa; they are orthogonal

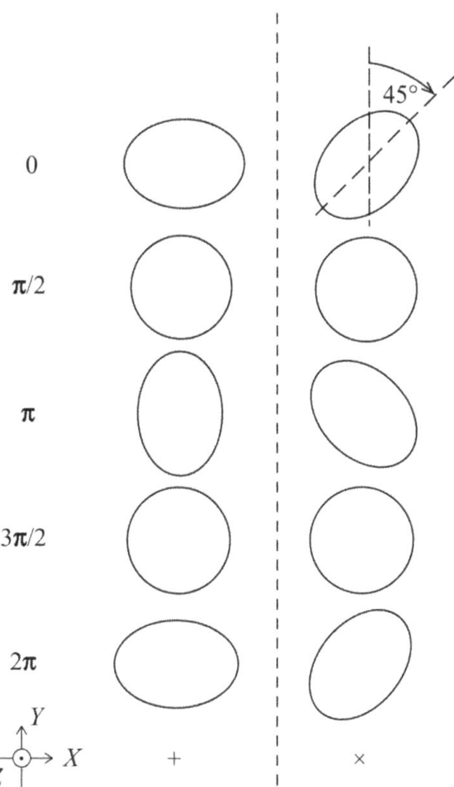

Figure 9.1. The effect of gravitational waves on a circle of test masses followed over one cycle. The observer is looking toward the source. Two orthogonal states of quadrupole radiation are illustrated.

polarization states. By analogy with electromagnetic waves the e_+ and $e_\times$ amplitudes can be added with a phase difference of $-\pi/2(+\pi/2)$ to obtain right- (left)-handed circularly polarized amplitudes. These are

$$h_{\mu\nu} = h\left[(e_+)_{\mu\nu}\cos(\omega t - kz) \pm (e_\times)_{\mu\nu}\sin(\omega t - kz)\right].$$

Their effects are shown in Figure 9.2 for the test mass arrangement described earlier. The arrows show the sense of rotation for these patterns. We can also draw diagrams complementary to Figures 9.1 and 9.2 that show tidal accelerations. Just two examples are given in Figure 9.3: for e_+ and $e_\times$ when the phase $\omega t - kx$ is zero. Not surprisingly these force patterns resemble the pattern of magnetic field lines across the aperture of a quadrupole magnet.

A fundamental difference between gravitational and electromagnetic radiation is that dipole radiation is absent in the gravitational case. To understand why this should be we start by making a comparison between potentials: one is an electrostatic potential due to charges q^a at r^a, and the other is a gravitational potential due to masses m_a at r^a. For simplicity we take the nearly Newtonian case of slowly moving masses/charges. At a vector distance **R** from the center of mass

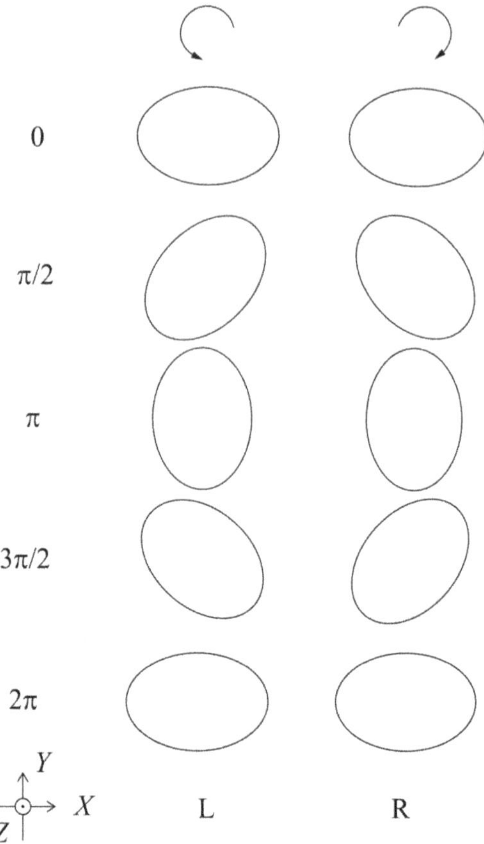

Figure 9.2. The effect of gravitational waves on a circle of test masses followed over one cycle. The observer is looking toward the source. Right- and left-handed circularly polarized patterns are illustrated.

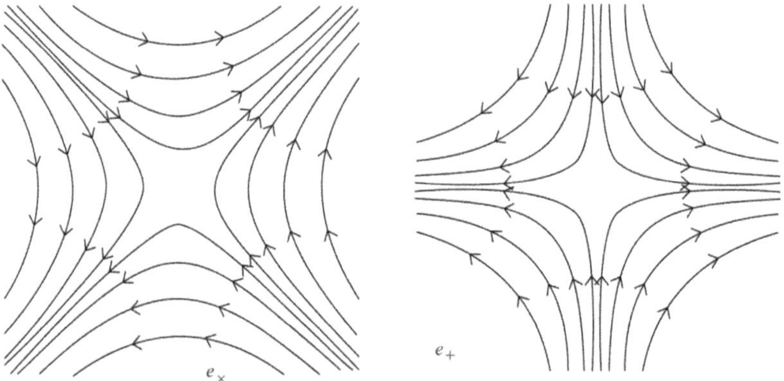

Figure 9.3. The patterns of tidal acceleration for the two orthogonal states of polarization, e_+ and $e_\times$, when the phase angle is zero.

$$4\pi\varepsilon_0\varphi_{\text{es}}(\mathbf{R}) = \sum q^\alpha/|\mathbf{R} - \boldsymbol{r}^\alpha|,$$

which can be expanded if $r \ll \mathbf{R}$ as

$$4\pi\varepsilon_0\varphi_{\text{es}} = \sum\frac{q^\alpha}{R} - \sum q^\alpha x_i^\alpha\left[\frac{\partial}{\partial X_i}\left(\frac{1}{R}\right)\right] + \cdots,$$

where x_i^α are components of r^α, and where X_i are components of $\mathbf{R}$. Similarly, the gravitational potential is

$$\frac{\varphi}{G} = \sum\frac{m^\alpha}{R} - \sum m^\alpha x_i^\alpha\left[\frac{\partial}{\partial X_i}\left(\frac{1}{R}\right)\right] + \cdots.$$

The radiation is proportional to $\mathrm{d}^2\varphi/\mathrm{d}t^2$ and the second term in the expansion is responsible for dipole radiation. In the gravitational case the dipole term contains a factor $\sum m^\alpha \mathrm{d}^2(x_i^\alpha)/\mathrm{d}t^2$, which equals the net force, and this vanishes for an isolated system. It is quite different for a charge distribution, which can easily have a non-zero oscillating electric dipole moment $\sum q^\alpha x_i^\alpha$. Similar considerations apply for magnetic dipole radiation, which is important when charges are in rapid motion. It too vanishes in the gravitational case. This makes the next term in the potential, containing the quadrupole moment of the source $\sum m^\alpha x_i^\alpha x_j^\alpha$, the important one in generating gravitational waves.

Gravitational waves carry energy which is how they deform spacetime. Therefore Equation (9.1) needs to be modified in the presence of such waves to take account of their stress–energy tensor $t_{\mu\nu}$:

$$t_{\mu\nu} = -\frac{c^4}{8\pi G}G^{(2)}_{\mu\nu}. \tag{9.14}$$

where $G^{(2)}$ only contains terms quadratic in $h_{\alpha\beta}$. This expression is evaluated in Appendix D. The energy flow per unit area per unit time is

$$F = \frac{c^3}{16\pi G}\langle\dot{h}_+^2 + \dot{h}_\times^2\rangle = \frac{c^3}{32\pi G}\langle\dot{h}_{ij}\dot{h}_{ij}\rangle, \tag{9.15}$$

where each dot over a quantity indicates taking a time derivative. It is misleading to assign energy to a point in spacetime because only the *relative* displacements are meaningful. Furthermore it is not even possible to specify whether the energy is in the peaks or valleys of the waves. For these reasons Equation (9.15) only applies when averages are taken over several cycles and wavelengths. This averaging is indicated by the angular brackets.

Solutions of Einstein's equation will be sought for sources that are nearly Newtonian, which means first that the curvature produced by the source and hence the strain are small, and second that the velocity v of the material within the source is very much less than c: $GM/rc^2 \ll 1$ and $v^2/c^2 \ll 1$. A linear approximation can be made to the metric in these circumstances. It turns out that the intensity of gravitational radiation predicted by this linearized theory generally only differs by small numerical factors from the results of more exact calculations (Davis et al. 1971). In Appendix E the contribution due quadrupole motion of the source is evaluated in Equation (E.4). At a distance r from the source

$$h_{ij}(t) = \frac{2G}{rc^4}\ddot{I}_{ij}^{\mathrm{TT}}\left(t - \frac{r}{c}\right) \tag{9.16}$$

where I_{ij}^{TT} is the transverse-traceless part of the quadrupole moment of the source (see Equation (E.3)). Equation (9.16) is a *retarded* solution: gravitational disturbances propagate at a velocity c, and hence the amplitude at r at time t is determined by the source behavior at an earlier time $t - r/c$. This argument will be omitted from here on in order to simplify the presentation. The energy flow in the gravitational wave given by Equation (E.6):

$$F = \frac{G}{8\pi r^2 c^5}\langle \dddot{I}_{ij}^{\mathrm{TT}} \dddot{I}_{ij}^{\mathrm{TT}} \rangle.$$

As before the angular brackets indicate the expectation value averaged over several cycles. The total energy flow through a sphere at distance r from the source is called its luminosity L. In Appendix E it is shown that

$$L = \left(\frac{G}{5c^5}\right)\langle \dddot{\mathcal{I}}_{ij} \dddot{\mathcal{I}}_{ij} \rangle, \tag{9.17}$$

where $\mathcal{I}_{ij}$ is the reduced quadrupole moment of the source:

$$\mathcal{I}_{ij} = \int (x_i x_j - \delta_{ij} x_k^2/3)\rho \, \mathrm{d}V$$

with ρ being the density in a volume $\mathrm{d}V$ at x_i. The range of integration is over the volume of the source. In the above equation δ_{ij} is the Kroneker delta, with value +1 when $i = j$ and zero otherwise.

9.3 PSR 1913+16

The pulsar PSR 1913+16 is one member of a binary pair located 6.4 kpc from the Earth in the constellation Aquila. Pulsars are neutron stars a few km in diameter, the outcome of the gravitational collapse described in Section 8.8. During the collapse to a neutron star the response is analogous to that of a spinning skater pulling her arms in; it ramps up the star's spin to about 1 kHz and the magnetic field to values as large as 10^8 T. Intense beams of electromagnetic radiation are emitted along the magnetic axis and if, as in the case of PSR 1913+16, the magnetic axis is offset from the spin axis, this beam whirls round like a lighthouse beam. If the Earth lies in the path traced out by the beam then a short burst at radio frequencies is observed at regular intervals. Pulsars are good time-keepers: for example, pulsations from PSR 1937+21 are stable to 10^{-19} s s^{-1}. Hulse & Taylor (1975), using the 305 m diameter Arecibo radio telescope, discovered PSR 1913+16 with a pulse period of 59 ms. They found that the pulse period was not constant, it changed over a 7.75 hour cycle by 80 μs. This meant that the pulsar was orbiting a companion so that its pulses were Doppler shifted during each 7.75 hour orbit. No radiation has been detected from the companion so it is likely to be another neutron star. Both the pulsar and its companion have masses close to 1.4 $M_\odot$ and their orbits would almost fit inside the Sun, enhancing GR effects far beyond those observed in the solar system. These GR effects can be measured remotely because the pulsar itself is a clock as precise as any atomic clock. The key observation made by Hulse and Taylor was that the orbits decay at a rate that is precisely that expected due to the loss of energy in gravitational radiation from the pair. Taylor and colleagues have timed the pulses over four decades and determined the orbital behavior in fine detail. In Figure 9.4 the data from the orbital collapse are compared to the prediction using GR. The predicted decay rate of the orbital period τ, is

$$\frac{\mathrm{d}\tau}{\mathrm{d}t} = -2.40263(5) \times 10^{-12} \text{ s s}^{-1}.$$

The fit to the data gives a rate 0.9983 ± 0.0016 times the GR prediction. Here we follow the steps leading to this prediction taking, for simplicity, the case that the binary pair have equal mass.

The quadrupole moment of the binary pair has an xx-component (see Exercises)

$$I_{xx} = 2Ma^2 \cos^2 \omega t = Ma^2(1 + \cos 2\omega t),$$

where ω is the angular frequency of rotation. Similarly

$$I_{yy} = Ma^2(1 - \cos 2\omega t).$$

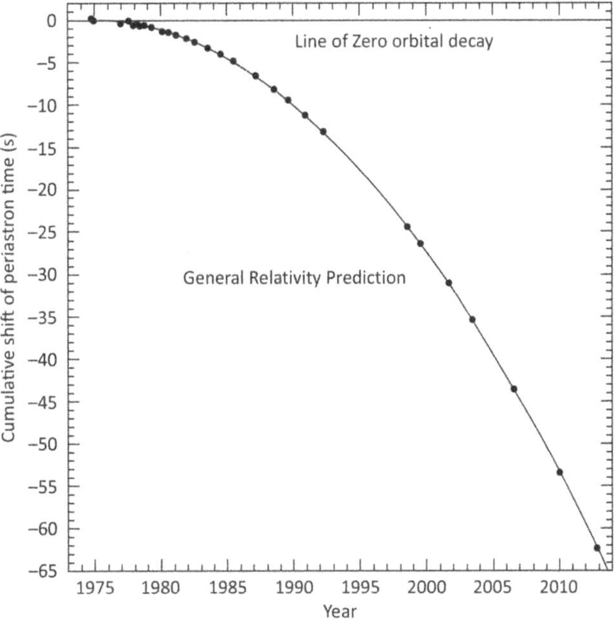

Figure 9.4. Orbital decay of PSR 1913+16 with time. The curve represents the orbital phase shift expected due to the emission of gravitational waves. The points are data with error bars that are smaller than the line width. Figure 3 from Weisberg and Huang (2016). Courtesy of Professor Weisberg on behalf of the copyright holders.

Working in two dimensions (x, y) the *traceless* or reduced quadrupole moments are

$$\mathcal{I}_{ij} = \int (x_i x_j - \delta_{ij} x_k^2/2)\rho \, \mathrm{d}V.$$

Thus

$$\mathcal{I}_{xx} = -\mathcal{I}_{yy} = Ma^2 \cos 2\omega t, \text{ and } \mathcal{I}_{xy} = \mathcal{I}_{yx} = Ma^2 \sin 2\omega t.$$

Substituting these values into Equation (E.7) gives a luminosity

$$L = \frac{G}{5c^5}(2\omega)^6(Ma^2)^2\langle 2 \sin^2 2\omega t + 2 \cos^2 2\omega t\rangle = \frac{G}{5c^5}(128\omega^6 M^2 a^4). \qquad (9.18)$$

In order to estimate the rate at which the orbit decays we need to compare this, the rate of energy loss, with the total energy of the binary pair:

$$E = Mv^2 - GM^2/2a.$$

Using the radial equation of motion for either star

$$Mv^2/a = GM^2/4a^2,$$

whence

$$\omega^2 = v^2/a^2 = GM/4a^3.$$

Now, substituting for v in the expression for total energy and then rewriting a in terms of ω gives

$$E = -\frac{GM^2}{4a} = -\frac{GM^2}{4}\left(\frac{4\omega^2}{GM}\right)^{1/3}.$$

Thus

$$\frac{\mathrm{d}E}{E} = \frac{2}{3}\frac{\mathrm{d}\omega}{\omega} = -\frac{2}{3}\frac{\mathrm{d}\tau}{\tau}.$$

Then the observable quantity, the orbital decay rate, is

$$\frac{(\mathrm{d}\tau/\mathrm{d}t)}{\tau} = -\frac{3}{2}\frac{\mathrm{d}E/\mathrm{d}t}{E} = \frac{3}{2}\frac{L}{E} = -\frac{768}{5}\frac{\omega^6 a^5}{c^5}.$$

Substituting for ω gives, finally,

$$\frac{\mathrm{d}\tau/\mathrm{d}t}{\tau} = -\frac{12}{5}\frac{G^3M^3}{c^5a^4}. \qquad (9.19)$$

Press and Thorne (1972) calculated the correction required for the case of elliptical orbits.[2] The precision with which the GR prediction has tracked the decay rate of

[2] With eccentricity e the right-hand side of Equation (9.19) is then multiplied by a factor

$$\frac{1 + 73e^2/24 + 37e^4/96}{(1 - e^2)^{7/2}}.$$

PSR 1913+16's orbit for forty years is convincing, but indirect evidence for the existence of gravitational waves. In eight other examples of binary pulsars studied by Weisberg and Huang similar agreement was found between the GR prediction and the observed orbital decay rates. These very precise quantitative comparisons with GR nicely complement the direct observations of gravitational waves by the LIGO/ Virgo collaboration; which is the next topic.

9.4 The LIGO and Virgo Interferometers

On 2015 September 14 the advanced LIGO Michelson interferometers (Abbott et al. 2016) at Hanford in Washington state and at Livingstone in Louisiana detected a burst of gravitational waves from the inspiralling and merger of two black holes occurring around 1.2 billion years ago.[3] The event is therefore labeled GW150914. Each rotation of the black holes about their center of mass generates two cycles of gravitational radiation. As they spiral inward the rotations become ever more rapid and the frequency of the waves rises in a characteristic *chirp* that terminates when the black holes merge. The time required for gravitational waves, or light, to travel directly between the two LIGO detectors is 10 ms: hence signals from a single source will arrive no more than 10 ms apart. In the case of GW150914 the time difference between the signals detected at the two sites was 7 ms. In the distant future the binary pair containing PSR 1913+16 will also merge.

The masses of the merging black holes in GW150914 were estimated to be around 35 $M_{\odot}$, and their merger to have released about three solar masses as gravitational wave energy. Reading down the left-hand column in Figure 9.1 illustrates how spacetime oscillates over a complete cycle of a gravitational wave. The wave direction of travel is perpendicular to the diagram in the Z direction. The shapes represent a string of freely suspended masses, initially forming a circle before the wave arrives. We can suppose the arms of the Michelson interferometer lie along the X and Y directions; then the arms lengthen and shorten in antiphase with each other as the wave passes. This produces an oscillating phase difference between light that travels along the two arms. The beams from the two arms exit together at the output port and form interference fringes there. Any oscillations in the relative length of the arms produced by an incident gravitational wave will cause the fringes to move in sympathy. It is this fringe motion that is detected, recorded and analyzed. Figure 9.5 shows the oscillations of spacetime at the two detectors over the crucial 150 ms of GW150914. In this interval more energy was radiated than our Galaxy will radiate in 10,000 years. As the wave passed the oscillation frequency rose at an increasing rate. This is the frequency *chirp* predicted for the evolution of gravitational waves from such mergers. At this point the two black holes had merged into a single black hole, which then *rang down* like a bell to a final equilibrium static state. The observed ringdown amplitude for GW150914 had a time dependence

[3] The material in this and the succeeding sections is substantially modified from similar material that appeared in "Quantum 20/20: Fundamentals, Entanglement, Gauge Fields, Condensates and Topology" published by Oxford Univ. Press in 2019 (Kenyon 2019). The author is indebted to Oxford Univ. Press for permission.

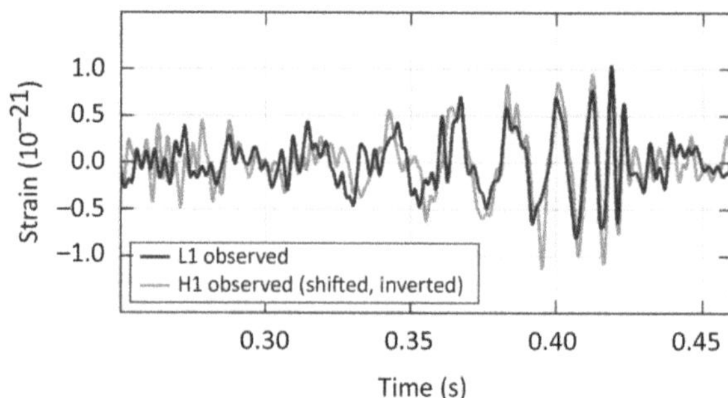

Figure 9.5. Strain versus time plot from the first gravitational wave signal observed by the advanced LIGO at Hanford (H1) and Livingstone (L1). LIGO Open Science Center at https://losc.ligo.org/events/GW150914.

$$\exp(-t/\tau)\cos(2\pi ft + \phi), \tag{9.20}$$

where $f \approx 260/M_f$ Hz and $\tau = 0.004M_f$ s, M_f being the mass of the final merged black hole, in units of 65 $M_\odot$. The masses of the merging black holes and the merged black hole were obtained by modeling the whole event, chirp plus ringdown, using general relativity and fitting this prediction to the observed oscillations by varying the masses. Figure 9.6 shows the details of the inspiral and coalescence found by this numerical modeling and fitting. Just think: spacetime stretched by only one part in 10^{21} as each wave crest passed, equivalent to an atomic width in the Earth–Sun separation. How this astonishing sensitivity was achieved is spelled out below.

The inspiral phase of GW150914 has been analyzed by making the approximation that the dynamics is nearly Newtonian (LIGO Scientific and VIRGO Collaborations et al. 2016). This leads to simple conclusions about the merger, which we now reproduce. Assuming the binary pair have masses m_1 and m_2, we define $\mu = m_1 m_2/(m_1 + m_2)$ and $M = m_1 + m_2$. The rate of change of the gravitational energy is

$$\dot{E} = \frac{G\mu M\dot{r}}{2r^2}, \tag{9.21}$$

where r is the pair separation at any moment and the dots over quantities signify differentiation with respect to time. Gravitational energy is radiated making $\dot{E}$ to be negative, r falls and the pair spiral in. We take the quadrupole moment of the pair to be $Q = \mu r^2/2$, oscillating at an angular frequency ω_g. During one complete orbit of the masses the moment goes through two complete cycles: thus ω_g is twice the orbital angular frequency of the masses ω_k. This gives

$$\dddot{Q}^2 = 32\mu^2 r^4 \omega_k^6. \tag{9.22}$$

For use later, we use Kepler's law to obtain expressions for r and $\dot{r}$:

$$r = [GM/\omega_k^2]^{1/3}, \tag{9.23}$$

$$\dot{r} = -(2/3)[r\dot{\omega}_k/\omega_k]. \tag{9.24}$$

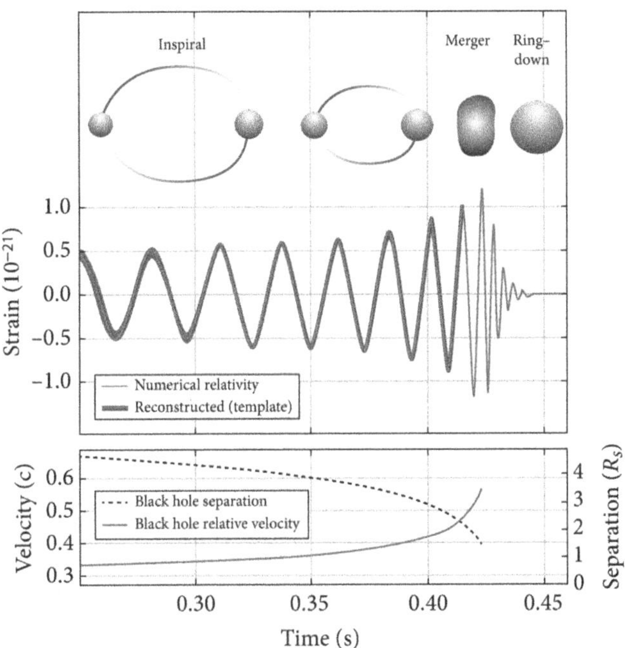

Figure 9.6. Estimated strain amplitude from GW150914 made with numerical relativity models of the black hole behavior as the holes coalesce. In the lower panel the separation of the black holes is given in units of the Schwarzschild radius $2GM/c^2$ and the relative velocity is divided by c. LIGO Open Science Center at https://losc.ligo.org/events/GW150914. The work is reported by the LIGO Scientific Collaboration and Virgo Collaboration (Abbott et al. 2016). Courtesy LIGO Collaboration. This figure is Figure 11.5 taken with permission from Kenyon (2019) published by Oxford Univ. Press in 2019.

Using Equation (E.7) the luminosity of the source is

$$L = \frac{32G}{5c^5}\mu^2 r^4 \omega_k^6. \tag{9.25}$$

Equating $\dot{E}$ and L from Equations (9.21) and (9.25) and substituting for r and $\dot{r}$ gives the result

$$\dot{\omega}_k{}^3 = (96/5)^3 (GM_c/c^3)^5 \omega_k^{11}, \tag{9.26}$$

which holds equally for ω_g in place of ω_k. $M_c = (\mu^3 M^2)^{1/5}$ is the *chirp mass*, useful in setting the mass scale in binary kinematics. This expression reveals that the frequency rises at an increasing rate, giving the chirp seen in Figures 9.5 and 9.6. Fitting this expression to the data gave a chirp mass of 30 $M_\odot$: then for equal masses $m_1 = m_2 = 35$ $M_\odot$. When the amplitude of oscillation reached its *peak* the frequency f_g is estimated to be 150 Hz, and the rotation frequency f_k to be 75 Hz. Correspondingly the separation of the masses was

$$R = (GM/\omega_k^2)^{1/3} = 350 \text{ km}. \tag{9.27}$$

The energy radiated in gravitational waves to that point was

$$GM\mu/2R \sim 3\ M_{\odot}c^2, \tag{9.28}$$

and the Schwarzschild radius of each star was

$$2Gm_1/c^2 = 103\ \text{km}. \tag{9.29}$$

This means that the black holes were, at that moment, only separated by not much more than the sum of their Schwarzschild radii. If they had been not black holes, but instead stars, for example white dwarfs which are tens of thousands of kilometers across, then the inspiral would have terminated far earlier.

In 2020, the two LIGO interferometers, and the Virgo interferometer in Italy were in operation and after this third data-taking period some 90 mergers had been detected. These included two neutron star mergers.

In parallel with the experimental development templates have been produced for the expected frequency and amplitude of the gravitational waves from black hole mergers: these are functions of the masses involved. Each template has three components that have to be smoothly connected: the inspiral, the merger (plunge) and the ringdown. Of these the inspiral is calculated using post-Newtonian approximation to GR, which includes higher orders in $\xi = GM/rc^2$ than the linear terms used in the analysis given earlier in this chapter. Both the merger and ringdown are more difficult to simulate because the term GM/rc^2 in the metric factor $(1 - GM/rc^2)$ begins to dominate, and hence a post-Newtonian approximation is expected to be inadequate. The calculations are therefore made with GR directly in a form that spacetime is presented as time-sliced: the calculation iterates from one time slice to the next. This stage of calculation of a template may require weeks on a processor farm. Fortunately the full simulations have revealed that high-order post-Newtonian approximations can be useful beyond the inspiral stage. During data-taking the outputs from the interferometers are matched continuously against templates to detect any possible signal. Noise from laser glitches, thunderstorms, and seismic shocks have other recognizably different frequency–amplitude profiles.

9.5 The Interferometers

We can now apply quantum measurement principles to understand how the design of the advanced LIGO detectors made it possible to achieve this amazing sensitivity. The motion of the mirrors under the action of gravitational waves is, so far as we know, purely classical: this motion modulates the electromagnetic field in the interferometer arms, and this is the point at which the quantum aspects of measurement impact.

Figure 9.7 shows the basic components of the advanced LIGO Michelson interferometers. The 40 kg mirrors are test masses used to transfer the gravitational wave motion to the electromagnetic field stored in the arms of the interferometer. Surprisingly, it is the quantum properties of the electromagnetic field that determine the precision achievable. A stabilized laser beam at 1064nm illuminates the 50/50 near lossless beam splitter. Each of the two 4 km long arms has, in addition to the far

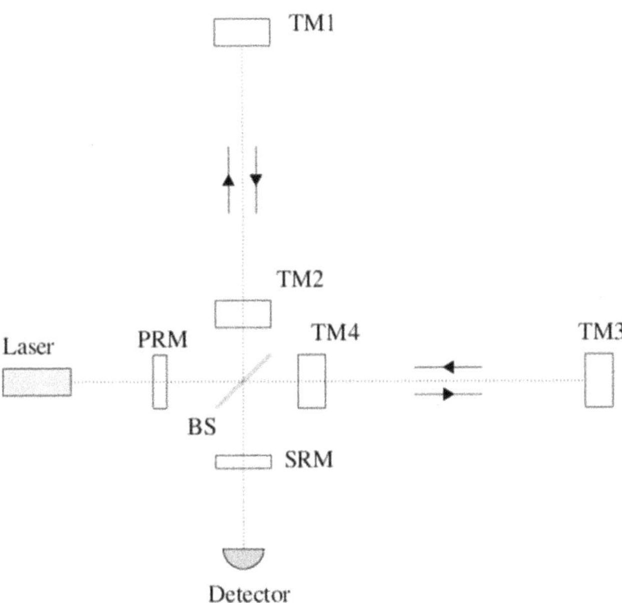

Figure 9.7. Sketch of components of the advanced LIGO detectors. The mirrors TM1, TM2, TM3, and TM4 form Fabry–Perot etalons in the arms of the Michelson interferometer. PRM is the power recycling mirror, SRM the signal recycling mirror. BS is the beam splitter. This figure is Figure 11.6 taken with permission from Kenyon (2019) published by Oxford Univ. Press in 2019.

mirror, a second similarly massive mirror placed close to the beam splitter. This second mirror converts each arm into a Fabry–Perot cavity and the laser is tuned to a cavity resonance. Radiation in resonance with the cavity passes to and fro about 300 times before its intensity falls significantly. As a result the phase resolution of the interferometer is improved by a similar factor. It is arranged that in the absence of gravitational waves the reflected beams arrive out of phase at the photodetector, in other words it views a dark fringe. An important consequence is that almost all the radiation entering through the input face of the beam splitter ends up exiting through this same face. In general the electric field at the detector is the real part of

$$E = E_{\text{in}}[\exp(i\phi_x) - \exp(i\phi_y]/2, \tag{9.30}$$

where $\phi_{x,y}$ are the phase changes in the two arms. When the interferometer is set on a dark fringe a gravitational wave traveling perpendicular to the plane of the interferometer produces an electric field at the detector

$$E_{\text{GW}} = E_{\text{in}}[\exp(i\phi_{\text{GW}}) - \exp(-i\phi_{\text{GW}})]/2 \approx i\phi_{\text{GW}}E_{\text{in}}, \tag{9.31}$$

where ϕ_{GW} is the oscillating phase difference induced by the gravitational wave. This signal is so small that in order to make a useful measurement a larger DC field is added to it. This is achieved by offsetting the output away from the dark fringe by a small angle in the absence of any gravitational wave. Then the total signal when a gravitational wave is detected is

$$E_T = E_{\rm GW} + E_{\rm DC}. \tag{9.32}$$

The power is then

$$P \propto E_{\rm DC}^2 + 2E_{\rm DC}E_{\rm GW}$$
$$= E_{\rm DC}[E_{\rm DC} + 2\phi_{\rm GW}E_{\rm in}]$$

where the negligible term $E_{\rm GW}^2$ has been ignored. This power is converted to a current in the photodetector and from the current the oscillating phase $\phi_{\rm GW}$ is extracted.

9.6 The Standard Quantum Limit

The role of the mirrors is to transfer the fluctuations of spacetime to the electromagnetic field in the interferometer.[4] Each precise measurement of a LIGO mirror position gives rise to uncertainty in its momentum, and because the measurement of the LIGO mirrors is continuous this momentum uncertainty gives a contribution to later mirror position measurement: a process known as quantum back action. Quantum fluctuations arising in the radiation field in the interferometer impose the ultimate limit on the precision of the phase measurement. There are two components: noise from statistical variation of the number of photons arriving at the detector, *shot noise*, and fluctuations of radiation pressure on the mirrors, known as the quantum radiation pressure noise (QRPN). This latter component is the back action.

First consider the shot noise. If the optical power is P at angular frequency ω and wave number k then during a measurement taking a time τ the integrated photon count at the detector, N is $P\tau/\hbar\omega$. The corresponding minimum phase uncertainty $\delta\phi_{\rm s}$ is $1/\sqrt{N}$. A phase uncertainty of 2π would give a path uncertainty of one wavelength λ, so the uncertainty in the path length is

$$\delta x_{\rm s} = \frac{\delta\phi_{\rm s}\lambda}{2\pi} = \sqrt{\frac{\lambda\hbar c}{2\pi P\tau}}. \tag{9.33}$$

It is the intensity (~power) spectrum of the noise that quantifies its impact. The shot noise has a *spectral density*, that is an intensity per unit frequency, of

$$S_{\rm s} = (\delta x_{\rm s})^2\tau = \frac{\lambda\hbar c}{2\pi P}, \tag{9.34}$$

with units $\rm m^2Hz^{-1}$. This imposes a limit on the detectable strain measured over a distance L of

$$h_{\rm s} = \frac{\sqrt{S_{\rm s}}}{L} = \frac{1}{L}\sqrt{\frac{\hbar c\lambda}{2\pi P}}. \tag{9.35}$$

[4] This matter is pursued more fully by Braginsky et al. (2003).

Next consider the QRPN. The force due to each photon striking a mirror of mass M during the measurement time τ is $2\hbar k/\tau$. With an average of N photons arriving at the mirror in time τ the fluctuation on this force is

$$\delta F = 2\sqrt{N}\hbar k/\tau. \tag{9.36}$$

When the freely suspended mirror responds to a gravitational wave of angular frequency $\Omega_{\rm GW}$ its equation of motion is

$$F\cos(\Omega_{\rm GW}t) = M\mathrm{d}^2x/\mathrm{d}t^2 = -M\Omega_{\rm GW}^2\, x\cos(\Omega_{\rm GW}t). \tag{9.37}$$

Then the arm length fluctuation due to the radiation pressure fluctuation is

$$\delta x_{\rm r} = \delta F/M\Omega_{\rm GW}^2 = \frac{2\sqrt{N}\hbar k}{\tau M\Omega_{\rm GW}^2}. \tag{9.38}$$

The corresponding spectral noise intensity is

$$\begin{aligned} S_{\rm r} = (\delta x_{\rm r})^2\tau &= N\tau\left[\frac{2\hbar k}{\tau M\Omega_{\rm GW}^2}\right]^2 \\ &= \left(\frac{P\tau^2}{\hbar\omega}\right)\left[\frac{2\hbar k}{\tau M\Omega_{\rm GW}^2}\right]^2 \\ &= \frac{8\pi\hbar P}{\lambda c}/(M^2\Omega_{\rm GW}^4). \end{aligned} \tag{9.39}$$

Expressed as a strain this gives

$$h_{\rm r} = \frac{\sqrt{S_{\rm r}}}{L} = \frac{1}{LM\Omega_{\rm GW}^2}\sqrt{\frac{8\pi\hbar P}{\lambda c}}. \tag{9.40}$$

Notice that the limiting strains due to the shot noise, Equation (9.35), and the QRPN, Equation (9.40), have inverted dependences on the laser power P. Thus the minimum detectable strain is obtained when these two contributions are equal:

$$\frac{\lambda\hbar c}{2\pi P} = \frac{8\pi\hbar P}{\lambda c M^2\Omega_{\rm GW}^4}, \tag{9.41}$$

hence we deduce that there is an optimum optical power

$$P = \frac{\lambda c M\Omega_{\rm GW}^2}{4\pi}. \tag{9.42}$$

Using this value for P the minimum total spectral noise power density,

$$S = S_{\rm s} + S_{\rm r} \geqslant \frac{4\hbar}{M\Omega_{\rm GW}^2}, \tag{9.43}$$

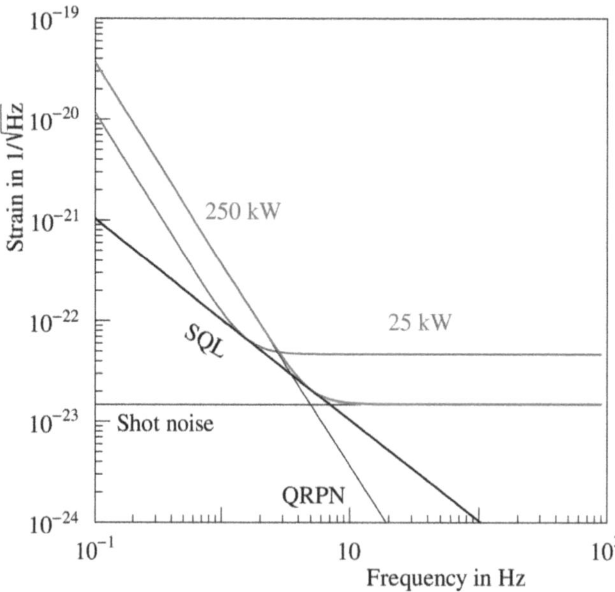

Figure 9.8. Limit on detectable strain with a simple Michelson interferometer having 10 km arms, 10 kg mirrors, and laser wavelength 1064 nm. Examples are shown for 50 kW and 500 kW circulating power.

with a corresponding strain

$$h = \frac{1}{L}\sqrt{\frac{4\hbar}{M\Omega_{\mathrm{GW}}^2}}. \tag{9.44}$$

We have here the *standard quantum limit* (SQL). Putting in the parameters of a simple Michelson interferometer with the dimensions similar but not identical to LIGO, the limiting strain detectable is plotted in Figure 9.8. The masses of the mirrors are 10 kg, the laser wavelength is 1064 nm, and the arms are 10 km long. Two cases are shown, for a circulating optical power in the arms of 50 kW and 500 kW. Raising the power lowers the shot noise and improves the sensitivity for detecting high frequency gravitational waves. However, the radiation pressure fluctuations increase with increasing power, which reduces the sensitivity to low frequency waves. The frequency at which the SQL is attained rises as the circulating optical power is increased.

Then our ideal detector with 100 kW power would have adequate sensitivity to detect the event at ~100 Hz observed by advanced LIGO. Evidently very high optical powers are required, much greater than the wattage available from the well-stabilized 1064 nm Nd:YAG lasers as used in LIGO.

9.6.1 Cavity Enhancement

Mirrors are inserted into both arms of the LIGO interferometers near the beam splitter making each arm into a 4 km long Fabry–Perot cavity. Fabry–Perot cavities are characterized by their *finesse, F*, $\pi\sqrt{R}/(1 - R)$, where R is the reflectance of the mirrors. Finesse is the ratio of fringe separation to the fringe width, which with the high reflectance mirrors used in advanced LIGO is ~300. This has the effect that the light at a cavity resonance traverses the cavity around 300 times before its intensity falls off appreciably. The phase resolution of the interferometer is improved by a similar factor and the energy in the radiation stored in the interferometer is similarly boosted, cutting down the shot noise. However a drawback is that gravitational waves with a period shorter than the lengthened storage time will be undetectable. In operation the laser is tuned close to a resonance of the cavity so as to benefit from maximum phase resolution.

Even with the Fabry–Perot cavities the power circulating inside the interferometer arms is still far less than the 100 kW required to detect the waves from astronomically distant black hole mergers. What saves the day is that there is the opportunity to recycle wasted power. The detection mode requires a dark fringe on the detector in the absence of any gravitational waves, meaning that negligible power enters the detector. Consequently almost all the radiation exits back through the entry port toward the laser. By placing a *power recycling mirror* (PRM) between the laser and the beam splitter, a new cavity is formed between this mirror and an effective mirror formed by the interferometer. The returning power is recycled in this cavity. In this way the power at the beam splitter is enhanced to over 200 kW. Standard mirrors would overheat in such beam intensities. Those in LIGO are highly homogeneous silica with coatings of such high reflectance that there is negligible absorption. The far mirrors have reflection coefficients of 0.999995 and the near mirrors a more modest 0.985 to facilitate entry and exit of beams.

The bandwidth of the composite interferometer-PRM cavity is only 1 Hz, which, with *mode cleaning*, renders the radiation entering the interferometer uniquely noise free. Mode cleaning is obtained by inserting a cavity before the power recycling mirror. This cavity is tuned to transmit only the principal TEM_{00} laser mode: this is the simplest mode with a rotationally symmetric Gaussian profile of 5 cm diameter. At the same time the mode cleaning cavity disperses all the other modes that make up the laser beam, thus eliminating most of the noise in the beam. Finally, yet another recycling mirror is placed between the detector and beam splitter to enhance the signal.

Figure 9.9 shows the noise floor achieved by the advanced LIGO detectors (Tse et al. 2019). At low frequencies seismic noise is suppressed by the mirror damping and the suspension. Individual spikes correspond to mechanical resonances of the components of the mirror assemblies. The phase amplification coming from the Fabry–Perot cavities is flat out to around 300 Hz and then falls off steadily with increasing frequency. For more details see Bond et al. (2016). The frequency at which this knee in the response occurs is roughly the inverse of the storage time in the Fabry–Perot etalons that are formed by the interferometer mirrors. There is an accompanying rise in the minimum detectable strain at higher frequencies in Figure 9.9. This effect was not included in the analysis used to produce Figure 9.8.

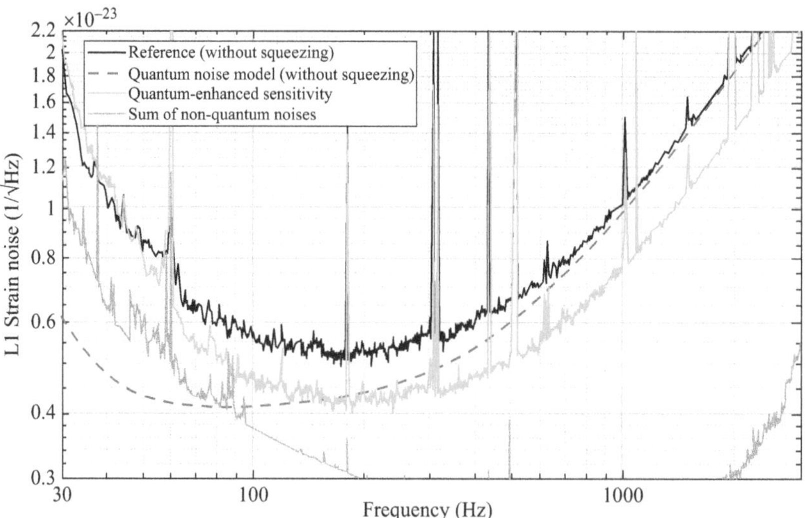

Figure 9.9. Strain noise level in advanced LIGO with unsqueezed (black) and squeezed (green) vacuum. Figure 1 from Tse et al. (2019). Made available through Creative Commons Attribution 4.0 International license.

9.7 Squeezing

Surprisingly the port where radiation exits to the detector has an impact on the noise. Through it the vacuum radiation field enters the interferometer and the vacuum fluctuations add to both the shot noise and the QRPN, introducing some correlation between them. It proved possible, by taking advantage of this correlation induced between the shot noise and the QRPN, to reduce the noise below the SQL (Unruh 1982; Braginsky et al. 1992; Kimble et al. 2001). The researchers have adapted LIGO by injecting a squeezed vacuum into the exit port, which by virtue of the correlation just noted, can be tuned to reduce the interferometer noise below the SQL over a range of frequencies. This strategy is now explained.

The quantum operator for the electric field in a single optical mode can be written[5]

$$\hat{E} = [\hat{a}\exp(-i\omega t) + \hat{a}^\dagger \exp(i\omega t)]/\sqrt{2}, \tag{9.45}$$

where $\hat{a}$ annihilates and $\hat{a}^\dagger$ creates a photon in the mode of angular frequency ω:

$$\hat{a}|n\rangle = \sqrt{n}\,|n-1\rangle, \tag{9.46}$$

$$\hat{a}^\dagger|n\rangle = \sqrt{n+1}\,|n+1\rangle. \tag{9.47}$$

Here $|n\rangle$ is the state vector describing a mode containing n photons. $\hat{a}$ and $\hat{a}^\dagger$ are conjugate operators with the property that $[\hat{a}, \hat{a}^\dagger] = 1$. Of interest here are the *quadrature operators*

[5] See for example Section 8.2 in Kenyon (2019) published by Oxford Univ. Press.

$$\hat{X}_1 = (\hat{a}^\dagger + \hat{a})/\sqrt{2} \text{ and } \hat{X}_2 = i(\hat{a}^\dagger - \hat{a})/\sqrt{2}, \tag{9.48}$$

in terms of which

$$\hat{E} = [\hat{X}_1 \cos(\omega t) + \hat{X}_2 \sin(\omega t)]. \tag{9.49}$$

It follows that

$$[\hat{X}_1, \hat{X}_2] = i, \tag{9.50}$$

so that $\hat{X}_1$ and $\hat{X}_2$ are also conjugate operators. Uncertainty in X_1 corresponds to an amplitude uncertainty, and impacts the QRPN; while uncertainty in X_2 corresponds to phase uncertainty, and impacts the shot noise. We can write the equivalent uncertainty relation between the variances of the observables ($\Delta X^2 = \bar{X^2} - \bar{X}^2$):

$$(\Delta X_1)^2(\Delta X_2)^2 \geqslant 1/4. \tag{9.51}$$

A fully coherent beam attains the minimum uncertainty limit with $(\Delta X_1)^2 = (\Delta X_2)^2 = 1/2$.

The mechanism employed in beating the SQL is to *squeeze the vacuum* lightly. This involves making a change in the content of a mode of the electromagnetic field of interest:

$$|0\rangle \to |\psi\rangle \equiv |0\rangle + s|2\rangle, \tag{9.52}$$

with $s \ll 1$ being the amplitude for injecting two coherent photons into that mode. Then

$$\begin{aligned} \langle\psi|\hat{X}_1^2|\psi\rangle &= \langle\psi|\,\{\hat{a}^\dagger + \hat{a}\}^2\,|\psi\rangle/2 \\ &= \{\langle 0| + s\langle 2|\}\,\{\hat{a}^\dagger\hat{a}^\dagger + \hat{a}\hat{a} + \hat{a}^\dagger\hat{a} + \hat{a}\hat{a}^\dagger\}\,\{|0\rangle + s|2\rangle\}/2 \\ &= (1 + 2\sqrt{2}s)/2 \end{aligned} \tag{9.53}$$

to order s. In making the second equality we have used the property of the annihilation operator that $\hat{a}|0\rangle = 0$. Also

$$\langle\psi|\hat{X}_1|\psi\rangle = [\langle 0| + s\langle 2|]\,[\hat{a}^\dagger + \hat{a}]\,[|0\rangle + s|2\rangle]/\sqrt{2} = 0. \tag{9.54}$$

Thus the variance

$$(\Delta X_1)^2 = \langle\psi|\hat{X}_1^2|\psi\rangle - |\langle\psi|\hat{X}_1|\psi\rangle|^2 = 1/2 + \sqrt{2}s. \tag{9.55}$$

Similarly it follows that

$$(\Delta X_2)^2 = 1/2 - \sqrt{2}s, \tag{9.56}$$

which demonstrates squeezing[6] in X_2 (impacts shot noise) while sacrificing the precision obtainable in X_1 (impacts QRPN). It requires a nonlinear process to

[6] The product of ΔX_1^2 and ΔX_2^2 appears to violate the uncertainty principle. If the expansion is continued to higher orders in s this violation disappears.

generate the required pairs of entangled photons appearing in Equation (9.52). The device used in LIGO is an *optical parametric oscillator* (OPO). A laser pumps a Fabry–Perot cavity containing a nonlinear crystal. The cavity mode is matched to that of the interferometer and the excitation produces pairs of entangled photons that are collinear and have frequency equal to that of the cavity resonance. These pairs of entangled photons are thus in the same mode and at the correct frequency to apply squeezing. The laser intensity is held *just below threshold* to excite oscillation in the crystal. With this setup the OPO input is the vacuum and the OPO output is the squeezed vacuum. The data in Figure 9.9 shows a 3 dB reduction in the noise spectrum above 50 Hz. This is the region in which the observed mergers were detected and shot noise dominates. There is a 14% increase in the distance at which mergers can be detected, which increases the detection rate by 40%.

9.8 GW170817 and the Velocity of Gravitational Waves

Once three detectors were in operation at the same time it became possible to use the difference between the arrival times at the three sites to triangulate, with varying precision, the direction of the source in the sky. One such merger GW170817, involving objects of masses 0.86 and 2.26 $M_{\odot}$. This was located more precisely thanks to the detection of a time-coincident gamma-ray burst, followed by prolonged emission over a wide spectral range from a single source. The spectral distribution and decay rate of the radiation observed were those expected following a binary neutron star merger. This was the process independently proposed to explain the production of most heavy elements beyond iron. Gold wedding rings and platinum bracelets predominantly contain matter processed in neutron star mergers! The merger of the two neutron stars and the associated gamma-ray burst lie in the direction of the galaxy NGC4993 whose distance from the Earth is 40 Mpc. This accords with the luminosity distance to GW170817 of 40 ± 14 Mpc estimated from the signal amplitude (Abbott et al. 2017). Now the time delay between the detections of the merger and gamma-ray burst was 1.74 s and we are at liberty to interpret this delay as due entirely to a difference between the velocities of gamma-rays and gravitational waves. Then taking the luminosity distance to lie at the lower limit of 26 Mpc, the fractional difference between these velocities would be at most $1.74c/26$ Mpc, roughly 10^{-15}. The expectation from GR that the velocities are identical looks safe.

9.9 Exercises

1. Calculate the quadrupole moment of a system of two stars, each of mass M in circular orbits of angular frequency ω at a separation $2a$. Take the center of mass as the origin and let the motion be in the x–y plane with the masses along the x axis at time zero. Calculate I_{xx}, $I_{xy} = I_{yx}$ and I_{yy} at time t. Then calculate the reduced quadrupole moments using the fact that only the time dependent part survives.
2. Calculate the luminosity of the gravitational wave emission from the binary pair described in the previous question, in terms of M, ω, and a.

3. Using the results from the previous question, determine the frequency and amplitude of gravitational waves from the binary pulsar system PSR 1913 +16 at the surface of the Earth. You can assume that the orbits are circular with a equal to 3.1961 lightsec, that the masses are each 1.414 $M_{\odot}$, that the period of rotation is 27,907 s, and that the pulsar is 6.4 kpc from the Earth.
4. LISA is a gravitational wave interferometer planned for launch into space around 2034. The laser light would follow a closed light path round an equilateral triangle of mirrors of side length 2.5×10^6 km. Suppose the laser used has 1 W power at wavelength 1 μm, and that 50 pW power is received in one pass along an arm. What is the minimum strain detectable?
5. Explain how the laws of conservation of momentum and angular momentum forbid emission of any dipole gravitational radiation.

Further Reading

Misner C W, Thorne K S and Wheeler J A 1971 *Gravitation* (San Francisco, CA: W. H. Freeman). Chapter 18 contains a thorough treatment of the formalism of gravitational waves. *Gravitation* established the notation generally used today for general relativity.

Bond A, Brown D, Friese A and Strain K A 2016 *Interferometric Techniques for Gravitational-Wave Detection: Living Reviews in Relativity* **19:3** (Berlin: Springer). http://www.springer.com/gp/livingreviews/relativity.

Abbott B P *et al.* 2020 *Prospects for Observing and Localizing Gravitational-wave Transients with Advanced LIGO, Advanced Virgo and KAGRA: Living Reviews in Relativity* **23:3** (Berlin: Springer). http://www.springer.com/gp/livingreviews/relativity.

The LIGO website, https://www.ligo.org/, is very useful.

Braginsky V. B. & Khalili F. Ya. 1990 *Quantum Measurement* ed. K. S. Thorne (Cambridge: Cambridge Univ. Press). This succinct introduction to quantum measurement was written by pioneers in the understanding of quantum measurement.

Scully M O and Zubairy M S 1997 *Quantum Optics* (Cambridge: Cambridge Univ. Press) contains a useful discussion of optical squeezing.

Ferrari V July 2010 *The Quadrupole Formalism Applied to Binary Systems, VESF School, Sesto val Pusteria.* A useful account of GR application to pulsars and merging binary systems. Available online.

References

Abbott, B. P., Abbott, R., Abbott, T. D., et al. 2016, PhRvL, 116, 061102

Abbott, B. P., Abbott, R., Abbott, T. D., et al. 2017, ApJ, 848, L13

Bond, C., Brown, D., Freise, A., & Strain, K. A. 2016, LRR, 19, 3

Braginsky, V. B., Gorodetsky, M. L., Khalili, F. Y., et al. 2003, PhRvD, 67, 082001

Braginsky, V. B., Khalili, F., & Thorne, K. S. 1992, Quantum Measurement (Cambridge: Cambridge Univ. Press)

Davis, M., Ruffini, R., Press, W. H., & Price, R. H. 1971, PhRvL, 27, 1466

Hulse, R. A., & Taylor, J. H. 1975, ApJ, 195, L51
Kenyon, I. R. 2019, Quantum 20/20: Fundamentals, Entanglement, Gauge Fields, Condensates and Topology (Oxford: Oxford Univ. Press)
Kimble, H. J., Levin, Y., Matsko, A. B., Thorne, K. S., & Vyatchanin, S. P. 2001, PhRvD, 65, 022002
LIGO Scientific and VIRGO CollaborationsAbbott, B. P., Abbott, R., et al. 2016, AnP, 529, 1600209
Press, W. H., & Thorne, K. S. 1972, ARA&A, 10, 335
Tse, M., Yu, H., Kijbunchoo, N., et al. 2019, PhRvL, 123, 231107
Unruh, W. G. 1982, in Quantum Optics, Experimental Gravity, and Measurement Theory, ed. P. Meystre, & M. O. Scully (New York: Plenum)
Weisberg, J. M., & Huang, Y. 2016, ApJ, 829, 55

AAS | IOP Astronomy

Introduction to General Relativity and Cosmology (Second Edition)

Ian R Kenyon

Chapter 10

Cosmic Dynamics

10.1 Introduction

To an excellent approximation the universe is flat. Consequently, this will be the case looked at in detail; less will be said about spacetimes with curvature. On the largest scale matter in the universe is approaching a homogeneous isotropic distribution. Within the framework of GR, Friedmann, Robertson, and Walker derived the metric for such homogeneous isotropic universes. Friedmann (again) and Le Maître applied Einstein's equation to these universes. The resulting three Friedmann–Le Maître equations are the tools used here for analyzing the properties of the universe. The inputs required when applying the Friedmann–Le Maître equations to our universe are the densities of matter, radiation, and dark energy; plus their equations of state, relating pressure to energy in each case. The derivations are carried through here and the Friedmann–Le Maître equations are applied to the case of the ΛCDM model.

10.2 Flat and Spatially-curved Universes

> The simple model that describes the universe on the large scale is that of a homogeneous, isotropic universe with a density equal to the mean density of the actual universe. Instead of galaxies and stars this model has a uniform fluid filling all space. This is our working hypothesis for how the universe appears on the very large scale. Our first step is to deduce the metric of such a model universe. Of course the motion of galaxies is distorted from that of particles in such an ideal fluid by the local variation of the gravitational attraction of other matter present.

doi:10.1088/2514-3433/acc3ffch10

We can imagine placing clocks at rest with respect to the cosmic fluid, and setting them to read the same reference time when the fluid density and temperature reach agreed values. Then using these synchronized clocks the physical state of the universe will depend on time in the same way everywhere. This time is called the *cosmic time*. Suppose now that a three-dimensional slice (hypersurface) is taken through spacetime at cosmic time t. This hypersurface will also be isotropic and homogeneous. Therefore there is a single Gaussian curvature $\kappa(t)$ characterizing it, and this depends solely on the cosmic time:

$$\kappa(t) = k/R^2(t) \tag{10.1}$$

where R^2 gives the magnitude and k the sign $(+1, 0, -1)$ of this curvature. A positively curved space with $k = +1$ is the three-dimensional analog of a spherical surface, i.e., a hypersphere. It can be embedded in a four-dimensional Euclidean space, just as a two-dimensional spherical surface is embedded in Euclidean three-space. Let x, y, z, and w be the Cartesian coordinates in Euclidean four space, with x, y, and z the spatial coordinates in the usual three-space. Then the hypersurface has the equation

$$x^2 + y^2 + z^2 + w^2 = r^2 + w^2 = R^2(t).$$

By differentiating at fixed time we obtain

$$w\,\mathrm{d}w = -r\,\mathrm{d}r \qquad \text{and} \qquad w^2\mathrm{d}w^2 = r^2\mathrm{d}r^2.$$

Then

$$\mathrm{d}w^2 = \frac{r^2\mathrm{d}r^2}{R^2 - r^2}.$$

The separation of nearby points on the hypersurface is given by

$$\mathrm{d}\ell^2 = \mathrm{d}r^2 + r^2\mathrm{d}\Omega^2 + \mathrm{d}w^2,$$

where (r, θ, φ) are the polar coordinates in the usual three-space and $\mathrm{d}\Omega^2 = \mathrm{d}\theta^2 + \sin^2\theta\,\mathrm{d}\varphi^2$. Eliminating $\mathrm{d}w^2$ we obtain the equation of the hypersurface

$$\mathrm{d}\ell^2 = \frac{R^2\,\mathrm{d}r^2}{R^2 - r^2} + r^2\,\mathrm{d}\Omega^2.$$

A change in angle θ produces a displacement $r\,\mathrm{d}\theta$, while a change in r in any direction gives a displacement of $R\,\mathrm{d}r/\sqrt{R^2 - r^2}$. These features show that the three-dimensional hypersurface is curved and isotropic. The choice of positive curvature means that it is a hypersphere. Note that the equation is the same for all points on the hypersphere, and so this space is also homogeneous. Changing the notation we introduce the dimensionless $\sigma = r/R$, so that the equation becomes

$$\mathrm{d}\ell^2 = R^2(t)\left(\frac{\mathrm{d}\sigma^2}{1 - \sigma^2} + \sigma^2\,\mathrm{d}\Omega^2\right),$$

where the time dependence of $R(t)$ is made explicit again. We now complete the metric equation by incorporating the cosmic time in a way consistent with SR. The invariant metric distance squared is

$$\mathrm{d}s^2 = c^2\,\mathrm{d}t^2 - R^2(t)\left(\frac{\mathrm{d}\sigma^2}{1-\sigma^2} + \sigma^2\,\mathrm{d}\Omega^2\right).$$

10.3 The Friedmann–Robertson–Walker Metric

Thus far we have restricted the hypersurface describing the shape of the universe to have positive curvature. More generally the curvature could be negative, or zero for a flat universe. The metric equation covering all three cases is

$$\mathrm{d}s^2 = c^2\,\mathrm{d}t^2 - R^2(t)\left(\frac{\mathrm{d}\sigma^2}{1-k\sigma^2} + \sigma^2\,\mathrm{d}\Omega^2\right) \tag{10.2}$$

where k can be -1, 0, or $+1$, consistent with Equation (10.1). $R^{-2}(t)$ is the magnitude of the Gaussian curvature; when $k = +1$, $R(t)$ is the hyper-radius of positively curved space which may vary with time. Equation (10.2) is the metric equation discovered independently by Friedman (1922), Robertson (1936) and Walker (1937) for isotropic and homogeneous spacetime.[1]

Taking some point A as origin and aligning the coordinates appropriately, we can make the coordinates of any other point B equal to $(\sigma, 0\ 0)$. The distance across the hypersurface to this second point is

$$d = R(t)\int_0^{\sigma} \frac{\mathrm{d}\sigma}{(1-k\sigma^2)^{1/2}} \tag{10.3}$$

which gives contrasting results according to the sign of k. Let us first take $k = +1$, then

$$d = R(t)\,\sin^{-1}\sigma.$$

If a sphere centered on A is drawn to intersect B the area of the sphere is

$$A = 4\pi R^2\sigma^2 = 4\pi R^2\,\sin^2(d/R).$$

As d increases from zero to $\pi R/2$ the area of the sphere increases steadily. Then as d increases further to $\pi\ R$ the area of the sphere falls to zero; by analogy with the case of a two-dimensional sphere we have reached the Antipodes of A. Finally when d reaches $2\pi\ R$, σ is zero and we are back at A. A hypersurface with positive curvature is a closed but unbounded surface. On the other hand, when $k = -1$ we have

$$d = R(t)\,\sinh^{-1}\sigma.$$

[1] Both Friedmann in Russia and Einstein in Switzerland belonged to clubs of young people sharing an interest in the latest in the physical sciences.

The area of the sphere through the other galaxy is

$$A = 4\pi R^2\sigma^2 = 4\pi R^2 \sinh^2(d/R)$$

which grows indefinitely as σ and d increase together. This type of universe is unbounded and open.

The remaining case is that of flat space. In a flat universe $k = 0$, and $R(t)$ becomes a reference length which we can set arbitrarily at some cosmic time: how this $R(t)$ then varies with time tells whether the flat universe is expanding, static or contracting.

A flat universe is open and unbounded. The FRW metric equation collapses to

$$\mathrm{d}s^2 = c^2\mathrm{d}t^2 - a^2(t)[\mathrm{d}r^2 + r^2\mathrm{d}\Omega^2]. \tag{10.4}$$

Here we have introduced notation that follows standard practice in cosmology: $a(t)$ varies with time and is made dimensionless; the length dimension is transferred to the *comoving length coordinate* r. As the universe expands r remains constant. The expansion is all in $a(t)$. In comparison with the Minkowski metric of SR the only difference is the appearance of the time varying scale factor $a(t)$.[2]

A universe's expansion from a singularity is easiest to picture for a universe of positive curvature. At each moment of cosmic time such a universe would be a hypersurface of some radius in four-dimensional Euclidean space. It would expand like a balloon, with the galaxies (once created), being analogous to dots on the balloon's surface. This balloon analogy brings out an important feature of an expanding universe like ours, namely that every point in the current universe originated at the singularity (Big Bang) and all have moved equally away from it. *Looking in any direction* we look back directly toward the Big Bang. However this is shrouded by the CMB that microwave dishes detect. The earlier phase of the universe's existence, containing a plasma of photons and matter, is not directly detectable using telescopes.

10.3.1 The Comoving Frame

Any point moving so that the spatial coordinates r, θ, and φ remain constant is simply in free fall in the cosmic fluid. r, θ, and φ are called *comoving coordinates* and the frame in free fall is called the comoving frame. Any real galaxy or cluster on which the net attraction of local matter is small would be comoving. Motion relative to the comoving frame due to local fluctuations in matter is called *secular motion*.

Figure 10.1 shows the radial comoving coordinate of three bodies in free fall in a uniformly expanding universe. The only motion of A, B, and C is that due to the expansion of the universe; their comoving coordinates are constant. Past light cones are shown at location B at times t_1 (blue) and t_2 (red). For light traveling radially in flat spacetime the Friedmann–Roberson–Walker Equation (10.4) reduces to

[2] a was already used in connection with Hubble's law in Section 1.4.

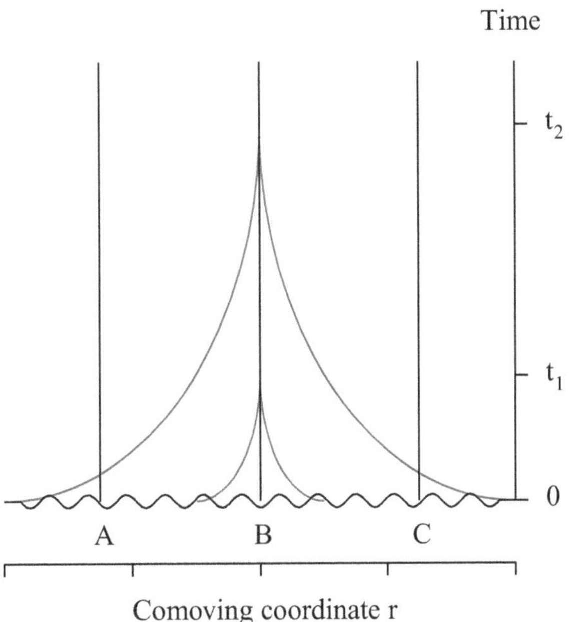

Figure 10.1. The development of the universe from a singularity. The past light cones are drawn for one location B at cosmic times t_1 and t_2.

$$c\ \mathrm{d}t = a(t)\mathrm{d}r.$$

Therefore the velocity of light in the comoving coordinate frame is $c/a(t)$. In an expanding universe $a(t)$ increases with time so that the edges of the light cones in the comoving frame are curved concave-up. Events lying within the past light cone at time t_1 can influence B at time t_1. Events early in the lives of A and C can affect B at time t_2 but not earlier at time t_1, before causal contact was made.

The wobbly line in the figure indicates an initial state. Note that light cones have been drawn for the case that $a(t)$ shrank to zero at $t = 0$. This makes the light cones become tangential at the initial point. That would allow instantaneous causal contact across the universe. Alternatively the comoving coordinates could continue for infinite time in the past. What actually happened is not known.

An observer in a flat, expanding, homogenous and isotropic universe sees all galaxies moving away radially: the observer's frame is a comoving frame and it is also a frame in free fall. For good measure this frame would coincide with the frame defined by the distant galaxies, and in it the CMB would appear isotropic. From the Earth we find that the rest frame defined by the distant galaxies and that defined by the CMB coincide, so this frame is a comoving frame: of course we move relative to the comoving frame with the vector sum of the Earth's motion around the Sun, of the Sun around the Galaxy and of the Galaxy itself. This vector sum is our secular motion.

It follows from the strong equivalence principle that because a comoving frame is in free fall, the laws of physics in such a frame should satisfy the postulates of SR. In the language of SR the comoving frame is an inertial frame. Other inertial frames can be obtained from the comoving frame by boosting to a frame with constant relative velocity. The comoving frame in free fall is unique because it is the only frame in which the CMB is seen to be isotropic. Consequently this frame is a preferred frame, something abhorrent to both classical and SR views.

As mentioned in Section 1.6, the existence of our local comoving frame is revealed by a Foucault pendulum, most vividly by one located at the North Pole. To us it appears to rotate through 360° in 24 hours. However, the plane of its swing remains in a fixed orientation relative to the most distant galaxies, and to the CMB. This makes specific the principle proposed by Mach in the 19th century: namely that the matter in the universe determines a reference (comoving) frame. It follows that any rotation with respect to that frame produces centrifugal and Coriolis forces.

10.4 The Friedmann–Le Maître equations

Einstein's equation provides the dynamical link between the metric for a homogeneous isotropic universe given in Equation (10.2) and the contents of the universe. This offers the possibility of inferring the life history and future of the universe, determining for example how Hubble's constant changes with time. The key equations were first derived by Alexander Friedmann in 1922 (Friedman, 1922) and then independently by Georges Le Maître in 1927 (Le Maître, 1927). They realized that the universe could have curvature and could expand or contract, whereas Einstein had envisaged a purely static universe.[3] For simplicity the universe is assumed to contain fluids with total pressure p and total rest density ρ. The other physical parameter is κ (in m^{-2}) the curvature of the universe. The derivations are carried through in Appendix F starting from the two independent components of Einstein's equation. Three useful equations are derived, which are, of course, not entirely independent. Despite their different appearance they are general relativistic versions of standard Newtonian equations. Some comments are made here to hopefully make the connection clearer.

The first of these three equations, Equation (F.14), often called *Friedmann's equation*, is

$$3\dot{a}^2/a^2 + 3\kappa c^2/a^2 = 8\pi G\varepsilon/c^2. \tag{10.5}$$

Here $a(t)$ is the time-dependent dimensionless scale factor that was introduced above: it expands with the universe.[4] From Equation (1.7) we recognize $\dot{a}/a$ as Hubble's constant, the expansion rate of the universe. The energy density, ε, can be expressed in terms of the equivalent mass density: $\varepsilon = \rho c^2$. It includes the contributions from matter and radiation that change with time and from the

[3] Einstein initially published a paper disproving Friedmann's solution, but then had to withdraw his disproof as incorrect. At that point he disowned the cosmological constant.

[4] For curved spacetime $\kappa = k/R^2$, where for a positive curved spacetime R is the radius of curvature in the comoving frame, and $k = \pm 1$ indicates the sign of curvature.

cosmological constant, which does not change. For the latter the energy density and pressure are given by Equations (F.12) and (F.13)

$$\varepsilon_\Lambda = \rho_\Lambda c^2 = -p_\Lambda = \frac{c^4\Lambda}{8\pi G}, \tag{10.6}$$

and, to reiterate Λ is invariant. Friedmann's equation is essentially an equation expressing the conservation of energy. The first term in $\dot{a}^2$ is the kinetic energy; that in ε is the gravitational energy; while the second term is the energy in the curvature of spacetime. If spacetime is flat the kinetic and gravitational energies sum to zero: the universe is created from nothing! The carry-home message is this: Friedmann's equation relates the expansion rate of the universe to the curvature and energy density of the universe.

A second equation, Equation (F.15), is the *acceleration equation*

$$\ddot{a}/a = -\frac{4\pi G}{3c^2}[\varepsilon + 3p]. \tag{10.7}$$

The RHS of the equation contains the gravitational force responsible for acceleration/deceleration of the universe's expansion. We see immediately that, as expected in GR, both energy ε and pressure p contribute. The gravitational attraction of matter acts to reduce the expansion rate. As we have seen earlier the cosmological constant, a.k.a. dark energy, has $p_\Lambda = -\varepsilon_\Lambda$, and it acts to accelerate the expansion.

The final equation, Equation (F.16), is the *fluid equation*

$$\dot{\varepsilon} = -(3\dot{a}/a)[p + \varepsilon]. \tag{10.8}$$

This relates the rate of change of the energy density to the expansion of the universe. Of the three equations, this is the only one for which a proof can be sketched using Newtonian mechanics. On the right-hand side of this equation the pressure and energy density of the cosmological constant mutually cancel. On the left-hand side its contribution is equally zero because the cosmological constant does not vary. In order to calculate the behavior of the universe from these equations the relationship between density and pressure must be provided, that is to say, the *equation of state* of the contents of the universe. We already know that for the cosmological constant. The other components are matter, whether baryonic or dark matter, and radiation, which can include the known neutrinos because they have negligible mass.

Notice that, in the case of a flat universe, all the terms containing a in the Friedmann–Le Maître equations are ratios making the size of a flat universe indeterminate. It is then convenient to choose, as we have done earlier, the scale factor at the current time t_0 to be equal to unity:

$$a(t_0) = 1.0.$$

This simplifies the redshift/scale factor relationship:

$$z = \frac{a(t_0) - a(t)}{a(t)} = 1/a - 1 \text{ and } a = 1/(1 + z). \tag{10.9}$$

Friedmann's equation shows that the critical energy density required to make the universe precisely flat, ε_c, is given by

$$3H^2 = 8\pi G\varepsilon_c/c^2.$$

Rearranging this in terms of the *critical density* $\rho_c = \varepsilon_c/c^2$ gives

$$\rho_c \equiv \frac{3H^2}{8\pi G}. \tag{10.10}$$

In a flat universe

$$\rho_r + \rho_m + \rho_\Lambda = \rho_c.$$

where the contributions are explicitly: ρ_m from baryonic matter plus dark matter, ρ_r from radiation, and ρ_Λ from the cosmological constant. Friedmann's equation for a flat universe can be re-expressed in terms of fractional contributions $\Omega_i = \rho_i/\rho_c$:

$$\Omega \equiv \Omega_r + \Omega_m + \Omega_\Lambda = 1,$$

When referring to the current era we add a subscript 0 to all parameters as well as to Hubble's constant (H_0). If the universe is not flat but curved then Friedmann's equation becomes

$$\Omega - 1 = \frac{\kappa c^2}{[aH]^2} = \Omega_\kappa. \tag{10.11}$$

Unless otherwise stated we analyze the case of a flat universe, which is consistent with all current measurements. Taking the current value of Hubble's constant as 70 km s^{-1} Mpc^{-1}

$$\rho_{c0} = 9.14 \times 10^{-27} \text{ kg m}^{-3},$$

or 1.35×10^{11} $M_\odot$ Mpc^{-3}. This is equivalent to 5.5 nucleon masses (mostly in non-baryonic matter) per cubic meter, which is tiny compared to the 10^{30} nucleons per cubic meter within the Earth.

10.5 Models of the Universe

The generally accepted model of the universe is an essentially flat FRW spacetime: the content is taken to be electromagnetic radiation, cold (non-relativistic) dark matter, baryonic matter and dark energy. This dark matter interacts gravitationally, but not through the other (strong, weak, electromagnetic) forces. What sort of elementary particles make up this dark matter is undetermined. The properties of dark energy are consistent with the properties of the cosmological constant proposed by Einstein. The cosmological constant Λ and the cold dark matter (CDM) are the origin of the name of this standard model of cosmology, ΛCDM.

How this model has been constructed and validated is described in the remainder of the book. In the following section we begin by considering simple models, each with only one component from those listed, and each useful in approximating the behavior of the universe on the large scale at some era.

10.6 Radiation, Matter and Λ Dominated Eras

At different epochs each of these components has, in turn, largely determined the way our universe developed. Hence we should examine how the universe would develop if each alone were present. Taking our universe to be flat, Friedmann's equation becomes

$$H^2 = \frac{\dot{a}^2}{a^2} = \frac{8\pi G\varepsilon}{3c^2}. \tag{10.12}$$

In this section, for simplicity, the subscript c is omitted from ε_c, and ε_{c0} becomes ε_0. If we are to use the fluid Equation (10.8) and the acceleration Equation (10.7) we need the *equations of state*, that is the relationships between pressure and energy for each component of the universe. To an adequate approximation all the equations of state can be presented in the form

$$p = w\varepsilon. \tag{10.13}$$

For non-relativistic dust $w = 0$ and hence this value serves for matter in its dilute overall state in the universe; $w = 1/3$ for radiation; and for the cosmological constant Equation (10.6) gives $w = -1$. This last choice assumes that the dark energy is fully equivalent to Einstein's Λ. Using the equation of state the fluid Equation (10.8) becomes

$$\frac{\dot{\varepsilon}}{\varepsilon} = -\frac{\dot{a}}{a}[3 + 3w].$$

Then

$$\varepsilon = \varepsilon_0 a^{-(3+3w)}.$$

When this relation is used in Friedmann's equation it gives

$$\dot{a}^2 = \frac{8\pi G\varepsilon_0}{3c^2}a^{-(1+3w)}. \tag{10.14}$$

We can solve this equation by substituting

$$a = (t/t_0)^u \text{ and } \dot{a} = (u/t_0)(t/t_0)^{u-1},$$

and equating the powers of t on the two sides. This gives

$$2u - 2 = -(1 + 3w)u,$$

whence $u = 2/(3 + 3w)$. Finally we have

$$a(t) = (t/t_0)^{2/(3+3w)}, \tag{10.15}$$

with

$$t_0 = \frac{1}{1+w}\sqrt{c^2/6\pi G\varepsilon_0}, \tag{10.16}$$

and Hubble's constant

$$H_0 = \frac{2}{(3+3w)t_0}. \tag{10.17}$$

We can now use the last three equations to determine how the universe would develop if radiation, matter or dark energy alone were present. The results give clues about the eras when either radiation, matter or dark energy dominated and how these fit together.

- If radiation dominates ($w = 1/3$) we get:

  $$a(t) = (t/t_0)^{1/2}, \tag{10.18}$$

  with

  $$t_0 = \frac{3}{4}\sqrt{c^2/6\pi G\varepsilon_0}, \tag{10.19}$$

  and Hubble's constant

  $$H = \frac{1}{2t}. \tag{10.20}$$

  The lifetime of the universe would be $1/2H_0$.
- With non-relativistic matter dominant ($w = 0$) we get:

  $$a(t) = (t/t_0)^{2/3}, \tag{10.21}$$

  with

  $$t_0 = \sqrt{c^2/6\pi G\varepsilon_0}, \tag{10.22}$$

  and Hubble's constant

  $$H = \frac{2}{3t}. \tag{10.23}$$

  The lifetime of the universe would be $2/[3H_0]$. Such a matter dominated universe is called an Einstein–De Sitter universe after its originators.
- Finally we consider an era in which dark energy (cosmological constant) dominates. In this case the energy is given by Equation (10.6)

  $$\varepsilon_\Lambda = \rho_\Lambda c^2 = \frac{\Lambda c^4}{8\pi G} = -p_\Lambda.$$

  The sign reversal of w shows how fundamentally dark energy differs from matter and radiation. Friedmann's equation, Equation (10.14) becomes

$$\dot{a}^2 = \frac{8\pi G}{3}\rho_\Lambda a^2.$$

Taking the square root of this equation,

$$H = \dot{a}/a = \sqrt{8\pi G\rho_\Lambda/3}.$$

ρ_Λ is constant so integrating this equation gives

$$a = \exp[H_0(t - t_0)]. \tag{10.24}$$

The eras in the life of the universe when radiation and matter dominate depend very much on how their energy densities change with time. Matter is non-relativistic for all but the earliest times, thus we can take its energy density to be inversely proportional to the volume of the universe,

$$\rho_{\rm m} \propto a^{-3}. \tag{10.25}$$

With radiation the energy density falls by an additional factor a to account for the stretching of the wavelength, and consequent fall in the individual photon frequency and energy,

$$\rho_{\rm r} \propto a^{-4}. \tag{10.26}$$

From the scale dependence of Equations (10.25) and (10.26) we see that at early enough times radiation would inevitably dominate and at some later time matter would take over. At the moment when the energy density of matter became equal to the energy density of radiation the value of a is

$$a_{\rm rm} = \rho_{\rm r0}/\rho_{\rm m0}, \tag{10.27}$$

where $\rho_{\rm r0}$ and $\rho_{\rm m0}$ are the current values of $\rho_{\rm r}$ and $\rho_{\rm m}$, respectively.

Dark energy increases as the volume of space expands; and because dark energy's gravitational force is repulsive this increases the expansion rate. The universe undergoes ever accelerating expansion. So it is guaranteed that eventually dark energy dominates over matter. The takeover from matter dominance occurs when a has the value

$$a_{\rm m\Lambda} = [\rho_{\rm m0}/\rho_\Lambda]^{1/3}. \tag{10.28}$$

Figure 10.2 compares the expansion of the universe predicted for radiation, matter and dark energy dominance; arbitrarily making them to match today's observed expansion rate. Putting $x = H_0(t - t_0)$, we have

- For a radiation dominated universe: $a = [1 + 2x]^{1/2}$,
- For a matter dominated universe: $a = [1 + 3x/2]^{2/3}$, and
- For dark energy dominant (Λ): $a = \exp x$.

These dependences are drawn in Figure 10.2. We shall see that the lifetime of the universe is approximately H_0^{-1} making the intercept on the time axis equal to -1.

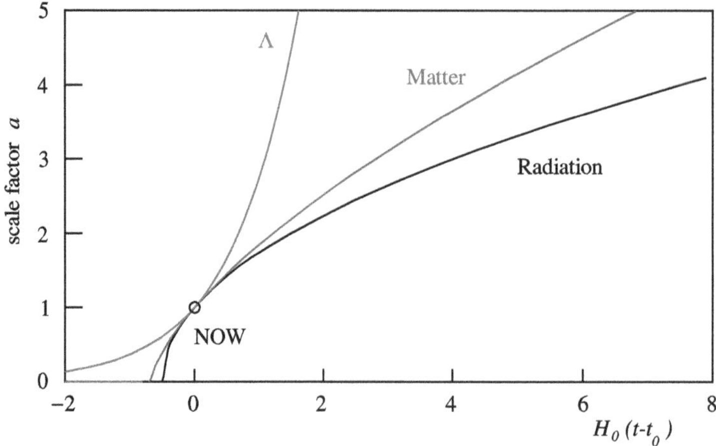

Figure 10.2. Predicted expansion of the universe for the cases of radiation or matter or dark energy dominant.

None of the single component universes matches ours particularly well, but a mix looks promising.

10.7 The ΛCDM Model

The ΛCDM model of the universe (Planck Collaboration et al., 2020) models the evolution according to general relativity of a universe containing radiation, baryonic matter, dark matter, and dark energy. Here we take the current value of the Hubble constant[5] H_0 to be 70 km s^{-1} Mpc^{-1}; and the current fractional contributions to the energy of the universe to be: dark energy $\Omega_{\Lambda 0} = 0.70$; matter (baryonic+dark) $\Omega_{m0} = 0.30$; baryonic matter $\Omega_{b0} = 0.045$; and radiation $\Omega_{r0} = 0.000\,09$. This makes the lifetime of the universe around 13.8 Gyr, very close to the crude estimate $1/H_0$ mentioned in Chapter 1. The lifetime of the oldest stars defines a lower limit to the lifetime of the universe: they give a limit of 8–9 Gyr, comfortably less than 13.8 Gyr. At various places in the following chapters fits to data are shown that have been made with the ΛCDM model. These are fits using essentially the same Ω values and the same other parameters in each case.

After an inflationary phase the universe is dominated first by radiation, then by matter and finally by dark energy, with the scale factor at the changeovers given by Equations (10.27) and (10.28) respectively. Table 10.1 gives the details of the changeovers for the ΛCDM model of the universe. The variation of the content of the universe in the ΛCDM model is shown in Figure 10.3 against a the scale factor. For reference later the matter contribution is separated into dark matter and baryonic matter, while the radiation is separated into the electromagnetic and neutrino contributions. In order to apply Friedmann's equation to the ΛCDM model universe we need to write the energy of all three components taking account

[5] There is some tension within the data, not currently resolved. Measurements of H_0 from the CMB data extrapolating forward to the present give a 2% lower value, while measurements from low redshift using standard candles, including supernovae SNe Ia, give a 2% higher value.

Table 10.1. Time, Redshift, Scale Parameter, and Temperature at Radiation/Matter Energy Equality and Matter/Dark Energy Equality

	Time (t) in Myrs	Redshift	$a(t)$	Radiation T (K)
Rad/matter	0.050	3443	2.90×10^{-4}	9384
Matter/dark energy	9900	0.320	0.758	3.60

Exercise 4 is worth doing to get $t_{\rm rm}$.

of their different dependence on a: $\varepsilon_0\Omega_{\rm r0}/a^4$, $\varepsilon_0\Omega_{\rm m0}/a^3$ and $\varepsilon_0\Omega_{\Lambda0}$. Then Equation (10.12) becomes

$$\frac{\dot{a}^2}{a^2} = \frac{8\pi G\varepsilon_0}{3c^2}(\Omega_{\rm r0}/a^4 + \Omega_{\rm m0}/a^3 + \Omega_{\Lambda0}). \tag{10.29}$$

We can make this more compact, multiplying through by a^2/H_0^2:

$$[\dot{a}/H_0]^2 = \Omega_{\rm r0}/a^2 + \Omega_{\rm m0}/a + \Omega_{\Lambda0}a^2.$$

Then taking the square root and rearranging gives

$$H_0\,{\rm d}t = [\Omega_{\rm r0}/a^2 + \Omega_{\rm m0}/a + \Omega_{\Lambda0}a^2]^{-1/2}\,{\rm d}a. \tag{10.30}$$

Finally this can be integrated to give

$$H_0 t = \int_0^a [\Omega_{\rm r0}/a^2 + \Omega_{\rm m0}/a + \Omega_{\Lambda0}a^2]^{-1/2}\,{\rm d}a. \tag{10.31}$$

This integration is carried through with the Ω values of the ΛCDM model to obtain the expansion of the universe with time. It will also be useful to express the Hubble constant at an earlier period in terms of H_0 its current value:

$$H = \dot{a}/a = H_0[\Omega_{\rm r0}/a^4 + \Omega_{\rm m0}/a^3 + \Omega_{\Lambda0}]^{1/2} \tag{10.32}$$

We now introduce the conventional *deceleration parameter q* using Equation (10.7)

$$q = -\frac{\ddot{a}a}{\dot{a}^2} = \frac{1}{H^2}\frac{4\pi G}{3c^2}[\varepsilon + 3p] = \frac{1}{2\rho_{\rm c}c^2}[\varepsilon + 3p], \tag{10.33}$$

where the last equality has used Equation (10.10). Next, writing the contributions of the components of the universe and then applying the equations of state $p_i = w_i\rho_i$ this becomes

$$q = \frac{1}{2\rho_{\rm c}c^2}\sum_i[\varepsilon_i + 3p_i)] = \frac{1}{2}\sum_i\Omega_i(1 + 3w_i). \tag{10.34}$$

At the current time the deceleration parameter

$$q_0 = \Omega_{\rm r0} + \Omega_{\rm m0}/2 - \Omega_{\Lambda0} \approx -0.55, \tag{10.35}$$

where the values $\Omega_{\rm m0} = 0.30$ and $\Omega_\Lambda = 0.70$ have been inserted for the approximation. This brings out the fact that already the effect of dark energy has brought about

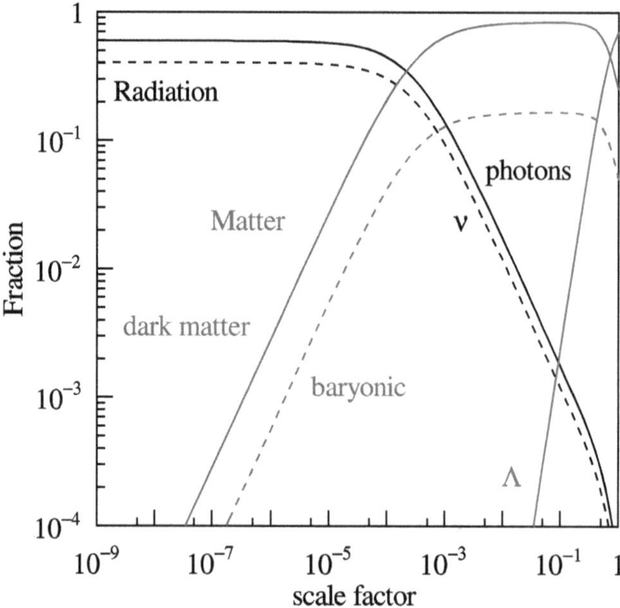

Figure 10.3. Fractional energy content of the universe in the ΛCDM model as a function of the scale factor. The redshift is $1/a - 1$, which is close to $1/a$ for most of the lifetime of the universe.

an acceleration of the expansion, and this will only increase as Ω_Λ increases toward unity and Ω_m declines correspondingly into the future.

10.8 Exercises

1. Show that for small lookback times Δt ($t = t_0 - \Delta t$) the scale parameter can be approximated by

$$a(t) = 1 - \Delta t H_0 - q_0 H_0^2 \Delta t^2/2.$$

Hence show that to the same approximation

$$z = H_0 \Delta t + (1 + q_0/2) H_0^2 \Delta t^2,$$

and that

$$H_0 \Delta t = z - (1 + q_0/2) z^2.$$

2. Show that in a flat universe in which the equation of state is $p = w\varepsilon$, then $q = (1 + 3w)/2$.
3. Estimate the baryon mass density of the Galaxy and its total mass density. How does this compare to the mean density of the universe?
4. In the early universe the Friedmann Equation (10.30) can be simplified, without much error, by ignoring Ω_Λ, which gives

$$H_0 \, \mathrm{d}t = \mathrm{d}a[\Omega_r 0/a^2 + \Omega_{m0}/a]^{-1/2}.$$

Now write $a_{\rm rm} = \Omega_{\rm r}0/\Omega_{\rm m0}$, which is the scale factor at which the energy density of matter overtakes the energy density of radiation. Then the Friedmann equation reduces to

$$\mathrm{d}t = \frac{a\,\mathrm{d}a}{\sqrt{\Omega_{\rm r}0}\;H_0}\left[1 + \frac{a}{a_{\rm rm}}\right]^{-1/2}.$$

Hence show that the time of matter–radiation equality was approximately

$$t_{\rm rm} = 0.391\,\frac{a_{\rm rm}^2}{\sqrt{\Omega_{\rm r0}}\;H_0} \approx 50,\,000\ \text{years}. \tag{10.36}$$

When integrating, try the substitution $x^2 = 1 + a/a_{\rm rm}$.

Further Reading

Liddle A 2015 *An Introduction to Modern Cosmology* (3rd ed.; New York: Wiley). This text provides a compact introduction.

Peebles P J E 2020 *Cosmology's Century: an inside history of our modern understanding of the universe* Princeton University Press. This book gives insights into the development of the subject by a Nobel laureate.

References

Friedman, A. 1922, ZPhy, 10, 377
Le Maître, G. 1927, ASSB, A47, 49
Planck CollaborationAghanim, N., Akrami, Y., et al. 2020, A&A, 641, A1
Robertson, H. P. 1936, ApJ, 83, 257
Walker, A. G. 1937, PLMS, s2-42, 90

AAS | IOP Astronomy

Ian R Kenyon

Chapter 11

Distances, Horizons and Measurements

11.1 Introduction

The acceptance of an expanding universe raises the question of how to measure distances when these are changing continually. If we could freeze the universe at a given moment of cosmic time, then distances between points would be easy to define and measure. That is not what we have, so how should we proceed? A connected question is whether light or gravitational waves could have traveled from one spacetime point to another. This range is crucial since it is the range of any causal influence. The limits of causal influence are generically called *horizons*.

11.2 Proper Distance and Horizons

When we observe a quasar at a redshift of 6 the light has been traveling for about ten billion years. During all that time the universe has been continuously expanding, and not at a constant rate. How should we define the distance light travels from the quasar to reach us? Each segment of the path will have expanded since the light passed over it, a lot if the segment is near the quasar source and less if near the Earth. The prescription used is to sum up the lengths of the segments scaled to the same cosmic time, for simplicity choosing the time at which we receive the light. During any short time interval $\mathrm{d}t$ at time t the light would travel a distance $c\,\mathrm{d}t$. The scale factor of the universe was then $a(t)$ so if we wish to know that segment's current length, at time the current time t_0, we need to multiply by the expansion since the light traversed it, that is by $1/a(t)$; taking as usual $a(t_0) = 1$. In order to get the total path length at the current cosmic time we integrate $c\,\mathrm{d}t/a(t)$ with respect to time from emission at the quasar, t_q until today t_0

$$d_\mathrm{P} = c\int_{t_\mathrm{q}}^{t_0} \frac{\mathrm{d}t}{a(t)}, \tag{11.1}$$

doi:10.1088/2514-3433/acc3ffch11

where d_P is known as the *proper distance* to the quasar at the current time. To get the proper distance when the light was emitted, simply multiply the result above by $a(t_q)$. The other distance frequently used is the *comoving distance* of Equation (10.4). For the interval between cosmic times t_1 and t_2, this is

$$d_C = c\int_{t_1}^{t_2} \frac{dt}{a(t)}, \tag{11.2}$$

which overlaps the definition of the current proper time. Another closely related quantity is the *conformal time*. The conformal time from the Big Bang up to the cosmic time t is

$$\eta = \int_0^t \frac{dt}{a(t)}. \tag{11.3}$$

Now turning to the horizons: the *particle horizon* is defined to be the boundary of the region containing all spacetime events that could have influenced the reference spacetime point. Thus the distance to the particle horizon is simply the distance that light, or equally gravitational effects, had traveled since the universe began until the reference time t. The proper distance to the particle horizon is therefore

$$d_P = a(t)c\int_0^t \frac{dt'}{a(t')}, \tag{11.4}$$

and the equivalent comoving distance of the particle horizon is

$$d_C = c\int_0^t \frac{dt'}{a(t')}. \tag{11.5}$$

Note that d_C can be re-expressed (dropping the primes on the integrated quantities) as

$$d_C = c\int_0^a [d(\ln a)/\dot{a}] = \int_0^a \left[\frac{c}{aH}\right] d(\ln a), \tag{11.6}$$

where $c/[aH]$ is called the *comoving Hubble radius* spanning the distance light travels in one *expansion time* of the universe, that is when the universe expands by a factor e. Another important boundary is the *event horizon*. For instance, the furthest events now occurring that will ever be visible from Earth constitute our event horizon. To reiterate, such horizons apply equally to gravitational and electromagnetic effects.

In many contexts reference is made to some structure *entering the horizon* or *leaving the horizon*. This is best explained by saying first what being within the horizon means. For a structure to be within the horizon requires that light from any part of the structure has had sufficient time since some reference time to reach all other parts of the structure. The reference time is usually the Big Bang. Within that sufficient time the whole structure could come in causal contact. When a structure enters the horizon all its parts come into causal contact. When a structure leaves the horizon all its parts are no longer in causal contact.

11.3 Measuring Distance

Having defined distance in a consistent manner, the practical question remains: how does one measure the distance to the stars and galaxies? The fundamental measurements explained here require that either the absolute luminosity of a source or its physical dimensions can be determined independently. We examine these two methods in turn. In the first case L, the absolute luminosity of the source, can be inferred independently. The energy S, arriving at Earth per unit area per unit time in a *static* universe,

$$S = L/[4\pi d_{\mathrm{P}}^2], \tag{11.7}$$

where d_{P} is the proper distance of the source. In an expanding universe the photon frequency (and energy) falls by a factor $(1 + z)$ and the frequency of their arrival falls by the same factor. Thus

$$S = L/[4\pi d_{\mathrm{P}}^2(1 + z)^2]. \tag{11.8}$$

The observer only has S and L available and so a *luminosity distance* d_{L} is defined by

$$S = L/[4\pi d_{\mathrm{L}}^2]. \tag{11.9}$$

Thus

$$d_{\mathrm{L}} = d_{\mathrm{P}}(1 + z). \tag{11.10}$$

Once the source's redshift is measured the proper distance d_{P} can be extracted.

In the second case, the diameter of the object viewed can be inferred independently. In an astronomical context the apparent angular diameter of any galaxy or star, $\mathrm{d}\theta$, is small: if its physical diameter is ρ then by definition the *angular diameter distance* is

$$d_{\mathrm{A}} = \rho/\mathrm{d}\theta. \tag{11.11}$$

The coordinate system has the observer at the origin with the light rays from the edges of the source traveling radially. Now

$$\rho = a(t)\, d_{\mathrm{P}}\, \mathrm{d}\theta \tag{11.12}$$

where t is the time at which light left the source. Using Equation (10.9) this gives

$$\rho = d_{\mathrm{P}}\, \mathrm{d}\theta/(1 + z). \tag{11.13}$$

Hence the angular diameter distance

$$d_{\mathrm{A}} = \rho/\mathrm{d}\theta = d_{\mathrm{P}}/(1 + z). \tag{11.14}$$

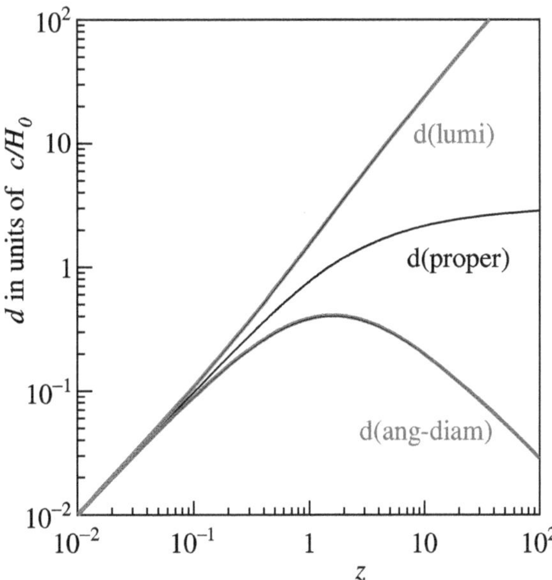

Figure 11.1. Proper distance, luminosity distance and angular diameter distance as a function of the redshift z of the source. The ΛCDM model of the universe was used taking $\Omega_{r0} = 0.000\,09$, $\Omega_{m0} = 0.30$ and $\Omega_{\Lambda 0} = 0.70$.

Combining Equations (11.10) and (11.14)

$$d_P = d_L/(1 + z) = (1 + z)d_A. \tag{11.15}$$

Figure 11.1 shows the proper, luminosity and angular diameter distances for the ΛCDM model of the universe with current energy fractions 0.00009 radiation, 0.30 matter, and 0.70 dark energy. The redshift dependence of d_A makes it turn over around $z = 1.0$. As a result the angular diameter distance eventually shrinks as the source recedes! Usually z is known well enough to remove the ambiguity.

The distances between ourselves and several comoving bodies, looking along one line of sight, is plotted against the cosmic time in Figure 11.2. These distances were obtained by integrating Equation (10.31) with the ΛCDM model parameters. A line parallel to the distance axis marks a constant cosmic time. Measuring from the time axis along such a line to the trajectory of a body gives the proper distance of that body from the Earth at that time. Note that the paths of these bodies have a point of inflection where the transition from matter dominance to Λ dominance occurs. The blue line is the path of light now reaching us. Light emitted from each of the bodies at the red dots would all reach the Earth simultaneously, and have the indicated redshifts. The proper distance at the current time from such a source at time τ is

$$d_P = \int_{a(\tau)}^{1} \left[\frac{c\,\mathrm{d}t}{a(t)}\right]. \tag{11.16}$$

This is the distance along the light path, compensating for the intervening expansion of the universe through the factor $1/a(t)$ at each step. The parameter space between

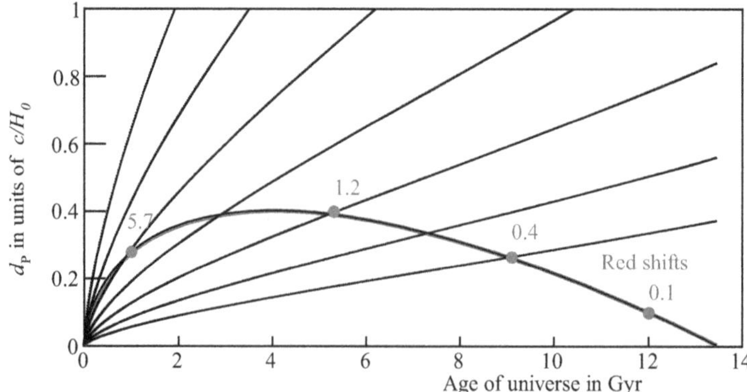

Figure 11.2. Several paths for matter at different comoving distances from the Earth. Their distances are plotted against cosmic time. The blue curve shows the path of light reaching the Earth at the present moment. Light emitted from the locations of the red dots arrives on Earth with the indicated redshifts. Adapted with permission from the ICHEP2020 talk by David Kirkby, University of California, Irvine. https://faculty.sites.uci.edu/dkirkby/.

the indicated path of light and the time axis encloses all the spacetime locations that were at some time in causal contact with us. Now because the universe is expanding isotropically, the figure will work equally well for any line of sight over the 4π solid angle. Thus the blue line encloses our past lightcone. Any radiation emitted from sources within the lightcone (closer to the time axis) would have reached Earth earlier: radiation emitted outside would either reach us later or not at all. Put another way, the blue line marks our particle horizon.

11.3.1 Cosmology Calculators

There are several useful websites that calculate cosmological parameters. Two of these are http://www.kempner.net and cosmocalc.icrar.org. First enter H_0, Ω_{m0} and $\Omega_{\Lambda 0}$ to define the universe. Then enter the redshift of interest: the corresponding age, lookback time, d_A, d_L, etc. are returned. A more professional package can be found at astropy.cosmology.

11.4 Exercises

1. Calculate the distance to sources at redshifts 0.1 and 1.0. Calculate also the temperature of the CMB, its peak wavelength and the age of the universe in both cases when the radiation was emitted from the source. By what factor would the luminosity and angular diameter estimates for the distance to the source have differed in each case? You can assume that H has its current value throughout the expansion.
2. Find $d_P(t_0)$ for the second example in the previous chapter in the same approximation as used there.
3. A source at redshift $z = 4$ is observed to change in intensity over a period of 5 years. What time interval does this correspond to at the source itself? Why

does this interval set an upper limit to the size of the source? Estimate this upper limit and its angular size viewed from the Earth.

4. Show that in a non-expanding universe the observed surface brightness of a source is independent of distance. Also show that in an expanding universe the brightness falls of like $(1 + z)^{-4}$.

Further Reading

Liddle A 2015 *An Introduction to Modern Cosmology* (3rd ed.; New York: Wiley). This text provides a compact introduction.

Peebles P J E 2020 *Cosmology's Century: An Inside History of Our Modern Understanding of the Universe* (Princeton, NJ: Princeton Univ. Press). This book gives insights into the development of the subject by a Nobel laureate.

Ian R Kenyon

Chapter 12

Cosmic Microwave Background

12.1 Introduction

This radiation was introduced in Section 1.6. It was first observed at Bell Labs by Penzias and Wilson with a microwave receiver previously used in satellite communication. They found that the instrument suffered from constant, unexpected noise with a characteristic equivalent temperature of around 3 K, and this appeared to be omnidirectional. They had no luck in eliminating it. The interpretation emerged that this *noise* is the radiation expected from the initial compact, high temperature, Big Bang with which the universe began. Penzias & Wilson (1965) earned a Nobel Prize for their discovery. The relic radiation from the Big Bang reaches us as nearly isotropic black body radiation, shown in Figure 1.5. The temperature is 2.7255 K, with the peak wavelength at 1.06 mm: earlier it was hotter by a factor $1 + z$ at redshift z.

The cosmic microwave background (CMB) originates from an initial compact source; yet, paradoxically, we detect it in every direction. Analogously, a two dimensional person living on the surface of a balloon expanding from a point would see radiation from its point origin in any direction in his/her two dimensions. We can visualize that situation easily, and can argue that viewed from a four dimensional space, three of which form our three dimensions, the paradox would be resolved. Figure 12.1 shows a section through our space with us at the center and the source of radiation apparently forming a sphere around us. Its radius is the distance that the CMB has traveled since decoupling from matter at about 365,000 years after the Big Bang. Before that there existed a plasma of charged baryonic matter in which radiation was continually scattered, so that direct observation of earlier eras is impossible: we detect the photons from the last such scattering. The CMB temperature only deviates from isotropy by less than parts in 10^4 ($\sim 10\,\mu$K): however a great deal has been learnt about the universe from studying how these deviations are

doi:10.1088/2514-3433/acc3ffch12

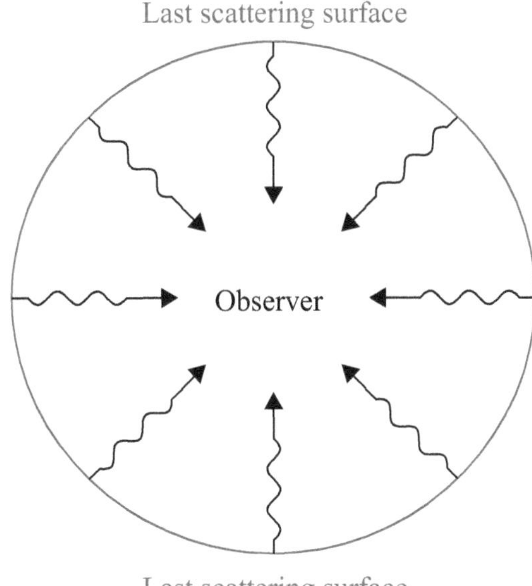

Figure 12.1. The observer receives the CMB streaming freely from the last scattering surface. According to the cosmological principle an observer anywhere in the universe would have a similar view of the CMB.

distributed across the sky. This success is built on the observations made with a sequence of impressively instrumented satellites, thus eliminating the substantial microwave absorption by the Earth's atmosphere at wavelengths shorter than 6 cm.

The latest satellite, Planck, was active from 2009 to 2013, having nine detectors across the frequency range from 30 to 900 GHz (10 cm–0.3 mm). Planck was positioned at the L2 point located along the Sun–Earth line and 1.5 Mkm beyond the Earth. Lying at L2, shielded from the Sun by the Earth, the cooling required to operate detectors sensitive to the CMB radiation at 2.7255 K is long-term manageable. This orbit is stable radially along the Sun–Earth direction. In the perpendicular direction (lying in the plane of the Earth's orbit) the orbit is weakly unstable. The time taken for any offset in this direction from L2 to double is one month, so that only small corrective rocket impulses are needed to hold station on L2.

The frequency distribution of photons in black body radiation at temperature T K was given by Planck (the physicist!):

$$n(f)\mathrm{d}f = \frac{8\pi f^2}{c^3[\exp(hf/k_\mathrm{B}T) - 1]}\mathrm{d}f, \tag{12.1}$$

where k_B is the Boltzmann constant, and h is Planck's constant. The energy distribution is $n(f)hf\,\mathrm{d}f$. These expressions were integrated over all f in Section 1.6 to give the total number density and total energy density in the black body radiation:

$$n = 2.03 \times 10^7\, T^3 \ \mathrm{m}^{-3}, \quad E = 7.56 \times 10^{-16}\, T^4 \ \mathrm{J\,m}^{-3}. \tag{12.2}$$

This translates today, when $T = 2.7255$ K, to 411 cm^{-3} (200,000 inside a cup), and an energy density of only 0.261 eV cm^{-3}. The mean photon energy is

$$E_{\text{mean}} = 2.70\, k_{\text{B}}T = 3.72 \times 10^{-23}\, T \text{ J}. \tag{12.3}$$

The angular correlations observed between the temperature fluctuations in the CMB across the sky are fundamental to understanding cosmology. Both the intensity and polarization correlations carry useful information. The distribution of perturbations of matter in the early universe, as preserved by the CMB, is complemented by measurements of the distribution of matter at low redshift. Knowing both has made it possible to ask whether the universe developed from those early perturbations in accordance with the known physical laws to give the present pattern of galaxies, galaxy clusters, and voids. We shall learn that it has. A consistent history connecting the era of the CMB, 365,000 years after the Big Bang, and now, 13,800,000,000 years later, has been constructed based on the ΛCDM model.

12.2 The Origin of the Cosmic Microwave Background

The early universe contained a hot plasma of baryons and photons in thermal equilibrium: the principal interactions maintaining the equilibrium were the scattering of photons by electrons (Thomson scattering) and Coulomb scattering of electrons off protons. As the universe expanded and cooled collisions in the plasma became less frequent, electrons could be captured by protons to form neutral hydrogen atoms, known as *recombination.* Eventually photons with enough energy to ionize atoms became rare, and the baryons in neutral atoms became invisible to the now lower energy photons. The content of the universe changed from an opaque plasma to a transparent gas. From that time onward the photons were *decoupled* from the baryons and traveled freely from the *last scattering surface.* These are the photons we now observe as the CMB, their wavelengths expanded with the universe's expansion, but preserving a picture of the universe at the *recombination era.*

A useful marker in this early life of the universe is the period when matter (baryonic plus dark) and radiation had equal energy density. Using Equation (10.29), it would have been when

$$\Omega_{\text{m0}}/a^3 = \Omega_{\text{r0}}/a^4, \quad \text{and} \quad a = \Omega_{\text{r0}}/\Omega_{\text{m0}}.$$

That is when

$$z = 1/a - 1 = \Omega_{\text{m0}}/\Omega_{\text{r0}} - 1 = 3443. \tag{12.4}$$

At that time, about 50,000 years after the Big Bang,[1] the temperature was 2.7255×3443 = 9384 K. The electron density, n_e, and proton density were $(1 + z)^3$ times the current baryon density of 0.25 m^{-3}, that is 1.02×10^{10} m^{-3}. Knowing the

[1] See Table 10.1 and Exercise 9.8.

cross-section for Thomson scattering, σ (6.65×10^{-29} m^2) we can calculate the mean free path between scatters:

$$\lambda = 1/(\sigma n_e) = 1.47 \times 10^{18}\ \mathrm{m} = 47.6\ \mathrm{pc},$$

and the time interval between scatters

$$\tau \approx \lambda/c \approx 5 \times 10^9\ \mathrm{s},$$

or 150 years. This illustrates how dilute the plasma already was at matter/radiation equality.

The mean photon energy $2.7 k_B T = 2.2$ eV - was well below $Q = 13.6$ eV the energy required to ionize hydrogen. However, because the photons outnumbered the baryons (explicitly nucleons) by a factor (the same as now) of 1.6×10^9 the photons in the high energy tail of the black body spectrum were numerous enough maintain the ionization. The buildup of hydrogen atoms, through the capture of the free electrons by the protons, took place between around 70,000 and 365,000 years after the Big Bang. The progress of recombination can be followed by applying the Saha ionization equation to the reactions in equilibrium:

$$\gamma + {}^1\mathrm{H} \rightleftharpoons \mathrm{e}^- + \mathrm{p}. \tag{12.5}$$

The Saha equation relates the numbers of the reacting particles in thermal equilibrium to the temperature T. Given that in equilibrium the number of hydrogen atoms per unit volume is n_H, of free protons is n_p, and of free electrons is n_e, then the Saha equation gives

$$\frac{n_H}{n_p n_e} = \left[\frac{m_e m_p}{m_{^1H}}\right]^{-3/2} \left[\frac{2\pi k_B T}{h^2}\right]^{-3/2} \exp\left[\frac{Q}{k_B T}\right]. \tag{12.6}$$

Using this equation we can find what we require, which is the fraction X of the atoms that remain ionized at temperature T. We have first

$$n_e = n_p = X(n_p + n_H) = X n_b,$$

where n_b is the number of baryons per unit volume. Inserting these values in Saha's equation, and rearranging, gives

$$n_p = \frac{1 - X}{X}\left[\frac{2\pi m_e k_B T}{h^2}\right]^{3/2} \exp\left[\frac{-Q}{k_B T}\right]. \tag{12.7}$$

The number of protons per unit volume n_p can be obtained from the number of photons per unit volume because the baryon/photon ratio is the same as the current value given in Equation (1.13) $\eta = 6 \times 10^{-10}$; and the number of photons is given by integrating[2] the black body spectrum in Equation (1.9) at temperature T:

[2] $\int_0^\infty x^2\, dx/[\exp x - 1] = 2.404$.

$$n_\gamma = 60.42\left[\frac{k_\mathrm{B}T}{hc}\right]^3. \tag{12.8}$$

Thus we have

$$n_\mathrm{p} = Xn_\mathrm{b} = X\eta n_\gamma = 60.42\eta X\left[\frac{k_\mathrm{B}T}{hc}\right]^3. \tag{12.9}$$

The right-hand sides of Equations (12.7) and (12.9) will be equal in thermal equilibrium. Applying this equality to the mid-point of recombination when $X = 0.5$ we get[3] an equilibrium temperature of 3755 K.

When a calculation is made taking account of the details of ionization and recombination the temperature at which the photons decouple from matter is found to be $T = 2974$ K. Correspondingly in the ΛCDM model the redshift is 1090 and the age of the universe is 365,000 years. Figure 12.2 shows the energy distribution in all the detectable background radiation arriving now at the Earth's upper atmosphere:

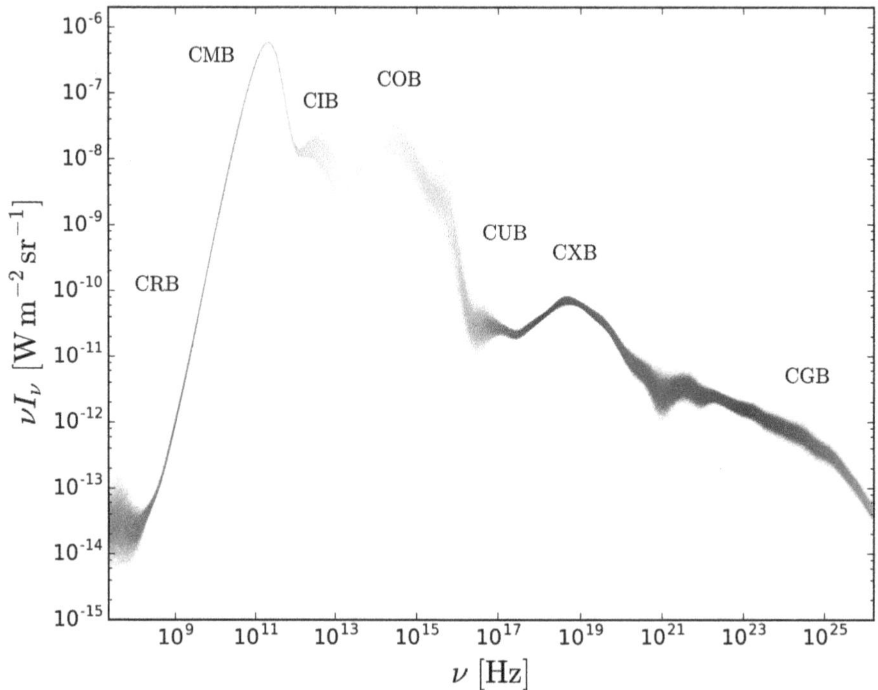

Figure 12.2. The background radiation of the universe. For example, CXB indicates the x-ray background, CUB the ultraviolet background, and so on. The line width indicates the experimental uncertainty. Compilation in Figure 9 of Hillet al. (2018).

[3] Taking a target value of 0.5 for X, the right-hand sides can be computed with trial values of T until they become equal.

for instance CXB indicates the x-ray background. Despite the tiny energy per photon the CMB dominates the energy flow. At recombination the photon flux in the CMB would have been $(1090)^2$ times greater and the energy flow would have been $(1090)^3$ times greater. The competing sources of radiation, stars and galaxies, would only form much later. This all fits in with the CMB being the radiation from the initial fireball, the Big Bang.

12.3 Thermal Fluctuations

The temperature distribution over the whole sky measured by the Planck mission is shown in Figure 1.6. This incorporates five scans of the whole sky, with typically an angular resolution of $10'$ and a temperature sensitivity[4] of 5 μK.

Studies of the CMB thermal fluctuations have been aimed at determining and analyzing the correlations between the temperature deviations across the whole celestial sphere, at all angular scales. Fortunately it turns out that two-point correlations alone provide a full statistical description of these fluctuations. The appropriate mathematical tool used by the collaborations is partial wave analysis, which captures features of a given angular size irrespective of where they are over the 4π solid angle of the heavens.

The temperature fluctuations shown in Figure 1.6 are presented using the polar coordinates (θ, ϕ) of the direction in the sky:

$$\frac{\delta T}{T}(\theta, \phi) = [T(\theta, \phi) - \langle T \rangle]/\langle T \rangle, \tag{12.10}$$

where $\langle T \rangle$ is the mean temperature

$$\langle T \rangle = (1/4\pi) \int T(\theta, \phi) \sin\theta \, \mathrm{d}\theta \, \mathrm{d}\phi = 2.7255 \text{ K}.$$

The temperature fluctuations across the sky can be projected onto spherical harmonic functions (met elsewhere in atomic wavefunctions)

$$\frac{\delta T}{T}(\theta, \phi) = \sum_{\ell=0}^{\infty} \sum_{m=-\ell}^{\ell} a_{\ell m} Y_{\ell m}(\theta, \phi). \tag{12.11}$$

The correlation between the temperature fluctuations at pairs of points separated by an angle θ is defined as the mean value of the product of these fluctuations

$$C(\theta) = \left\langle \frac{\delta T}{T}(\hat{a}) \frac{\delta T}{T}(\hat{b}) \right\rangle, \tag{12.12}$$

where θ is the angle between the directions defined by the unit vectors $\hat{a}$ and $\hat{b}$ pointing at the two points in the sky, that is to say $\hat{a} \cdot \hat{b} = \cos\theta$. The mean is taken moving $\hat{a}$ over the whole sky, and for all orientations ϕ of $\hat{b}$ around $\hat{a}$. Averaging

[4] The three low-frequency detectors at 30–70 GHz were high electron mobility transistors cooled to 20 K and the six high-frequency detectors at 100–857 GHz were bolometers cooled to 0.1 K.

over the azimuthal angle is reasonable because there is no preferred direction across the sky. Now using the expansion in spherical harmonics this reduces to

$$C(\theta) = (1/4\pi)\sum_{\ell=0}^{\infty}(2\ell + 1)C_\ell P_\ell(\cos\theta) \tag{12.13}$$

where P_ℓ are Legendre polynomials: $P_0(x) = 1$, $P_1(x) = x$ and $P_2(x) = (3x^2 - 1)/2$, etc. The coefficient

$$C_l = \frac{1}{2\ell + 1}\sum_{m=-\ell}^{+\ell} |a_{\ell m}|^2. \tag{12.14}$$

This is the average of the $2\ell + 1$ samplings, $|a_{\ell m}|^2$, for a given value of ℓ. The rms deviation of C_ℓ, ΔC_ℓ, called the *cosmic variance*, is given by

$$\frac{\Delta C_\ell}{C_\ell} = \sqrt{\left[\frac{2}{2\ell + 1}\right]},$$

and is biggest for the low ℓ multipoles with their fewer samples. The $\ell = 0$ multipole vanishes because the mean deviation is zero.

The secular motion of the Earth relative to the CMB Doppler shifts the CMB so that its apparent temperature becomes (see Equation (1.15) and surrounding text in Section 1.6)

$$T(\theta) = \langle T\rangle(1 + \beta\cos\theta),$$

where θ is the angle between the Earth's instantaneous velocity, βc, and whatever reference axis is chosen. Thus the Doppler shift induces an artificial dipole moment ($\ell = 1$) in the CMB temperature distribution. In all, the Earth's motion has components from the motion of the Galaxy group, of the Galaxy in the group, and of the Sun round the Galaxy, as well as its motion round the Sun. Overall $\beta \sim 3 \times 10^{-3}$, and this produces a few milli-Kelvin effect, which is ~100 times the intrinsic temperature perturbations in the CMB. Thus a first essential step in analyzing the CMB is to compensate for this Doppler shift before carrying out the analysis described in this section.

A first quantity of cosmological interest is the angular power spectrum of the temperature fluctuations

$$\mathcal{D}_\ell^{\mathrm{TT}} = \frac{\ell(\ell + 1)}{2\pi}C_\ell\langle T\rangle^2. \tag{12.15}$$

In Figure 12.3 the measurements of the temperature correlations made by the Planck Collaboration are plotted against ℓ. In Section 12.8 we meet a complementary measurement, that of the polarization of the CMB: the corresponding correlations in polarization are shown in Figure 12.6. The superposed curves in both figures are the outcome of fits to the data calculated using the ΛCDM model. These detailed calculations, which help determine the values of cosmological parameters, are outside our scope here. Instead the main features of the temperature and

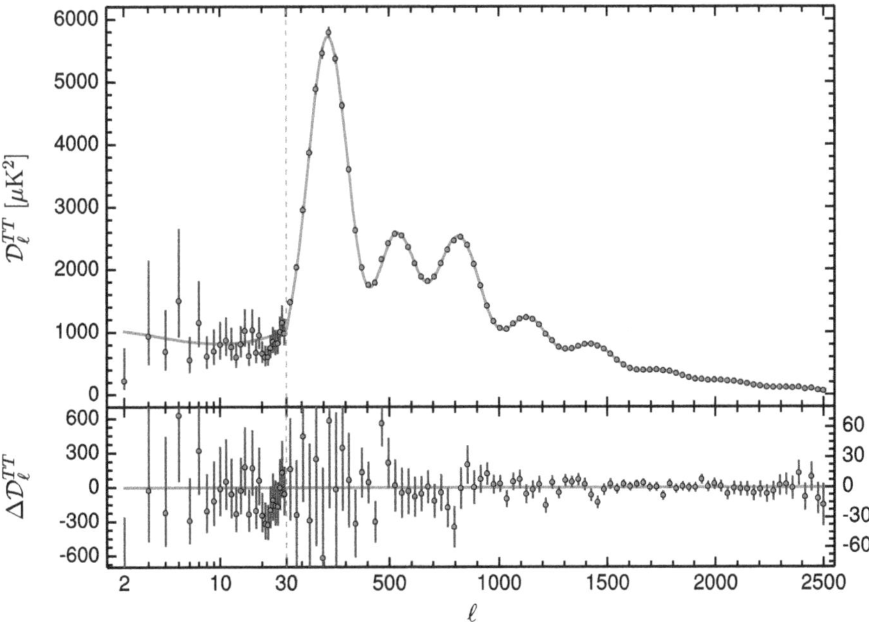

Figure 12.3. Temperature–temperature correlations of the CMB measured by the Planck Collaboration. The experimental errors are displayed in the lower panel. Adapted from Figure 1 in Planck Collaboration et al. (2020). Reproduced with permission © ESO.

polarization angular spectra are discussed and inferences made about the impact they have on the cosmological parameters. Of these parameters, we have already met Ω_{r0}, Ω_{m0}, Ω_{b0}, $\Omega_{\Lambda 0}$ and H_0. In the later universe radiation from the stars reionizes the hydrogen gas it contains and the CMB undergoes further Thomson scattering from the free electrons thus created. We meet a measurable parameter used to quantify this effect: τ the *optical thickness* of the universe.

12.4 Interpreting the Power Spectrum

The striking features visible in the CMB correlation power spectrum in Figure 12.3 are the peaks with roughly equal spacing in ℓ, extending up to $\ell \sim 2000$: they have varying heights, but show an overall decline in contrast with increasing ℓ. The angular scale expressed in terms of ℓ is simply $\theta \approx 180°/\ell$. This is because the Legendre polynomials $P_\ell(\cos\theta)$ used to project out the moments each contain a leading term $[\cos\theta]^\ell$ so that they capture features in the CMB of angular extent $180°/\ell$. Hence a peak in the power spectrum at ℓ shows that at recombination the fluctuations with angular extent $180°/\ell$ were prominent. The origin of such prominent features is understood to lie in the quantum fluctuations in the early universe that are the natural accompaniment of inflation. These can be viewed as curvature fluctuations away from flatness, or equivalently as gravitational potential hills and valleys. Photons escaping a gravitational potential well at the last scattering lose energy and are colder; those leaving a potential hill gain energy and are hotter.

This gives the thermal variation across the last scattering surface, resulting in the correlations that Planck measured.

Dark matter particles had become non-relativistic much earlier, as we shall prove. Lacking any interactions other than gravitational, dark matter simply accumulated in the gravitational valleys and was depleted on the hills left after inflation. This process steadily enhanced the contrast between the hills and valleys. Relativistic baryonic matter however formed a plasma in thermal equilibrium with the photons, coupled by Thomson scattering of photons from electrons and Coulomb scattering of electrons off protons. This plasma reacted to the gravitational potential with baryonic matter flowing into valleys and becoming compressed there. The photons then exerted a restoring pressure that produced an outflow from the potential wells. Together, the inertia of the dust-like baryons and the spring provided by the radiation produced plasma oscillations in the existing gravitational potential wells. These oscillations would have traveled at the speed of sound in the plasma.[5] Photons dominated the baryons in number and in their total energy so that the speed of sound was

$$c_s \approx c/\sqrt{3},$$

not very different from the speed of light.

Plasma oscillations that completed an exact half cycle at recombination would have reached maximal compression at recombination and hence been hotter, and contribute to the first and largest peak in the power spectrum in Figure 12.3. More generally the odd numbered peaks in $\mathcal{D}_\ell^{\mathrm{TT}}$ are the result of 1/2, 3/2, 5/2, ... cycles of oscillation being completed at recombination: they correspond to maximal compression at recombination and a hotter plasma. When 1, 2, 3, ... full cycles are completed at recombination there is maximal rarefaction and a colder plasma: this gives the even order peaks in $\mathcal{D}_\ell^{\mathrm{TT}}$. Even or odd, these peaks, the result of longitudinal sound waves in the photon-baryon plasma, are called *acoustic peaks*.

During the compression of the plasma the gravitational attraction of the baryons acts in the same sense as the gravitational attraction of the dark matter, by contrast they are in opposition at rarefaction. Hence the relative heights of the even numbered acoustic peaks with respect to the odd peaks is determined by the baryon to dark matter ratio. The fits to the CMB spectra by the Planck Collaboration give mean values $\Omega_{b0} = 0.049$ and $\Omega_{m0} = 0.31$. It follows that the ratio of the number of protons to the number of photons now, and equally ever since decoupling, is 6.0×10^{-10}.

The observed angular size of the acoustic peaks is dependent on the intrinsic curvature of the universe. A positive curvature would lead to perceived angular sizes being smaller, and hence acoustic peaks would be displaced to larger ℓ values than for a flat universe. Conversely a negative curvature would displace the peaks to smaller ℓ values. The position of the peaks therefore can be used to determine the sum $\Omega_{m0} + \Omega_{\Lambda 0}$, which would be unity in a flat universe (ignoring the tiny Ω_{r0}). We

[5] In Chapter 15 the speed of plasma waves is examined in more detail in connection with structure growth later in the universe.

can show in a simple manner that the location of the first peak is where it would be expected in the ΛCDM model with its flat universe.

The distance the plasma waves would have traveled up to the time of recombination is the *sound horizon* distance

$$r_s = a_{\rm rec} \int_0^{t_{\rm rec}} \frac{c_s {\rm d}t}{a(t)}, \tag{12.16}$$

where the subscript rec signifies the time at recombination when the components of the plasma decouple. Using Equation (10.31) with the parameters of ΛCDM model, $t_{\rm rec}$ is 365,000 years and

$$r_s = 0.150 \text{ Mpc}. \tag{12.17}$$

This distance, the extent of the sound horizon over the last scattering surface, is important in providing a well-defined length scale (a ruler) at recombination. We shall see in Chapter 15 how a comparison is made with a length defined by the oscillations inherited by the baryons in the universe at low redshift. The comparison provides a crucial test for the ΛCDM model of the universe. (Note that we will need to tweak Equation (12.16) slightly in the following chapter.) The sound horizon today subtends an angle at the Earth of

$$\theta = r_s/d_{\rm A}, \tag{12.18}$$

where $d_{\rm A}$ is the angular diameter distance to the last scattering surface. $d_{\rm A}$ evaluated using the ΛCDM model is

$$d_{\rm A} = \left[\int_0^{t_0} c{\rm d}t/a(t)\right]\frac{1}{1 + z_{\rm rec}} = 13.0 \text{ Mpc}. \tag{12.19}$$

Here the time before recombination has been included because it is negligible compared to the lifetime of the universe today. Thus the angle subtended by plasma waves just reaching maximum compression for the first time at recombination is

$$\theta \approx 0.150/13.0 = 0.0115, \tag{12.20}$$

that is 0.69°. Projecting this onto partial waves gives $\ell \approx 180°/0.69° = 260$. This is broadly consistent with the measurements shown in Figure 12.3: the first peak in the power spectrum lies at $\ell \approx 220$, while the spacing between peaks is around 300.

The curves shown in Figures 12.3 and 12.6 are the result of fitting the ΛCDM model to the data, by varying parameters including $\Omega_{\rm m0}$ and $\Omega_{\Lambda 0}$. Their sum is found to lie close to unity providing strong evidence for the flatness of the universe. As a measure of the sensitivity, note that if the sum $\Omega_{\rm m0} + \Omega_{\Lambda 0}$ were reduced to 0.1 then the first peak would move to $\ell = 800$. In more detail, Figure 12.4 shows the restriction imposed by such fits to the CMB spectrum (Suzuki et al., 2012) on the permissible range of values of Ω_{m0} and $\Omega_{\Lambda 0}$. Those consistent with the CMB data lie within the orange-red boundaries at confidence levels of 68%, 95% and 99.7%. The line $\Omega_{m0} + \Omega_{\Lambda 0} = 1.0$ traces out the combinations that would make the universe flat. The other limitations shown come from measurements of baryon acoustic

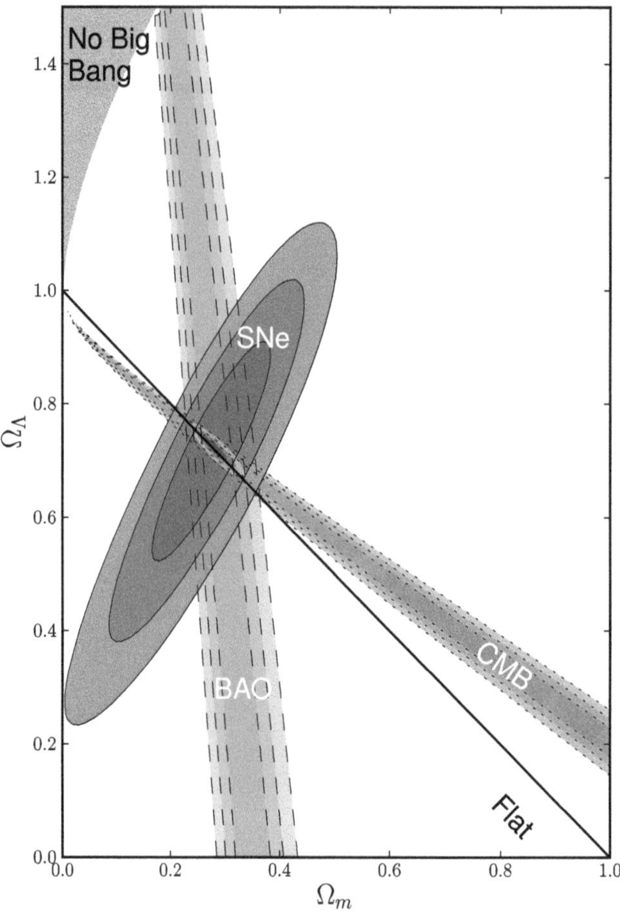

Figure 12.4. Current values of Ω_Λ ($\Omega_{\Lambda 0}$) versus Ω_m (Ω_{m0}). Figure 5 from (Suzuki et al., 2012). Courtesy Professor Suzuki for the copyright holders.

oscillations (Chapter 15), and measurements of the current acceleration of the universe using SNe Ia (Chapter 17). The cross-over region satisfying all constraints identifies the parameters of the ΛCDM model. Taking account of these three constraints the conclusion is that the universe is close to being flat, if not exactly flat.

The total matter content of the universe (examined in detail in Chapters 14, 15 and 16) only accounts for 30% of the critical energy required in a flat universe. The only viable interpretation of this missing 70% is that it is due to dark energy which has a uniform density, constant in time, across the whole universe, with only a gravitational interaction. These properties match, within our current knowledge, with Einstein's cosmological constant, introduced in Chapter 6. To reiterate: the values used in this book are $\Omega_{m0} = 0.30$ and $\Omega_{\Lambda 0} = 0.70$, with the photon density $\Omega_{r0} = 0.000\,09$ given directly by the density of the CMB photons observed today. Very similar values are met in all recent published fits to comprehensive cosmological data using the ΛCDM model.

The last scattering of the CMB photons is spread over time and during that period there is the opportunity for scatters to smooth out the temperature fluctuations. This effect, known as *Silk damping*, is limited to the thickness of the last scattering layer. The effect grows in importance as the potential wells' physical scale decreases, and thus with increasing ℓ. Making the baryon density larger would shorten the photon mean free path and spread the damping to larger ℓ values. This as well as the relative heights of the even to odd peaks constrains Ω_{b0}.

The power in the fluctuations with a given ℓ value is observed to have a Gaussian distribution: this is what is expected to result from the model for inflation discussed in the next chapter. One consequence of a Gaussian distribution is that the two-point correlations analyzed above account for all correlations. All the information content of the correlations is contained in the power spectra shown in Figures 12.3 and 12.6. In addition the fluctuations observed in the CMB are consistent with being adiabatic. That is to say, the fractional change induced in number density is the same for all the particle species: this is a further prediction made with the model of inflation presented in the next chapter.

12.5 The Sachs–Wolfe Plateau

The plateau in the CMB power correlation spectra in Figures 12.3 and 12.6 at $\ell < 30$ is less striking than the peak structure but provides direct information about the primordial gravitational fluctuations left by inflation. The low ℓ long wavelength modes only complete a small fraction of an oscillation before the last scattering. Thus they remain close to the pristine state determined by the quantum fluctuation of the gravitational potential during inflation. The analysis, given here, of these low ℓ modes, is due to Sachs & Wolfe (1967).

They consider photons at recombination at a location where the gravitational potential departs by Φ from the average potential. There will be a gravitational spectral shift, as discussed in Chapter 2, for photons on, thereafter, reaching the average potential. The fractional frequency shift of the photons is

$$\delta f/f = \Phi/c^2,$$

while time intervals change in a complementary way

$$\delta t/t = -\Phi/c^2.$$

The fractional energy shift, and hence the fractional shift in the temperature of the photons, are thus

$$\delta T/T = \delta E/E = \Phi/c^2.$$

We must take into account that the scale factor also depends on the time. The universe is matter dominated at recombination so that $a \propto t^{2/3}$ and hence

$$\delta a/a = (2/3)\delta t/t = -(2/3)\Phi/c^2.$$

Now the black body radiation temperature $T \propto 1/a$ and it follows that black body temperature undergoes a shift

$$\delta T/T = -\delta a/a = (2/3)\Phi/c^2.$$

Relative to this modified black body temperature the spectral shift of the photons is reduced to

$$\delta T/T = \Phi/c^2 - (2/3)\Phi/c^2 = (1/3)\Phi/c^2.$$

The temperature fluctuations calculated for a super-horizon mode are therefore independent of the dimensions of the potential fluctuation and this gives a plateau in the power spectrum at $\ell < 30$, the *Sachs–Wolfe plateau.*

12.6 The Three-dimensional Power Spectrum

Thus far the perturbations left by inflation have been viewed in terms of the angular pattern across the sky. However, the perturbations are necessarily three dimensional. These perturbations were imprinted equally on the baryons as on the CMB and are the progenitors of the distribution of matter in today's galaxies and clusters of galaxies. What is then of interest is the physical size of the CMB perturbations and how the overall energy is distributed. The cosmological principle rules out any particular importance in where the perturbations are located: what counts is the correlations. In order to evolve the perturbations and compare with the current distribution of matter their distribution is presented using the Fourier transform of the spatial distribution. The perturbation power spectrum in **k**, the Fourier transform of the spectrum in the coordinate **r**, is parametrized by a power law distribution (Planck Collaboration et al., 2020), namely

$$P(k)\mathrm{d}k = 2\pi^2 A_s k^{-3}\left[\frac{k}{k_p}\right]^{n_s-1}\mathrm{d}k. \tag{12.21}$$

k_p is an arbitrary *pivot scale*, chosen by the Planck Collaboration to be 0.05 Mpc^{-1}. It follows that

$$A_s = k_p^3 P(k_p)/(2\pi^2). \tag{12.22}$$

The correlation coefficients in the temperature fluctuations are projected out in k-space rather than coordinate space. This gives[6] for the pristine modes that Sachs and Wolfe analyzed

$$C_\ell^{\mathrm{SW}} = \frac{4\pi}{25}A_s k_p^{1-n_s}\int_0^\infty k^{n_s-2} j_\ell^2(k)\mathrm{d}k \tag{12.23}$$

where j_ℓ is the ℓ th spherical Bessel function. If n_s is unity the coefficient for the spectrum shown in Figure 12.3 is simply

$$\ell(\ell+1)C_\ell^{\mathrm{SW}} = 8A_s/25, \tag{12.24}$$

[6] Section 9.9 in the second edition of "Modern Cosmology" by Scott Dodelson and Fabian Schmidt, Academic, New York, 2021.

giving the observed near constancy of the Sachs–Wolfe plateau. Fitting the data gives $A_s = 2.1 \times 10^{-9}$, making the fluctuations $\sqrt{A_s}$ ~few 10^{-5}. This is the *most direct* measure of the strength of the curvature fluctuations produced during inflation.

By fitting the measured power spectrum to data the Planck Collaboration found the value $n_s = 0.965$, equal to unity within the uncertainty in the determination. Two physicists, Harrison and Zeldovich, had independently proposed that the power spectrum for the perturbations had n_s equal to unity. The distribution is then *scale invariant*, that is to say each logarithmic interval in k then contains equal power. Put another way, it is fractal. The model of inflation presented in the following chapter accounts for n_s being close to, but marginally less than unity.[7]

12.7 Optical Depth

Matter in the universe became neutral at recombination; later, within a few hundred million years, matter began to condense into stars. The radiation emitted by the first stars reionized the gas around them, and this process has continued up to the present time. As a result CMB photons scatter from the free electrons produced by reionization. An *optical depth* τ is defined to quantify the effect: the fraction of the photons that arrive now, unscattered, is then

$$f = \exp(-\tau).$$

Consequently, if the CMB temperature in some direction is $T(1 + \Theta)$ where T is the mean CMB temperature, we receive a fraction $\exp(-\tau)$ unscattered at this temperature. The remaining fraction $[1 - \exp(-\tau)]$ that we receive has been scattered in from other directions at the average temperature T of the CMB. Then the mean temperature of the radiation in this direction is

$$T(1 + \Theta)\exp(-\tau) + T[1 - \exp(-\tau)] = T[1 + \Theta\exp(-\tau)].$$

Hence the observed fluctuation in temperature is reduced by a factor $\exp(-\tau)$. τ was measured by the Planck Collaboration to be around 0.054.

12.8 Polarization

The CMB is weakly polarized and the correlations in polarization across the sky have been measured by the Planck Collaboration. They simultaneously fitted the temperature correlations, the polarization correlations, and the temperature–polarization correlations using the ΛCDM model, varying parameters including Ω_{r0}, Ω_{m0}, and $\Omega_{\Lambda 0}$. Two of these precise fits are shown in Figures 12.3 and 12.6.

Polarization of the CMB was produced at the last scattering surface, in the final Thomson scatter of each photon, before it began traveling freely across the universe. Asymmetries in the CMB temperature distribution are converted in this last scatter to polarization of the CMB. The explanation for this conversion takes a few steps: remember that photons are polarized perpendicular to the direction of travel. It

[7] An attractive mathematical property of the Harrison–Zeldovich power spectrum is that it is the only power law distribution that avoids a divergence at both very small and very large values of k.

starts by noting that the Thomson cross-section depends on the relative alignment of the polarizations (**e**) of the incoming and outgoing photons. This happens because the incoming photon excites the electron to oscillate along the direction of its polarization, and then the electron emits a photon with that same polarization. As a result the Thomson differential cross-section is

$$\mathrm{d}\sigma_{\mathrm{T}}/\mathrm{d}\Omega = r_{\mathrm{e}}^2|\mathbf{e}_{\mathrm{in}} \cdot \mathbf{e}_{\mathrm{out}}|^2, \tag{12.25}$$

where r_{e} is the classical radius of the electron, $2.82 \times 10^{-15}\ \mathrm{m}^2$. Thus the scattering is strong when the polarizations of incoming and outgoing photons are aligned, and null when they are orthogonal. The polarization of the electric field measured by the detectors on board Planck is of course exactly the polarization of these outgoing photons. Figure 12.5 illustrates what happens at the last scattering in the simplest arrangement in which incoming photons enter in the horizontal plane and scatter through 90° upward. The lower panel shows the possibilities for scattering where a dipole temperature asymmetry exists: that is where a rotation of π around the direction of the outgoing scattered photon reverses the sign of the temperature perturbation. The upper panel shows the possibilities for scattering where there is a quadrupole temperature asymmetry: that is where a rotation of $\pi/2$ around the direction of the outgoing photon reverses the sign of the temperature perturbation. The incident photons from the hotter regions are drawn as incoming arrows using full lines in the figure. They are more numerous than the incident photons from the colder

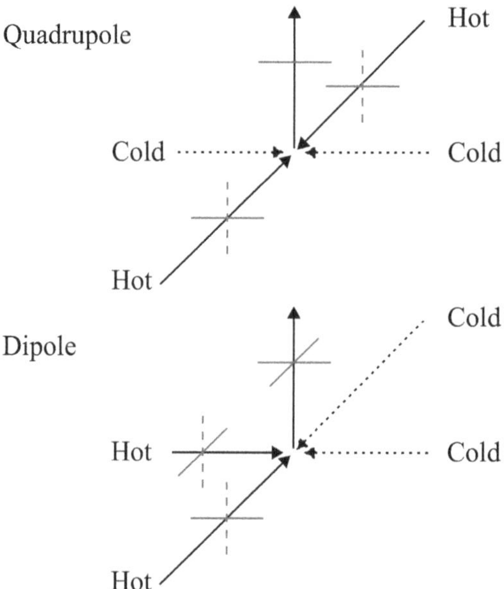

Figure 12.5. The last scattering in the presence of quadrupole and dipole temperature asymmetries. Only the polarizations of the more numerous photons from the hotter regions are drawn. The polarization of a photon that is scattered in the direction of the outgoing ray is drawn as a solid blue line: the polarization that cannot be scattered in that direction is shown as a broken blue line.

regions: these latter are drawn with dotted arrowed lines. The two orthogonal possible polarizations for the incoming photons are only indicated for the more numerous hotter photons. The polarizations that lead to scattering along the outgoing ray are drawn with full lines; polarizations not scattered along the outgoing ray are indicated using broken lines. In the lower panel, illustrating the case of a dipole asymmetry the more numerous outgoing hotter photons can have either of the orthogonal polarizations. On the other hand in the upper panel, for the case of a quadrupole temperature asymmetry, the more numerous hotter scattered photons are linearly polarized. Thus, finally, we see that quadrupole asymmetries in the temperature perturbations at recombination produce corresponding net polarization patterns in the CMB.

During recombination the CMB polarization was being continually reset by Thomson scattering. It is therefore only a layer of the universe of thickness equal to the mean free path of the photons that influences the polarization. Consequently the strength of polarization correlations is much weaker than the thermal correlations. The polarization across the sky is again expressed in terms of spherical harmonics

$$E(\theta, \phi) = \sum_{\ell=0}^{\infty} \sum_{m=-\ell}^{\ell} a_\ell^m Y_\ell^m(\theta, \phi). \tag{12.26}$$

The correlations between polarizations are defined in a similar way to those for temperature asymmetries

$$C^{\mathrm{EE}}(\theta) = \langle E(\hat{a})E(\hat{b})\rangle. \tag{12.27}$$

Projecting out the angular momentum states gives the coefficient C_l^{EE}, which is plotted in Figure 12.6. The acoustic peaks in the polarization correlations are in antiphase to those in the temperature correlations seen in Figure 12.3. This is expected because the polarization is proportional to the velocity of the plasma flow, while the temperature maps the plasma density.

In principle the primordial quantum fluctuations left by inflation could have been scalar, that is density fluctuations, or tensor fluctuations due to gravitational waves excited during inflation. These types of fluctuation produce very different and distinctive patterns of polarization in the area of the sky around hot and cold spots. In the case of scalar fluctuations the electric field is either radial or tangential, the *E-mode*. In contrast the tensor pattern would have the field at 45°, left or right of these directions, swirling around the hot/cold spot. This is the *B-mode*, in appearance analogous to magnetic field lines. The observed polarization is almost entirely E-mode, thus requiring scalar primordial perturbations.

It is worth repeating that Figures 12.3 and 12.6 illustrate how well the ΛCDM model can fit all the detailed features of the power and polarization spectra. There are two key properties of the CMB that imply analogous features in baryonic matter, and these features should be apparent in baryonic structures in the low redshift universe. First the near scale invariant (Harrison–Zeldovich) power spectrum of perturbations; second the angular correlation peak marking the sound horizon at recombination. As we shall see fits made with the ΛCDM model provide a consistent match to the observed evolution.

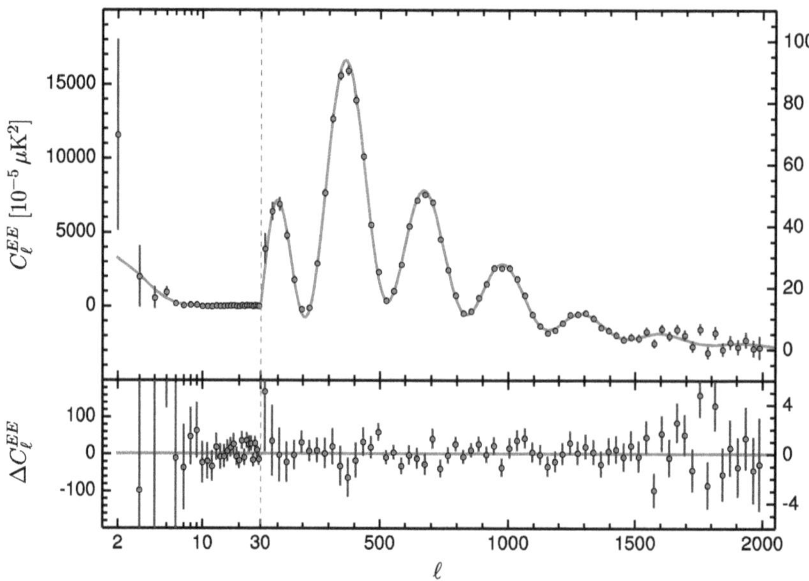

Figure 12.6. Polarization–polarization correlations of the CMB measured by the Planck Collaboration. The experimental errors are displayed in the lower panels. Adapted from Figure 1 in Planck Collaboration et al. (2020) Reproduced with permission © ESO.

12.9 The Horizon Distance

One defect that could undermine our analysis of the CMB must now be tackled. The horizon distance at decoupling is the distance over which causal contact could have extended by that time. We can readily calculate this distance by re-using Equation (12.16) with c replacing c_s. Taking the result from the ΛCDM model, quoted a few lines on, the horizon distance is $\sqrt{3} \times 0.150 = 0.260$ Mpc. Then proceeding as with Equation (12.18), the angular size of the horizon distance seen on Earth is 0.260/13.0 = 0.02 rad or 1.2°. This result poses a serious challenge to the interpretation of the CMB made in this chapter. On the one hand we observe that the CMB is uniform to better than one part in ten thousand over the whole sky. And yet only areas subtending a degree or so could ever have got into causal contact at decoupling. How can that be? In the next chapter we find out.

12.10 Neutrino Decoupling

The neutrinos decoupled earlier than the photons and form a *cosmic neutrino background* (CNB). There is little prospect for direct detection because these neutral particles only interact through the weak force. The reaction that had maintained thermal equilibrium between the neutrinos (ν), anti-neutrinos ($\bar{\nu}$) and baryonic matter is the weak interaction process

$$\nu + \bar{\nu} \rightleftharpoons e^- + e^+. \tag{12.28}$$

This is only possible energetically if the neutrinos have energies in the center-of-mass frame greater than the electron mass $\times c^2$ ($\epsilon = 511$ keV). The corresponding temperature of the universe is ϵ/k_B or 6×10^9 K. At lower temperatures the neutrinos decouple from matter.

Between neutrino and photon decoupling the photons picked up energy when the electrons and positrons annihilated

$$e^- + e^+ \rightleftharpoons \gamma + \gamma; \tag{12.29}$$

this energy is not accessible to the neutrinos. The resulting difference between the temperatures of the CNB and the CMB can be predicted precisely:

$$T_\nu = (4/11)^{1/3} T_\gamma, \tag{12.30}$$

which makes the present temperature of the cosmic microwave neutrinos 1.94 K. The ratio of the number of CNB neutrinos plus anti-neutrinos to the number of CMB photons is again precisely predicted at 3/11, so there are 1.12×10^8 m^{-3} today. Hence you and I are each being traversed instantaneously by well over ten million CNB neutrinos! The corresponding CMB photons incident mostly get absorbed by water molecules in the atmosphere.

Experiments in particle physics show that the neutrino masses are of order 0.2 eV c^{-2}. Consequently the neutrinos formed a relativistic gas in the early universe, and exerted pressure. Therefore they must be included with the photons in the relativistic radiation when calculating the properties of the universe for this period. Figure 10.3 separates the CNB and the CMB in showing how the contents of the universe have changed with time. Big Bang nucleosynthesis, discussed in Chapter 14, provides the most direct information on the number density of neutrinos.

12.11 Exercises

1. Make an argument in terms of angular momentum conservation, which rules out polarization of the CMB, originating from higher multipole components of the temperature asymmetry than the quadrupole asymmetry.
2. Make an estimate of the optical depth along the path of the CMB, assuming that reionization is complete at $z = 10$. Also assume that the universe develops as if matter dominated so that $H = H_0/a^{3/2}$.
3. Use Equation (12.1) to show that for photon energies, E much larger than $k_B T$ the black body spectrum reduces to

$$n(E)\mathrm{d}E = \frac{8\pi E^2}{c^3 h^3} \exp[-E/k_B T]\,\mathrm{d}E.$$

Show that the number with energy greater than E per meter cubed is approximately

$$8\pi \left[\frac{k_B T}{hc}\right]^3 \left[\frac{E}{k_B T}\right]^2 \exp[-E/k_B T].$$

This requires integration by parts, keeping only the term in $\left[\frac{E}{k_{\rm B}T}\right]^2$. What fraction is this of the black body spectrum? At 5700 K what fraction of the black body photons has energy greater than 13.6 eV?

4. Calculate the redshift and the time after the Big Bang when the CMB was at room temperature. Would it have suffered rescattering at that era?
5. At $z = 1090$ what was the mean free path for Thomson scattering and what was the interval between collisions?

Further Reading

Liddle A 2015 *An Introduction to Modern Cosmology* (3rd ed.; New York: Wiley) (Oxford: Oxford Univ. Press). This text provides a compact introduction.

Samtleben D, Staggs S and Winstein B 2007 *The Cosmic Microwave Background for Pedestrians* in *Annual Reviews of Nuclear Science* 57 245. This is a very helpful 38 page-long article for the non-expert.

Donaldson S and Schmidt F 2021 *Modern Cosmology* (2nd ed.; Amsterdam: Elsevier). This gives a detailed expert analysis using the Boltzmann and Einstein equations.

References

Hill, R., Masui, K. W., & Scott, D. 2018, ApSpe, 72, 663
Penzias, A. A., & Wilson, R. W. 1965, ApJ, 142, 419
Planck Collaboration,, Akrami, Y., & Arroja, F. 2020, A&A, 641, A10
Sachs, R. K., & Wolfe, A. M. 1967, ApJ, 147, 73
Suzuki, N., Rubin, D., Lidman, C., et al. 2012, ApJ, 746, 85

AAS | IOP Astronomy

Introduction to General Relativity and Cosmology (Second Edition)

Ian R Kenyon

Chapter 13

Inflation in the Early Universe

13.1 The Horizon and Flatness Problems

The hypothesis that the CMB originated from the initial Big Bang poses a quandary known as the *horizon problem*. Recall that the event horizon is the distance light could have traveled since the universe began. In Section 12.9 we found that the event horizon at decoupling covers of order one degree of the sky as presently seen from the Earth. It follows that regions 180° apart in the sky were never in causal contact before recombination. And yet the temperature of the CMB across the whole sky is the same to better than one part in 10^4: which implies that at some earlier time there was thermal, and hence causal contact across the universe. A second quandary is that posed by the flatness of the universe. In fits of the ΛCDM model to existing data, the curvature parameter in Friedmann's equation, Equation (10.11), $\Omega_k < 0.002$. We now show that this means that the universe began not only very small but astonishingly close to perfect flatness.

We start by rearranging Friedmann's equation, Equation (10.5),

$$\left[\frac{3H^2}{8\pi G} - \frac{\varepsilon}{c^2}\right] = -\frac{3\kappa c^2}{8\pi G a^2}.$$

Using Equation (10.10) this becomes

$$a^2(\rho_c - \rho) = -\frac{3\kappa c^2}{8\pi G}.$$

Here ρc^2 is the total energy density of matter, radiation and dark energy: $\rho_c c^2$ is its value if the universe is flat. Using $\Omega = \rho/\rho_c$, this becomes

$$a^2\rho(\Omega^{-1} - 1) = -\frac{3\kappa c^2}{8\pi G}, \tag{13.1}$$

doi:10.1088/2514-3433/acc3ffch13

where the right-hand side is a constant $-C$ because it is made up of quantities that do not change. Thus

$$|\Omega^{-1} - 1| = C/[a^2\rho]. \tag{13.2}$$

Now the early universe was dominated by radiation and for that period, using $\rho \propto a^{-4}$ and Equation (10.18),

$$a^2\rho \propto a^{-2} \propto t^{-1}, \tag{13.3}$$

making

$$|\Omega^{-1} - 1| \propto a^2 \propto t \tag{13.4}$$

so that $\Omega \to 1$ as $t \to 0$. For example at 10^{-36} s after the origin of the universe, $|\Omega^{-1} - 1| \equiv \Omega_k/\Omega$ would have needed to be of order 13.8 Gyr/10^{-36} s = 10^{54} times smaller than it is now. Then extrapolating from the current upper limit, $\Omega_k = 0.002$, back to 10^{-36} s gives an upper limit,

$$\Omega_k(t_{\rm infl}) = 2 \times 10^{-57}. \tag{13.5}$$

The inference drawn here that the universe started with vanishing curvature poses the *flatness problem*.

This puzzle has a bright side. If the universe had started out slightly overdense then gravitational attraction would have soon led to the collapse of the universe: alternatively if the universe started out slightly underdense then the expansion would have been rapid enough to prevent any structures forming. We are here only because ours is the Goldilocks universe.

A third problem may exist if tentative theories extending the standard model of particle physics turn out to be valid. These predict that there should be a dense population of magnetic monopoles in the universe. These would be readily detectable because they would produce ionization in matter about 137 times stronger than that produced by an electron.[1] None have been seen, which could be the *monopole problem*.

13.2 Inflation

The commonly accepted solution for all three problems was suggested by Guth (Guth & Steinhardt, 1984) in 1981, with later contributions from Linde (Linde, 1987) and Steinhardt (Guth & Steinhardt, 1984). The solution is that the universe inflated almost instantaneously shortly after its creation and by a huge factor: space expanded much faster than the speed of light. As a result regions that were in causal contact expanded to a size larger than the part of the universe we observe. This explains why the CMB is close to homogeneous of over the whole sky. As a bonus the expansion flattened out any existing curvature of spacetime in the universe. The cause of this inflationary epoch is thought to be a state change of the empty universe, the vacuum, dropping to a lower energy vacuum. This change released an enormous

[1] The ratio is predicted to be the inverse of the fine structure constant.

amount of energy into matter and radiation: an event which we recognize as being the Big Bang.

First we look further into how inflation solves the three problems, and then describe the commonly inferred version of the phase transition. In Figure 13.1 the broken line shows how the horizon would increase with time in the absence of inflation, while the solid line shows how the universe now visible to us would have grown. The figure shows again that the points on opposite sides of the sky we see today would never have been in causal contact in the remote past. Figure 13.2 shows the corresponding development with inflation by a factor of, for example, 10^{50} at 10^{-36} s. With inflation the region that was in causal contact in the past has become much larger than the universe now visible to us. This explains why the CMB is so uniform across the whole sky. Note that the figures were drawn choosing a scale factor $a \propto t^{1/2}$, appropriate for a radiation dominated universe: other choices would not affect the conclusions reached here. Inflation by 10^{50} would flatten the curvature of the universe by the same factor, leaving a curvature undetectable using current measurement techniques.

The vast change in volume following inflation would so dilute the numbers of existing monopoles, that detection of even a single monopole in the near future is highly unlikely. Equally all other matter would be diluted to the same extent leaving a virtually empty universe. This invites a question: why is the universe now populated with matter, despite further gentle expansion? As previously noted, it is thought that at the end of inflation the universe fell into a lower energy vacuum state. The compensating huge energy release appeared as matter and radiation: this event becomes the Big Bang.

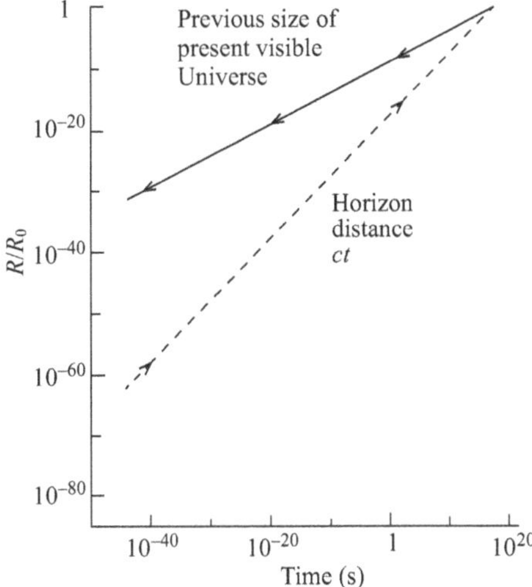

Figure 13.1. The broken line indicates how the distance to the horizon increased through the life of the universe. The solid line extrapolates the size of that part of the universe, which we now see back into the past, without inflation.

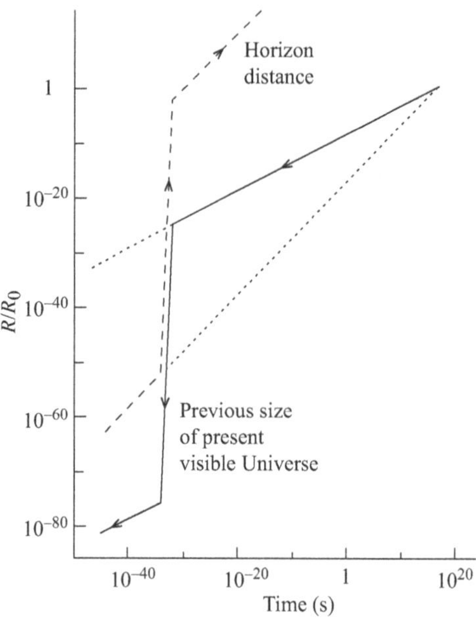

Figure 13.2. The broken line indicates how the distance to the horizon increased through the life of the universe. The solid line extrapolates the size of that part of the universe, which we now see back into the past, taking account of inflation.

13.3 The Vacuum Transition

The process driving inflation is the change of the vacuum state of the universe to one of lower energy. The phase transition that occurred was analogous to the phase transition when steam condenses, with an accompanying release of latent heat, but on a more dramatic scale. That the universe could undergo such a phase transition can be inferred from elementary particle physics as follows.

The coupling strengths of the strong and electroweak forces vary logarithmically with energy. When the dependence is extrapolated above the highest accelerator energies (10,000 GeV) as shown in Figure 13.3 it appears that the strengths converge at an energy per particle of around 10^{15} GeV, which translates to a temperature of 10^{28} K across the universe. At higher temperatures, that is earlier on, grand unified theory (GUT) predicts that there would have been a single *grand unified* interaction. The change from a single unified force to distinct forces would provide the looked-for phase change. The time at which the forces became distinct can be roughly estimated by assuming radiation dominated thereafter. Then $t \propto a^2 \propto T^{-2}$ and

$$t_{\text{GUT}} = t_0 \left[\frac{2.7255}{10^{28}}\right]^2 \approx 10^{-38}\text{ s}.$$

Using the ΛCDM model yields an improved estimate for the time at which the forces became distinct of 10^{-36} s.

In a similar transition, when the energy per particle in the universe fell to below ~1 TeV, the weak and electromagnetic forces became distinguishable. That phase

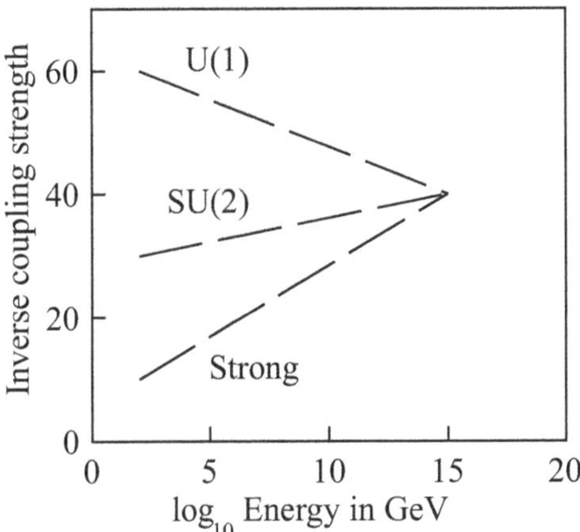

Figure 13.3. Long range extrapolation of the coupling strengths of the electroweak and strong forces, according to the standard model of particle physics.

transition is now well understood, and some reasonable inferences can be drawn about the grand unification transition. At the electroweak phase transition the vacuum state of the universe changed. A neutral scalar field extending uniformly over the whole universe, the Higgs field, dropped into a lower energy state with a release of energy into baryonic matter. It is argued that at the grand unification transition an analogous neutral scalar field underwent a similar transition. This field is generally called the *inflaton* field. Although a rigorous quantum field theory has been developed to describe the electroweak transition, a parallel theory does not exist yet to analyze what happens in the early universe. In such cases where the mathematical underpinning is not available an *effective field theory* is built nonetheless: this approach proved highly successful in modeling reality in the cases of the superfluid and superconducting transitions.[2] The approach here is to present at an elementary level the corresponding analysis of the inflationary epoch in terms of this inflaton field. Following inflation by the factor $\sim 10^{30}$ any pre-existing matter and radiation is so diluted that the universe is essentially empty, apart from the inflaton field. This field is assumed to be a scalar field like the Higgs field, and equally, like the Higgs field, uniform and isotropic.

Our analysis of inflation starts with the acceleration Equation (10.7) in the era of interest

$$\ddot{a}/a = -\frac{4\pi G}{3c^2}[\varepsilon_\phi + 3p_\phi],$$

where ε_ϕ and p_ϕ are the energy density and pressure of the inflaton field. If there is to be inflation, then $\ddot{a} > 0$, and in the equation of state of the inflaton $p_\phi = w\varepsilon_\phi$, w must be less than $-1/3$. The simplest case is to have $w = -1$, which, we shall see, fits the

[2] The success of the earlier developed effective field theory for superconductivity led Anderson to propose how the eventual rigorous theory of the electroweak transition could be constructed.

data closely. It also duplicates the effect of a cosmological constant. Taking $p_\phi = -\varepsilon_\phi$ we can integrate the above equation, giving

$$a(t + \Delta t) = a(t)\exp(H\Delta t), \tag{13.6}$$

where

$$H = \sqrt{\left[\frac{8\pi G\varepsilon_\phi}{3c^2}\right]}. \tag{13.7}$$

If H remains constant during inflation the number of e-foldings of the universe's dimensions during an inflationary period of duration Δt is

$$N = H\Delta t. \tag{13.8}$$

Using Equation (13.5) we see that any initial regions of extreme curvature with Ω_k approaching unity were smoothed out by inflation if

$$\exp(2H\Delta t) > 10^{57}. \tag{13.9}$$

Applying Equation (13.8), the flattening and smoothing of the universe requires 65 or more e-foldings. The 10^{57} expansion in the scale factor during inflation caused the temperature to fall by the same large factor. At this point the universe was both very cold and virtually empty. The burst of inflation then ended with the release of the energy of the inflaton field into an energetic soup of matter and radiation. This process reheated and repopulated the universe with matter and radiation. *This was effectively the Big Bang.*

The comoving Hubble radius

$$\frac{c}{aH} = \frac{c}{\dot{a}}$$

introduced after Equation (11.6) measures the range of causal effects at a given moment. Figure 13.4 shows how this distance changed as a function of the scale parameter a during and after inflation, on a log-log scale. During inflation H was constant so that a grew exponentially, and so therefore did $\dot{a}$. As a result the horizon for causal contact shrank correspondingly rapidly. After inflation, while radiation dominated, $H \propto a^{-2}$ so that now the horizon grew steadily and its growth continues to the present. The path of a typical quantum fluctuation (mode) responsible for the thermal correlations in the photons in the CMB is drawn in Figure 13.4. During inflation it would first *exit the horizon*, and then at some later time *re-enter the horizon.*

13.4 Plasma Wave Coherence

When a mode of the photon perturbation left the horizon causal contact was lost across this mode and its amplitude froze. As a result, when the mode re-entered the horizon the rate of change of its amplitude, $\dot{\delta}$, was instantaneously zero. This fixed the phase of the mode and ensured that its amplitude was a cosine function of the

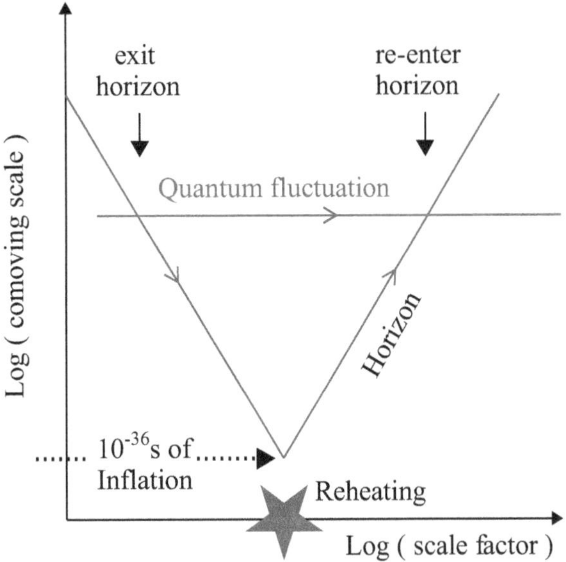

Figure 13.4. Evolution of the comoving Hubble radius during and after inflation. A typical fluctuation is shown, first leaving and then re-entering the horizon. The energy release and reheating after inflation are discussed later.

time measured from the moment of re-entering the horizon. Among all the modes, those modes whose amplitude completed one half cycle of oscillation between re-entering the horizon and recombination were in phase with one another at recombination and added coherently to produce the first peak in the CMB temperature correlation power spectrum. If instead these waves had re-entered with random phases there would be cancellation between them at recombination and hence no spectral peak in the CMB power spectrum. This coherence was essential in producing all the acoustic peaks seen in these spectra (Dodelson, 2003). Figure 13.5 shows how the photon perturbations in plasma waves changed after inflation between the moment of re-entering the horizon and recombination. The black line shows an example of a wave that completed one half cycle at recombination; such waves added coherently to give the first peak in the CMB temperature correlation power spectrum Referring back to Equation (12.16) we see now that the integral used to evaluate the sound horizon distance must run from the moment at which this perturbation re-enters the horizon up to the moment of recombination:

$$r_s = a_{\rm rec} \int_{t_{reentry}}^{t_{\rm rec}} \frac{c_s {\rm d}t}{a(t)}. \tag{13.10}$$

Going back to Figure 13.5, the blue line shows a wave that completed one full cycle at recombination and contributed to the second peak. Waves such as that shown in red did not change much before recombination because their wavelengths were much larger than the horizon at recombination: these are called *superhorizon* modes. We saw that their contributions at recombination formed the Sachs–Wolfe plateau.

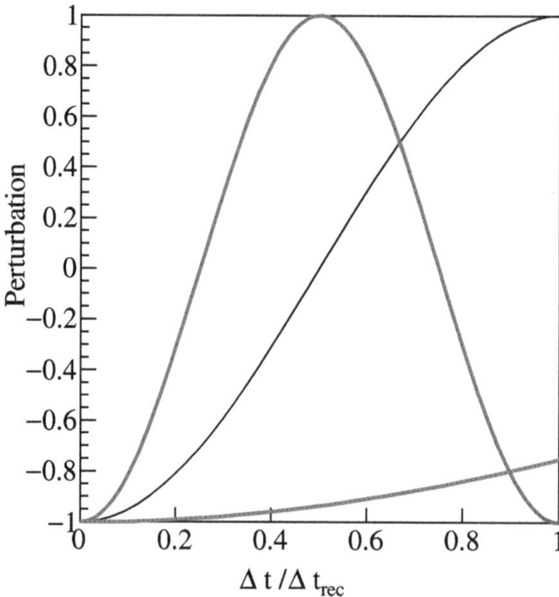

Figure 13.5. Notional development of photon perturbations between reentering the horizon and recombination. Δt is the time since reentering the horizon and $\Delta t_{\rm rec}$ is the time between reentry and recombination. Waves drawn in black contribute to the first (compression) acoustic peak; those in blue contribute to the second (rarefaction) acoustic peak. An example of a superhorizon mode is drawn in red.

Other key features of the inhomogeneities observed in the CMB, the scale invariance, Gaussianity and adiabaticity, also emerge naturally in the model of inflation presented here.

13.5 Slow Roll Inflation

In terms of the inflaton field the CMB anisotropies are due to an effect common to all quantum fields including the electromagnetic field. This is the existence of quantum fluctuations of the vacuum in which locally the field and its energy fluctuate from the null value anticipated classically. In the case of the inflaton field this generates local curvature fluctuations in spacetime that get inflated to large scales. Four properties expected of such fluctuations, and seen in the CMB, offer strong support to this interpretation. These are the isotropy, Gaussian nature, near perfect (Harrison–Zeldovich) scale invariance, and adiabatic nature of the fluctuations; properties already discussed in Chapter 12. Isotropy, independence of direction, is a necessity if the inflaton field is a scalar field like the Higgs field. Next, the fluctuations of a quantum field are random. It follows from the central limit theorem that the temperature–temperature fluctuations of any given ℓ value will have a Gaussian distribution in amplitude: such a distribution in turn requires that the two point correlations encompass all the correlations as seen in the CMB. Lastly, because inflation arises from a single scalar field the fractional change in density in quantum fluctuations is the same for all particle species. The fluctuations are then called adiabatic.

An alternative, associated with more complex fields, would be to have isocurvature fluctuations. In these the changes in particle densities compensate to maintain a constant total density. One result of isocurvature fluctuations is that instead of the partial cancellation seen in the Sachs–Wolfe plateau the contributions add and the prediction for the correlations at large angles would be six times larger. This rules out isocurvature fluctuations.

We now apply classical field theory in order to learn more about the transition. The energy density of a scalar field like the inflaton field is

$$\varepsilon_\phi = \dot{\phi}^2/2 + V(\phi), \tag{13.11}$$

where ϕ is the field amplitude. $\dot{\phi}^2/2$ is the density of kinetic energy and $V(\phi)$ is the potential energy density. This potential energy density is the result of the interaction that the field at one point has with the surrounding field. The field exerts a pressure

$$p_\phi = \dot{\phi}^2/2 - V(\phi). \tag{13.12}$$

The equation of state of the field is $p_\phi = w\varepsilon_\phi$, so that, as with dark energy, w must be negative to give inflation, and close to -1 to give scale invariance. That means that the potential must dominate:

$$V(\phi) \gg \dot{\phi}^2. \tag{13.13}$$

Figure 13.6 shows a suitable form for the potential energy of the vacuum of the universe as a function of the amplitude of the inflaton field. This potential gives rise to what is called *slow roll* inflation.

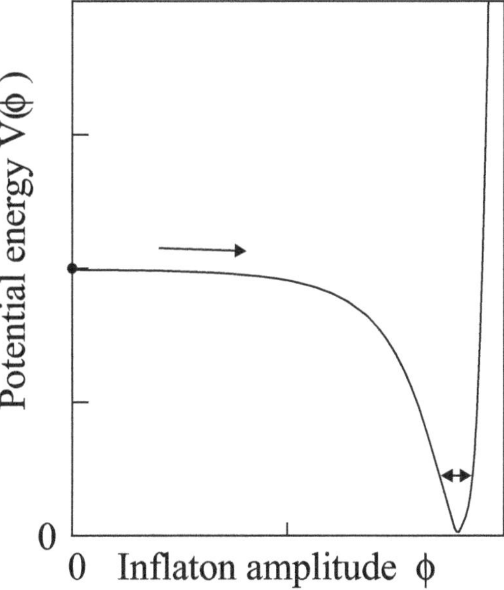

Figure 13.6. Slow roll inflation: variation of the inflaton field's potential energy as a function of the field's amplitude. The solid circle marks the initial metastable state before inflation.

Initially the universe would have been in a metastable false vacuum with vanishing amplitude of the inflaton field: at the point marked by the black ball. Like a ball poised on a shallow slope the state of the universe rolled down the slope ending in the true current vacuum at the minimum in the potential well. In the steep fall into the minimum the inflaton field picked up kinetic energy so that Equation (13.13) no longer held. Then the inflaton field performed damped SHM in the well. This damping took the form of the conversion of the energy of the inflaton field to a soup of matter and radiation that became the matter and radiation that now fill the universe. Consequently the universe was reheated back to its temperature before inflation.

The analysis of the slow roll starts with the fluid Equation (10.8)

$$\dot{\varepsilon} = -3H(p + \varepsilon).$$

Substituting for the energy density and pressure using Equations (13.11) and (13.12) this becomes

$$\ddot{\phi}\dot{\phi} + \dot{V}(\phi) = -3H\dot{\phi}^2. \tag{13.14}$$

Writing V' for $\mathrm{d}V/\mathrm{d}\phi$, this can be written

$$\ddot{\phi}\dot{\phi} + V'\dot{\phi} = -3H\dot{\phi}^2, \tag{13.15}$$

whence canceling $\dot{\phi}$ we have the equation of motion of the inflaton field

$$\ddot{\phi} = -V' - 3H\dot{\phi}. \tag{13.16}$$

In this equation the potential gradient (V' is negative) provides the inflationary force, and this is opposed by a damping term on the right-hand side, called *Hubble friction*. The condition for prolonged inflation is that the terms on the right-hand side almost cancel, the potential gradient term being larger. If the energy of the inflaton field is the dominant energy in the universe, the Hubble constant

$$H = \sqrt{\left[\frac{8\pi GV}{3c^2}\right]}. \tag{13.17}$$

Then over the range where the slope of the inflaton field is constant and sufficiently shallow both terms on the right-hand side of Equation (13.16) remain nearly constant and there is constant acceleration down the slope. The near constant value of V appearing in Equation (13.17) makes H in turn nearly constant, producing nearly exponential inflation.

$$a = a_{\text{initial}} \exp[H\,(t - t_{\text{initial}})]. \tag{13.18}$$

Evidently the process is self-similar: after a time $1/H$ the growth repeats itself on a scale e-times larger. As a result the quantum fluctuations in the inflaton field would have close to the scale invariant form suggested by Harrison–Zeldovich. The inflaton fluctuations induce corresponding fluctuations in the curvature/gravitational potential of the universe. Whence come the CMB thermal fluctuations. In more detail H cannot be precisely constant and the slow roll must have a finite rate: each effect gives rise to a slight deviation from the Harrison–Zeldovich scale invariance. Both effects reduce the spectral power n_s appearing in Equation (12.21) below unity. We saw in Chapter 12 that the CMB power spectrum measured by the Planck Collaboration is indeed slightly tilted from the Harrison–Zeldovich form: n_s is 0.965 rather than being exactly unity.

To conclude, inflation solves the problems of flatness and homogeneity in the universe at the time the CMB decouples. If there are monopoles then they would be dispersed to undetectable levels in the universe. The proposed origin for the fluctuations in quantum field fluctuations ensures that the fluctuations are Gaussian, while the quantum field can be designed to give near scale invariant and adiabatic fluctuations, which accounts naturally for these observed properties of the CMB.

Thus far we have looked at two major events in the early universe: at 10^{-36} s when grand unification ended in inflation, and at 365,000 years after the Big Bang, when radiation and matter decoupled. In between the temperature of the universe fell through several thresholds. The one we understand best, when the nuclei of the lightest elements in the universe condensed out of baryonic matter, is the main topic of the next chapter.

13.6 Exercises

1. Estimate the value of the inflaton energy density at the start of inflation.
2. Explain why fluctuations that leave the Hubble horizon are frozen until they re-enter this horizon.
3. How important is dark energy during the inflation at the grand unification energy scale?

Further Reading

Liddle A 2015 *An Introduction to Modern Cosmology* (3rd ed.; New York: Wiley) (Oxford: Oxford Univ. Press). This text provides a compact introduction.

Dodelson S and Schmidt T 2021 *Modern Cosmology* (2nd ed.; New York: Academic). Chapters 7 and 9. A text for anyone who wants to penetrate more deeply into the subtleties of inflation and the CMB. It is rigorous, clear and fully up-to-date.

References

Dodelson, S. 2003, AIPCP, 689, 184

Guth, A. H., & Steinhardt, P. J. 1984, SciAm, 250, 116

Linde, A. 1987, PhT, 40, 61

Ian R Kenyon

Chapter 14

Big Bang Nucleosynthesis

14.1 Timeline

We can infer much about the history of baryonic matter in the universe following inflation from what we know about the properties of elementary particles. This era culminated between one second and five minutes after the Big Bang, with baryonic matter condensing into light nuclei; a process that is called *Big Bang nucleosynthesis* (BBN). It was then that the proportions of hydrogen, helium, and lithium in the universe were established. A small fraction of this primordial matter was later reprocessed into the heavier elements in stars and supernovae. Table 14.1 summarizes the important steps in the evolution of the early universe up to the BBN and beyond. Before describing BBN in detail, the evolution of baryonic matter between the end of inflation and BBN will be sketched.

The existence of the first phase change affecting the baryons after inflation, the electroweak transition, is integral to the standard model of elementary particle physics. At that transition the weak and electromagnetic forces became distinct and their overall symmetry was lost. The discovery of the predicted scalar Higgs boson in 2011 at the Large Hadron Collider (LHC) at CERN was ample confirmation of the essential soundness of our understanding of electroweak physics. The Higgs boson is the quantum of a scalar field filling all space uniformly. At the electroweak transition the Higgs field changed to a lower energy state and the universe dropped from one vacuum state to one of lower energy. The Higgs' mass $m_{\mathrm{H}} = 125$ GeV c^{-2} sets the energy of the transition; collisions in which the available energy falls below 125 GeV can no longer produce Higgs bosons. The model for inflation as described in the previous chapter is analogous to the better understood electroweak transition.

After the electroweak transition the temperature of the universe continued high enough that baryonic matter remained in the form of a quark-gluon plasma. At particle energies below 1 GeV the quarks condensed into *hadrons*: they become

doi:10.1088/2514-3433/acc3ffch14

Table 14.1. Timeline for Radiation and Baryonic Matter up to the Decoupling of the CMB from Matter

	Time	Temperature	Energy	Comment
Planck era	5.4×10^{-44} s	$1.4\ 10^{32}$ K	$1.2\ 10^{19}$ GeV	Quantum gravity required
GUT era				
Inflation	10^{-36} s	10^{28} K	10^{15} GeV	Strong force decouples from electroweak
Electroweak era. Antimatter common.				
Electroweak transition	10^{-12} s	10^{15} K	10^{2} GeV	Higgs field enters new vacuum state
Quark-gluon plasma.				
Quark to Hadron transition	10^{-5} s	1.7×10^{12} K	0.150 GeV	π-Meson mass scale
Hadron era.				
Neutrino decoupling	1 s	10^{10} K	1 MeV	Now the CNB at 1.95 K
BBN	$1 \rightarrow 600$ s	$10^{10} \rightarrow 10^{8}$ K	$1 \rightarrow 0.01$ MeV	
Radiation era; e^{---} and e^{+} mostly annihilate to photons.				
Radiation era ends	50,000 yrs	9400 K	0.8 eV	z=3400
Matter era.				
CMB decouples	365,000 yrs	2970 K	0.27 eV	z=1090

bound in the lowest-energy stable three-quark structures (protons and neutrons). The quarks are bound together by the interchange of gluons, the massless carriers of the strong, or color force, in a mechanism analogous to how the interchange of photons binds electrons in atoms. The lightest hadrons are the π-mesons formed from a bound quark–antiquark pair. The π-meson mass is an indicator of the energy scale of the condensation to hadrons, in round numbers ~150 MeV. The π-mesons have a short lifetime, 2.6×10^{-8} s, and would have decayed soon after the transition.

At some period in the life of the universe there occurred an event posing a critical unanswered question about our understanding of particle physics and cosmology. It is presumed that initially, after inflation, there was a plasma containing equal proportions of matter and antimatter. However we live in a universe containing, as far as we can tell, matter alone. One obvious question is this: why didn't the equal cohorts of matter and antimatter simply annihilate to leave electromagnetic radiation and nothing else? If they had, we would not be here. The very fact that we are here requires that there must be a difference between the interactions of matter and antimatter. Taking the photons in the CMB to be the outcome of these annihilations, only one in ~10^{10} of the baryons survived, giving the baryon/photon ratio that is preserved until today.

14.2 Big Bang Nucleosynthesis

Up to roughly one second after the Big Bang, energies exceeded the binding energy of light nuclei (~1 MeV, $T = 1\ \mathrm{MeV}/k_B \sim 10^{10}$ K) so that any nucleus heavier than ^{1}H would be ripped apart by a collision. In the following five minutes or so deuterium, helium, and a relatively few lithium nuclei were formed: these became the building materials for the first stars. This is called Big Bang nucleosynthesis to distinguish it from the nuclear processing in stars and supernovae that much later produced heavier elements: the *metals* as the cosmologists call them.

Predictions for the proportions of the light elements produced are made using well-known nuclear reaction cross-sections, and should be particularly reliable. Nuclear processes taking place later in stars and in stellar explosions converted nuclear matter from the primordial light nuclei into the nuclei of heavier elements. These are the elements oxygen, nitrogen, carbon, silicon, and so on, from which we and our Earth are made. However, only a small fraction of matter was converted, so that the mix of light nuclei as calculated should agree substantially with the mix observed today. This prediction has been tested by measuring the spectra of local and distant sources, and correcting for the expected later reprocessing. Abundances of the isotopes of the light elements vary by a factor of 10^{10}, which makes for a demanding test. Despite the yawning gap between 300 s and 13.8 Gyr, good agreement is found: our model of the universe passes this fundamental test.

14.3 The Neutron Decay

The masses of the proton and the neutron are $m_p = 938.272$ and $m_n = 939.565$ MeV c^{-2}, respectively; and their difference 1.293 MeV c^{-2} is much larger than the electron mass $m_e = 0.511$ MeV c^{-2}. Thus free neutrons decay into a proton, an electron and a near massless electron anti-neutrino $\bar{\nu}_e$:

$$\mathrm{n} \rightarrow \mathrm{p} + \mathrm{e}^- + \bar{\nu}_e \tag{14.1}$$

with a mean life of 880 s, ~15 min. As long as the temperature of the universe remained high enough the weak processes involving neutrinos and the antiparticle of the electron, e^+:

$$\mathrm{e}^- + \mathrm{p} \rightleftharpoons \nu_e + \mathrm{n},$$
$$\bar{\nu}_e + \mathrm{p} \rightleftharpoons \mathrm{e}^+ + \mathrm{n}$$

maintained thermal equilibrium between protons and neutrons. The number ratio in thermal equilibrium at temperature T is given by the Boltzmann factor

$$n/p = \exp[-Q/(k_B T)], \tag{14.2}$$

where Q is 1.293 MeV. This ratio froze out when the rate at which the weak reactions occurred fell below the expansion rate of the universe. We now calculate when this happened.

Let the cross-sections for these weak processes be σ; that is to say σ is the equivalent area of the target nucleon. Also let n_ν be the number density of neutrinos. Neutrinos have masses less than 1 eV c^{-2} so that they are relativistic in this era, like photons. In one second they travel a distance c, hence there are

$$\Gamma = \sigma c n_\nu$$

interactions per second. Also, like photons, their number density n_ν is proportional to a^{-3}, hence

$$n_\nu \propto a^{-3} \propto T^3.$$

The weak cross-sections have a temperature dependence

$$\sigma \propto G_F^2 T^2,$$

where G_F is the coupling constant for the weak interaction. Combining the last three equations, the interaction rate

$$\Gamma \propto G_F^2 T^5.$$

Space was expanding at the Hubble rate $H \propto \sqrt{G}T^2$ and when $\Gamma \sim H$ the weak reactions could no longer maintain equilibrium, the neutron/proton ratio was frozen. The freeze-out temperature is

$$T_{fr} \sim (G/G_F^4)^{1/6} \sim 10^{10}\ \mathrm{K}.$$

The energy per particle was thus ~0.8 MeV, and the time around one second after the Big Bang. At freeze-out the Boltzmann factor was

$$\frac{n_n}{n_p} = \exp\left[\frac{-(m_n - m_p)c^2}{k_B T}\right] = \exp[-1.29/0.8] = 1/5. \tag{14.3}$$

After freeze-out the remaining neutrons could either decay or become bound into light stable nuclei. Given the short neutron lifetime of 15 min it became a race to make light nuclides; about as long as a 5000 m race. It was certainly successful because helium makes up about a quarter of the mass of all baryonic matter today.

14.4 Deuterium Formation

The first step in forming nuclei from protons and neutrons was to form deuterons (^{2}H or D):

$$\mathrm{p} + \mathrm{n} \rightleftharpoons {}^2\mathrm{H} + \gamma. \tag{14.4}$$

In the forward process the binding energy of the deuteron, $B_D = 2.22$ MeV, is released. The reverse process is another ionization process, so we can use the Saha equation again to calculate the time it takes for the neutrons to get absorbed into deuterons. In this case the appropriate Saha equation is

$$\frac{n_D}{n_p n_n} = \frac{3}{4}\left[\frac{m_n m_p}{m_D}\right]^{-3/2}\left[\frac{2\pi k_B T}{h^2}\right]^{-3/2} \exp\left[\frac{B_D}{k_B T}\right].$$

The prefactor 3/4 is required because the Saha equation relates the number of particles in eigenstates. Protons and neutrons have two spin states ($\pm(1/2)\hbar$) so that their numbers each scale down by a factor 2, while the deuteron has three

($[+1, 0, -1]\hbar$) giving a factor 3.[1] To a good approximation $m_n = m_p = 0.5m_D$, so the Saha equation collapses to

$$\frac{n_D}{n_p n_n} = \frac{3}{4}\left[\frac{\pi m_n k_B T}{h^2}\right]^{-3/2} \exp\left[\frac{B_D}{k_B T}\right]. \tag{14.5}$$

This equation is tractable. First we know the ratio of baryons to photons at the present day, 6.0×10^{-10}, and can take this to be approximately the proton to photon ratio today, and therefore equally during BBN. Then we can re-use Equation (12.8) to relate the number of photons now to the number when the universe was at temperature T:

$$n_\gamma = 60.42\left[\frac{k_B T}{hc}\right]^3.$$

Putting these numbers together

$$n_p = 3.62\ 10^{-8}\left[\frac{k_B T}{hc}\right]^3.$$

Using this value of n_p Equation (14.5) becomes

$$\frac{n_D}{n_n} = 2.72\ 10^{-8}\left[\frac{\pi m_n c^2}{k_B T}\right]^{-3/2} \exp\left[\frac{B_D}{k_B T}\right].$$

When all the neutrons are fused into deuterons this reduces to

$$1 = 2.72\ 10^{-8}\left[\frac{2952\ \text{MeV}}{k_B T}\right]^{-3/2} \exp\left[\frac{2.22\ \text{MeV}}{k_B T}\right].$$

This gives a temperature of 0.066 MeV in energy units, making the temperature $T_n = 7.7 \times 10^8$ K. The time t_n that this all took place was in the radiation era so that using Equation (10.18)

$$t \propto a^2 \propto 1/T^2.$$

We can use the radiation/matter equality from Table 10.7 as a reference point: the temperature T_{rm} was 9384 K and the time t_{rm} was 50,000 yr. Then

$$t_n = t_{rm}\left[\frac{T_{rm}}{T_n}\right]^2,$$

which gives 224 s. The neutron mean life is 880 s and some refinement of the calculation is required to take account of the neutron decays during the formation of deuterium. In 224 s a further 23% of free neutrons would decay so that the estimated ratio given in Equation (14.3) would drop to

[1] In the case of decoupling Equation (12.6) the prefactor is unity.

$$\frac{n_{\rm n}}{n_{\rm p}} = 1/6 \tag{14.6}$$

by the time BBN is completed.

14.5 ^{4}He Formation

Nuclear reactions between the deuterons, protons and neutrons produce more complex nuclei, which can in turn take part in further reactions. The principal reactions leading to light stable nuclei are these

$$^2\text{H} + ^2\text{H} \rightarrow (^3\text{H} + \text{p}) \text{ or } (^3\text{He} + \text{n}),$$

$$\text{p(n)} + ^2\text{H} \rightarrow \ ^3\text{He}(^3\text{H}) + \gamma,$$

$$(^3\text{H} + \text{p}) \text{ or } (^3\text{He} + \text{n}) \rightarrow \ ^4\text{He} + \gamma.$$

Tritium, ^{3}H, is included, although unstable, because its lifetime is 24 years. ^{4}He is very well bound by 7.1 MeV per nucleon, while ^{3}H and ^{3}He have binding energies of 2.8 and 2.6 MeV per nucleon respectively. As a result most of the neutrons end up in ^{4}He. Building beyond ^{4}He is difficult because, as Figure 14.1 shows that there are no stable isotopes with atomic numbers 5 and 8: the nuclides with these atomic numbers quickly decay into the deeply bound ^{4}He. First, the atomic number 5 nuclides' decay modes are

$$^5\text{Li} \rightarrow \text{e}^+ + ^5\text{He} \quad \text{and} \quad ^5\text{He} \rightarrow \text{n} + ^4\text{He},$$

with lifetimes around 10^{-22} s. Second, the nuclides with atomic number 8 decay to a pair of ^{4}He nuclei. ^{4}He is a sink for neutrons. Assuming that all the reactions ended

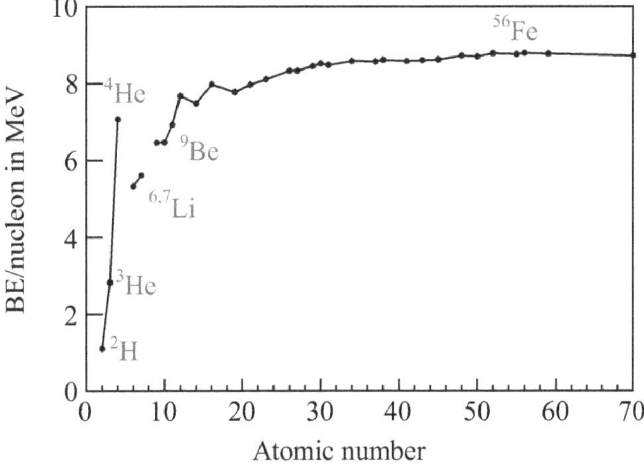

Figure 14.1. Binding energies per nucleon for stable nuclides against atomic number. Above atomic number 56 a slow continuous decline in binding energy per nucleon with atomic number commences: a few nuclides are shown to indicate this.

in ^{4}He then the fraction, by mass, of baryonic matter converted to ^{4}He in BBN would be

$$Y = 2n_n/[n_n + n_p] = 0.28,$$

which amounts to an upper limit on ^{4}He production in BBN. Very little ^{6}Li and ^{7}Li are produced in BBN: the production of nuclei with higher atomic numbers only took place much later, in stars and stellar explosions.

14.6 Primordial Abundances: Prediction and Measurement

Calculations of the nuclide yields after BBN are made nowadays with standard codes originated by Wagoner (969). The calculations make use of the ΛCDM model of cosmology, the standard model of particle physics, and measured nuclear reaction rates (Cyburt et al., 2016). Dark matter and dark energy do not exert any effect. Consequently it is a parameter-free calculation which should immediately fit the data. However the primordial yields, that is the yields directly after BBN, have to be deduced from what we observe in galaxies and stars which came into existence at least a billion years later. The degree of reprocessing in stars and in stellar explosions has to be compensated for.

With a binding energy per nucleon of only 1.1 MeV deuterium is easily destroyed in stellar processes. Hence the best estimates for the primordial abundance of deuterium are obtained from studying the spectra of the earliest sources. As we now explain the measurements, surprisingly, use absorption spectra. Quasars, described in Section 8.10, are bright sources visible from the early universe emitting continuous spectra. Their spectra are punctuated by the absorption lines due to clouds of deuterium and hydrogen lying in their path to the Earth. The depths of absorption lines at the Lyα transitions in ^{1}H and ^{2}H (121.567 nm and 121.435 nm, respectively) are proportional to the abundances of the two isotopes. Conveniently, the redshift of the lines fixes the time at which the gas clouds existed. Hence, both the era and the relative abundance of the isotopes are measurable. The ratio of abundances is then extrapolated to the redshift of the BBN.

^{4}He is made in stellar processes as well as in BBN. Consequently, measurements of the abundance of ^{4}He at later eras have to be extrapolated back to the era of BBN. Measurements are made on ^{4}He and ^{1}H emission lines in around 100 different regions of interstellar singly ionized hydrogen, *HII regions*, beyond our Galaxy.[2] The primordial abundance of ^{7}Li is only a few parts in 10^{10} compared to hydrogen. To complicate matters, ^{7}Li is also produced in stars and in reactions initiated by cosmic rays.

The predicted and observed abundances of the light nuclides after BBN are shown in Figure 14.2 as a function of the baryon to photon ratio, η (Tanabashiet al., 2018). Apart from Y, the mass fraction of ^{4}He, the ratios are between numbers of the named nuclides and the number of protons. The ratios span ten orders of magnitude, making it a severe test for the predictions to match the data. The predicted

[2] The ^{1}H emission lines come from the excited minority neutral atoms.

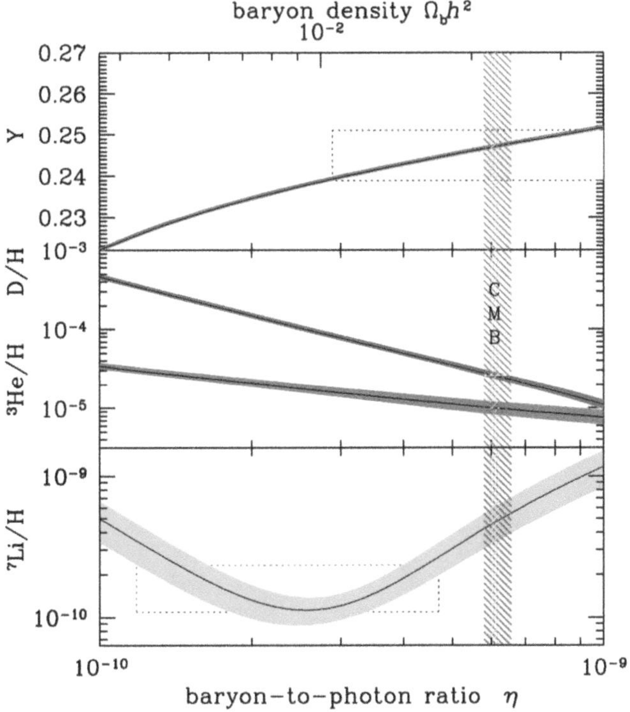

Figure 14.2. Predicted light element abundances relative to hydrogen as a function of the baryon to photon number ratio. The broader (red) vertical band shows the range of the baryon to photon ratio consistent with the D/H measurements. The narrower (blue) band is the expectation from the CMB. The yellow bands show the results of measurements for ^{4}He and ^{7}Li. Figure 23.1 from Tanabashiet al. (2018). Courtesy of the American Physical Society.

deuterium production rate depends strongly on η. Increasing the number of protons causes nucleosynthesis to commence earlier and this increases the conversion to ^{4}He. At the same time the yield of unconverted deuterons and ^{3}He must fall off sharply.[3] The measurements with error estimates are indicated by the colored boxes. The broader (red) vertical band indicates the range of values of η consistent with measured D/H ratio and lies in the range consistent with the less well determined helium ratio. In the case of ^{7}Li there is tension with the other data: the cause is not understood, but could be due to uncertainties in the measurement of ^{7}Li. The narrower (blue) vertical band is the range of η derived from fits to the CMB, which essentially means from the location and magnitude of the acoustic peaks. It is very impressive that the independent determinations of η from the BBN (at 300 s) and CMB (at 365,000 years) agree. This argues for the overall consistency of the picture we have of the development of the universe. The value found for η confirms that baryonic matter can only account for a fraction of the total matter in the universe.

[3] The proportion of ^{3}He has only been determined locally so that no comparison is possible with the predicted ^{3}He/H ratio (sloping red line).

Summarizing: the virtually parameter-free predictions are in good agreement with the ^{4}He and deuterium data and the CBM value for η. The most important conclusion to draw is that the universe did begin with a hot Big Bang. The CMB and BBN emerge as twin pillars underpinning the ΛCDM model of the universe.

Elements beyond helium in the table of elements are produced by nuclear fusion in stars, as well as in nuclear reactions during stellar explosions; cosmologists call these elements metals. Three mass fractions are defined: X the fraction of hydrogen, Y the fraction of helium and Z, the *metallicity* or fraction of metals. In the universe in general $X = 0.75$, $Y = 0.25$ while the metallicity varies with history of the body considered: the reference for metallicity is the solar system with $Z_\odot \sim 0.015$. Iron, which has the largest binding energy per nucleon, is copiously produced, and its abundance relative to hydrogen, (Fe/H), is also used to quantify the metallicity. The ratio of this parameter relative to its value in the solar system is given the symbol [Fe/H]:

$$[\mathrm{Fe/H}] = \log\{(\mathrm{Fe/H})/(\mathrm{Fe/H})_\odot\}.$$

During the period following inflation while the baryon eras listed in Table 14.1, including BBN, were unfolding the dark matter was accumulating in the gravitational potential valleys that had appeared as a result of quantum fluctuations during inflation. Baryonic matter, on being released from the oscillating baryon-photon plasma at decoupling, fell swiftly into these gravitational wells, now enhanced by the accumulation of dark matter in them. The stage was set for baryonic structure formation. This is our next topic.

14.7 Exercises

1. What is the number of ^{4}He nuclei formed per second in the Sun? Take the luminosity of the Sun to be 3.8×10^{26} W, the proton mass to be 938.272 MeV c^{-2}, the ^{4}He mass to be 3727.380 MeV c^{-2} and the electron mass to be 0.511 MeV c^{-2}.
2. What fraction of the hydrogen in the Sun is converted to ^{4}He in 1 Gyr?
3. How many solar neutrinos pass through you per second?
4. How many of these interact inside you? Assume the body to be simply water and the cross-section for any neutrino interaction with a nucleon or an electron to be approximately 10^{-47} m^2, for neutrinos of energy around 1 MeV.

Further Reading

Liddle A 2015 *An Introduction to Modern Cosmology* (3rd ed.; New York: Wiley) (Oxford: Oxford Univ. Press). This text provides a compact introduction.

Coles P and Lucchin F 2002 *Cosmology: The Origin and Evolution of Cosmic Structure* (2nd ed.; New York: Wiley). Clear account of the BBN and associated topics.

References

Cyburt, R. H., Fields, B. D., Olive, K. A., & Yeh, T.-H. 2016, RvMP, 88, 015004
Tanabashi, M., Hagiwara, K., Hikasa, K., et al. 2018, PhRvD, 98, 030001
Wagoner, R. V. 1969, ApJS, 18, 247

AAS | IOP Astronomy

Introduction to General Relativity and Cosmology
(Second Edition)

Ian R Kenyon

Chapter 15

Structure Origins

15.1 Introduction

The density fluctuations of radiation and of baryonic matter at decoupling were miniscule, $\sim 10^{-5}$. Since then the clumping together of matter has increased enormously. We may compare local matter densities today to the critical density. The density is about 10^2 higher in galaxy clusters; around 10^5 higher in galaxies; and in stars like the Sun it is 10^{29} times higher. Most of the matter in and around a galaxy or galaxy cluster is dark matter, called its *dark matter halo*. Such a halo envelops the visible structure, rather than standing apart like a saint's halo.

This chapter, and the final two chapters, cover the evolution from the tiny perturbations existing in the universe after inflation to the stars, galaxies and clusters of galaxies we see today. The early phase dominated by the growth of dark matter fluctuations is described in this chapter.

The initial density fluctuations gravitationally attracted more matter, the potential wells deepened, and equally the depleted regions lost matter. Countering this the expansion of the universe acted to disperse matter: the *Hubble drag*. Gravitational collapse in a static and an expanding universe are tackled in the first sections here. Following this the growth of the majority dark matter perturbations is followed until they reached equilibrium. Until decoupling the baryons oscillated within the gravitational wells formed by accumulations of dark matter. Once released, the baryonic matter then fell into these gravitational potential wells and enhanced their growth.

If there had only been dark matter present it would simply have become progressively hotter until its kinetic motion prevented any further gravitational contraction, the state of *virial equilibrium*. Baryonic matter differs from dark matter by coupling to the electromagnetic field. Ions, atoms, molecules, and electrons can all radiate photons which carry off energy. Thus, concentrations of baryons in the

doi:10.1088/2514-3433/acc3ffch15

gravitational potential wells were able to cool by emitting radiation that escaped the wells. Baryonic matter could thus continue to contract and so build the dense structures present today: the stars, the galaxies, and the clusters of galaxies. These are still embedded in their more massive dark matter halos.

In the previous chapter we saw that seconds after the Big Bang baryonic matter was already non-relativistic. We know less, directly, about the state of dark matter particles because they have never been detected. Dark matter only interacts gravitationally, its interactions are said to be *collisionless*. Both baryonic and dark matter were non-relativistic throughout the era of structure formation described here. We show that if dark matter particles had been so light as to be relativistic their streaming like photons would have dispersed the overdense accumulations of matter that, in fact, grew denser and became nurseries of galaxies of stars. There is a hierarchy of structures with stars forming before galaxies, and galaxies forming before clusters of galaxies. We shall show in this chapter that this hierarchical ordering provides critical evidence that dark matter is non-relativistic: that it is *cold dark matter* (CDM), rather than relativistic and *hot dark matter* (HDM). The ΛCDM model will be seen to make predictions that account for the large-scale structures existing today.

The oscillations of the baryon–photon plasma were necessarily imprinted on baryonic matter as well as on the CMB at decoupling. At the end of this chapter the detection of these *baryon acoustic oscillations (BAO)* in the large-scale structure of the universe is described. This provided a critical quantitative test of the ΛCDM model of the evolution of the universe from decoupling at 365,000 years after the Big Bang to the present, 13.5 Gyr later.

15.2 Gravitational Instability

Following the same lines as before matter is treated on the cosmic scale as a fluid. In the case of dark matter a collisionless fluid, one without internal pressure ($w = 0$). The free fall time to reach the center of an isolated uniform sphere of fluid of density ρ under the gravitational attraction alone would be

$$t_{\text{free}} \approx \sqrt{1/[G\rho]}. \tag{15.1}$$

This isn't the whole story describing collapse. The obvious neglected item is the conversion of gravitational potential energy into kinetic energy during the collapse. The increased average speed of the fluid particles would on its own lead to expansion. Evidently the kinetic energy builds up and resists the gravitational contraction, and this resistance acts whether the matter is baryonic *or* dark matter. How the competition works out depends on how fast the pressure waves travel compared to the free fall velocity. If the volume is tiny the pressure waves will win and resist contraction. If the volume is very large gravitation produces contraction, which pushes up the density, which reduces the free fall time, which make contraction unstoppable. The limiting size and mass at which the two forces are in balance are named after the originator of this idea affecting our understanding of the formation of structures in the universe: the Jeans length and the Jeans mass.

In baryonic matter pressure waves travel at the speed of sound c_s, so that the time for a pressure wave to cross a sphere of diameter d

$$t_s \approx d/c_s. \tag{15.2}$$

For a sphere that is overdense and on the edge of stability $t_s \approx t_{free}$. The length at which the pressure wave crossing time equals the free fall time is called the *Jeans length*, λ_J: exact calculation[1] gives

$$\lambda_J = c_s\sqrt{\pi/[G\rho]}. \tag{15.3}$$

There is a corresponding limiting *Jeans mass*

$$M_J = \frac{4\pi}{3}\rho\left[\frac{\lambda_J}{2}\right]^3 = \frac{\pi c_s^3}{6}\sqrt{\left[\frac{\pi^3}{G^3\rho}\right]}. \tag{15.4}$$

If the extra mass due to the overdensity of a region exceeds the Jeans mass it will collapse.[2] Given the right conditions the collapse may go as far as concentrating matter in an overdense region that becomes a galaxy and the stars within it. The speed of sound is seen to be a critical determinant of the response of a local overdense region; whether it collapses or supports oscillations. The speed of sound is given by

$$c_s = c\sqrt{\frac{\partial p}{\partial \varepsilon}} = c\sqrt{\frac{\dot{p}}{\dot{\varepsilon}}}, \tag{15.5}$$

where p is the pressure and ε the energy density, and a dot above indicates the time derivative of the parameter. Applying the fluid Equation (10.8) to the plasma

$$\dot{\varepsilon} = -3H(p + \varepsilon) = -H(4\varepsilon_r + 3\varepsilon_b). \tag{15.6}$$

Also for radiation

$$\dot{p}_r = \dot{\varepsilon}_r/3, \tag{15.7}$$

and

$$\frac{\dot{\varepsilon}_r}{-H} = -a\frac{d\varepsilon_r}{da} = 4\varepsilon_r. \tag{15.8}$$

Collecting terms, the speed of sound in the plasma is

$$c_s = \frac{c}{\sqrt{3}}\sqrt{\frac{4\varepsilon_r}{4\varepsilon_r + 3\varepsilon_b}}. \tag{15.9}$$

[1] Peebles P J E 1993 *Principles of Physical Cosmology* (Princeton, NJ: Princeton Univ. Press) 116.

[2] This and succeeding results in the following chapters are somewhat fluid. The choice of the λ_J to be the diameter rather than the radius is sometimes made, shifting the Jeans mass by a factor 8.

Over the life of the plasma

$$c_s(\text{plasma}) \approx c/\sqrt{3}, \tag{15.10}$$

because the photons overwhelm the roughly 10^{10} times fewer baryons. Hence the Jeans length in the plasma is very similar to the Hubble length

$$c/H = c\sqrt{\frac{3}{8\pi G\rho}}.$$

After decoupling the non-relativistic baryons can be treated as an ideal gas, then

$$c_s(\text{baryonic}) = \sqrt{\left[\frac{k_B T}{\mu m_H}\right]} = \sqrt{\left[\frac{k_B T}{\mu \varepsilon_H}\right]}\, c, \tag{15.11}$$

where m_H is the mass of atomic hydrogen and ε_H its rest energy. μ is the molecular weight of the gas. At decoupling the temperature T_{dec} was 2974 K. The kinetic energy of particles at that time, $(3/2)k_B T_{dec} = 0.40$ eV, while the binding energy of molecular hydrogen 4.52 eV. Any hydrogen molecules once formed thus become less and less likely to be dissociated the more the universe cools. Here for simplicity we therefore take $\mu = 2$. Hence the speed of sound in baryonic matter soon after decoupling became

$$c_s(\text{baryonic}) = 3.50 \times 10^3 \text{ m s}^{-1} = 1.17 \times 10^{-5}\, c. \tag{15.12}$$

We now calculate the Jeans masses for baryonic matter just before and after decoupling. Directly before decoupling the baryons' mass density was

$$\rho_b = \rho_c \Omega_{b0}(1090)^3 = 5.33 \times 10^{-19} \text{ kg m}^{-3}.$$

Inserting this density and a speed of sound $c/\sqrt{3}$ into Equation (15.3) the Jeans length for baryons is 5.15×10^{22} m or 1.66 Mpc. Using Equation (15.4) the Jeans mass is 3.81×10^{49} kg or $1.90 \times 10^{19}\, M_\odot$. Immediately after decoupling the speed of sound dropped sharply and the Jeans length fell by a factor

$$c_s(\text{plasma})/c_s(\text{baryonic}) = 4.93\; 10^4, \tag{15.13}$$

to 33.7 pc. In turn the Jeans mass tumbled precipitously by this factor cubed, to only $1.59 \times 10^5\, M_\odot$. This was a dramatic change: it meant that after decoupling it now became possible for a typical galaxy such as our own Galaxy (mass $1.5 \times 10^{12}\, M_\odot$) to collapse gravitationally. During the matter era after decoupling, the temperature varied as a^{-1}, the density of matter varied like a^{-3}: as a result the Jeans mass for baryonic matter in this era, given by Equation (15.4), would appear to stabilize at $5.2 \times 10^4\, M_\odot$. However we have still to take account of the fact that baryons in a collapsing body can lose energy by emitting radiation that escapes from the body, taking its energy with it. We shall see that stars are born in giant molecular clouds, whose temperatures are maintained in the region of 10 K. These clouds are in turbulent motion altering the local density. Wherever the density is increased sufficiently the Jeans mass falls and fragmentation occurs.

The analysis of gravitational collapse is easily extended to cover dark matter. In the case of dark matter the dispersion of the velocity distribution σ replaces the speed of sound in the formulae. This follows because any fluctuation in a collisionless fluid can disperse over a distance s in a time s/σ. Dark matter *streaming* would disperse the dark matter structures smaller than the corresponding Jeans length. The evidence from the size distribution of existing structures in the universe is used in Section 15.6 to argue that dark matter became non-relativistic very early in the life of the universe.

15.3 Instability in an Expanding Universe

Gravitational self-attraction has always acted to collapse any overdense region, making it denser, while the expansion of the universe acted to expand and diffuse the matter it contains. Here we deduce the evolution under these influences of density fluctuations left in the universe after the inflationary era. The parameter used to quantify the local overdensity is δ: then if the mean matter density is ρ the local density is $\rho(1 + \delta)$. By the time of decoupling δ had only grown to $\sim 10^{-5}$, thus in the early universe

$$\delta \ll 1.0.$$

This limitation simplifies the calculation of the growth of fluctuations in that era. All second and higher order terms in δ and its derivatives may be neglected, making this a *linear regime*. Because the overdensities and velocities in the linear regime were small we can analyze it using Newtonian kinematics.

After reentering the horizon following inflation the dark matter and baryonic matter behaved very differently up to the time baryonic matter decoupled from the CMB. As we have seen the baryons were trapped in a plasma that oscillated within the gravitational potential wells formed by the dark matter overdensities. Here we follow the history of growth of the dark matter overdensities, which became the nurseries for the baryonic structures we see today. Facets of the analysis can be applied to baryonic matter after decoupling.

The regions of overdensity of matter in the universe are assumed to be uniform and spherical in shape: this is called the *spherical top hat* model. Obviously the actual fluctuations will have structure and less symmetry, but this simple picture enables us to capture the essentials of instability. Birkhoff's theorem (Birkhoff & Langer, 1923) tells us that we can treat such a spherically symmetric region as gravitationally independent of the rest of the universe. The requirement of uniform overdensity in the top hat model rules out pressure variation, so the fluid is effectively pressureless. This is a useful approximation for describing the evolution of cold dark matter. The effect of including pressure on the final result will be noted.

First note that M, the mass enclosed in the top hat, does not change as the universe expands:

$$M = \frac{4\pi}{3} r^3 \rho(1 + \delta),$$

so that to first order in δ and taking into account that $\rho \propto a^{-3}$

$$r = Ka/(1 + \delta)^{1/3} = Ka(1 - \delta/3), \tag{15.14}$$

where K is constant. Differentiating this twice with respect to time gives

$$\ddot{r} = K[\ddot{a}(1 - \delta/3) - 2\dot{a}\dot{\delta}/3 - a\ddot{\delta}/3]. \tag{15.15}$$

Dividing this equation by r using Equation (15.14) and ignoring terms in the second order like $\delta\dot{\delta}$,

$$\frac{\ddot{r}}{r} = \frac{\ddot{a}}{a} - \frac{2}{3}\frac{\dot{a}}{a}\dot{\delta} - \frac{1}{3}\ddot{\delta}. \tag{15.16}$$

At the surface of the spherical top hat the acceleration inward is given by

$$\ddot{r} = -GM/r^2 = -\frac{4\pi}{3}Gr\rho(1 + \delta).$$

This gives

$$\frac{\ddot{r}}{r} = -\frac{4\pi}{3}G\rho(1 + \delta),$$

while the acceleration Equation (10.7) for non-relativistic matter reduces to

$$\frac{\ddot{a}}{a} = -\frac{4\pi}{3}G\rho.$$

Making these substitutions in Equation (15.16) produces

$$-\frac{4\pi}{3}G\rho(1 + \delta) = -\frac{4\pi}{3}G\rho - \frac{2}{3}\frac{\dot{a}}{a}\dot{\delta} - \frac{1}{3}\ddot{\delta}.$$

Multiplying by 3 and putting the Hubble constant in place of $\dot{a}/a$ gives our final compact result[3]

$$\ddot{\delta} = 4\pi G\rho\delta - 2H\dot{\delta}. \tag{15.17}$$

The second term on the right-hand side of Equation (15.17) is the damping of the growth of density fluctuations due to the expansion of the universe: it is known as the *Hubble drag*. Then using Equation (10.12) to replace the density in Equation (15.17) gives

$$\ddot{\delta} + 2H\dot{\delta} - (3/2)H^2\,\Omega_{\mathrm{m}}\delta = 0. \tag{15.18}$$

This equation yields a time dependence of growth that is very different in the radiation- and matter-dominated eras. We treat the radiation-dominated era first.

---

[3] The effect of pressure would add a term $[c_s^2/a^2]\nabla^2\delta$ to the right-hand side. See Chapter 5 in Peebles P J E 1993 *Principles of Physical Cosmology* (Princeton, NJ: Princeton Univ. Press).

During this era the free fall time $t_{ff} \propto (G\rho_m)^{-1/2}$ was much longer than the Hubble time, $t_H \propto (G\rho_r)^{-1/2}$, which is the expansion time of the universe. The universe was expanding so fast that the growth of the matter overdensities stalled, which is called the Meszaros effect. At sufficiently early times in the radiation era we can drop the final term in Equation (15.18) and take $H = 1/(2t)$:

$$\ddot{\delta} = -\dot{\delta}/t.$$

Thus an approximate solution is

$$\delta = A + B \ln t = C + D \ln a,$$

with constants A, B, C, and D. More generally, changing variables in Equation (15.18) from t to $y = a/a_{rm}$ gives, after some manipulation,

$$\delta'' + \frac{2 + 3y}{2y(1 + y)}\delta' - \frac{3}{2y(1 + y)}\delta = 0, \tag{15.19}$$

where primes indicate differentiation with respect to y. There is a growing solution, which also has limited growth

$$\delta \propto 1 + (3/2)a/a_{rm}, \tag{15.20}$$

only approaching linear growth with a at the transition to matter dominance.

Moving on to the matter-dominated era we can use Equation (10.23) to replace Hubble's constant in Equation (15.18), giving

$$\ddot{\delta} + 4\dot{\delta}/(3t) - 2\delta/(3t^2) = 0. \tag{15.21}$$

Substituting a power law solution in the form [δ = constant times t^n] gives two solutions, which you may check: $n = -1$ or $n = 2/3$. The former solution dies away leaving

$$\delta = \text{constant } t^{2/3}. \tag{15.22}$$

In a matter-dominated universe using Equation (10.21) gives

$$\delta \propto t^{2/3} \propto a = 1/(1 + z). \tag{15.23}$$

Accordingly the fluctuations would have grown by a factor of 1091 between decoupling and the present. This would only boost the fluctuations in matter from 10^{-5} at decoupling to 10^{-2} at the present time. Vastly more compaction is required to reach even the overdensity 10^2 of galaxy clusters. Because dark matter was immune to the plasma oscillations of radiation and baryonic matter, dark matter overdensities commenced growing from much earlier than decoupling. Without this boost there would be no viable explanation for how today's structures developed in the available time.

Figure 15.1 sketches the behavior of fluctuations in dark matter, baryonic matter, and radiation up to and beyond decoupling. After decoupling baryonic matter fell into the gravitational potential wells provided by dark matter and thereafter tracked dark matter. The density fluctuations continued to grow beyond the figure's range:

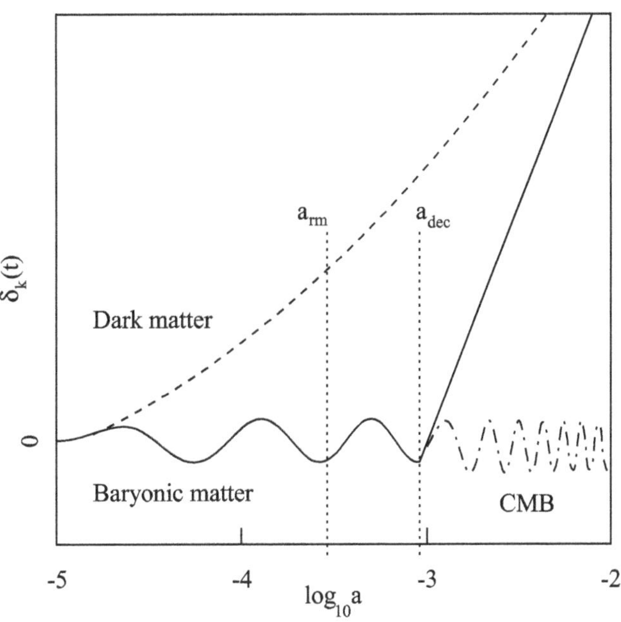

Figure 15.1. Sketch of the early growth of fluctuations showing a representative mode: the solid line for baryonic matter both in the plasma and after decoupling, the broken line for dark matter, and the dot-dash line for the CMB. After decoupling the growth of baryonic matter structures tracks that of dark matter. The dotted vertical lines mark the times of matter–radiation equality and of decoupling.

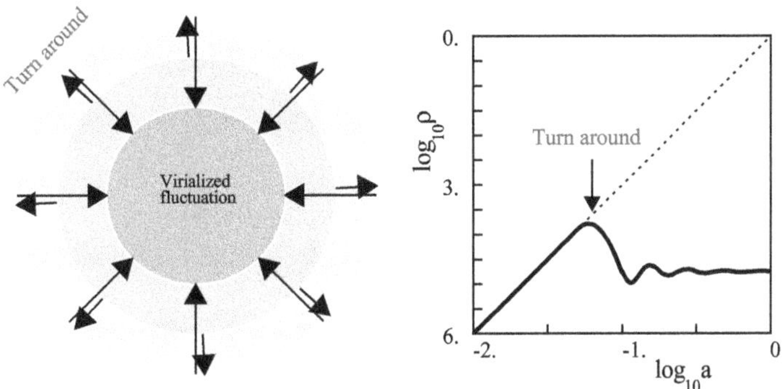

Figure 15.2. A sketch of the progress of a uniform spherical overdensity fluctuation that exceeds the Jeans mass (using the top hat model). The fluctuation expands until turn-around, and then collapses and virializes at half the turn-around radius. In the graph log ρ is the log of the density (relative to the mean density of the universe today) plotted against the log of the scale factor, beyond decoupling. The solid curve is for a virialized region; the broken curve is for the universe.

Figure 15.2 shows the subsequent evolution of an overdense fluctuation whose mass became greater than the Jeans mass. The latter figure shows how the fluctuation contracted violently before finally settling into equilibrium. This phase of structure formation, still driven by dark matter, is the next topic.

15.4 Collapse and Virialization

Once the local overdensity δ had become large, let us say above unity, we can no longer treat the fluctuation as small. We therefore go back to Friedmann's Equation (10.5), but omit the dark energy, as being unimportant at early enough times

$$\dot{a}^2 = 8\pi G\varepsilon a^2/[3c^2] - \kappa c^2. \tag{15.24}$$

The local imbalance between the expansion and the gravitational attraction of matter resulted in a local positive curvature κ. As before we analyze the simple case that the overdense region was spherical and uniformly overdense in an otherwise uniform universe: the spherical top hat model. Again, as before, Birkhoff's theorem tells us that we can treat such a spherically symmetric region as gravitationally independent of the rest of the universe. Now define

$$A^2 = 8\pi G\varepsilon a^3/[3c^2].$$

During virialization matter dominates: εa^3 will be constant, and therefore A will be constant. Substituting A into Friedmann's equation gives:

$$\dot{a}^2 = A^2/a - \kappa c^2. \tag{15.25}$$

It is convenient to choose coordinates such that κ is arbitrarily set to unity. This leaves the outcome of virialization unaffected. Then the equation to be solved becomes

$$\dot{a}^2 = A^2/a - c^2. \tag{15.26}$$

Its parametric solutions are

$$\begin{aligned} a &= \frac{A^2}{2c^2}[1 - \cos\phi], \\ t &= \frac{A^2}{2c^3}[\phi - \sin\phi]. \end{aligned} \tag{15.27}$$

The top hat region continues to expand, but less rapidly than the background. It reaches its maximum size, at *turn-around*, when $\phi = \pi$. The time is then $t_{ta} = \pi A^2/[2c^3]$ and the scale factor $a_{ta} = A^2/c^2$. In the surrounding flat universe Equation (15.25) simplifies to

$$\dot{a} = A/\sqrt{a}. \tag{15.28}$$

Integrating this equation up to turn-around gives the scale factor of the universe at that point, t_{ta},

$$a_{univ} = [3At_{ta}/2]^{2/3} = [3\pi/4]^{2/3}A^2/c^2.$$

The density contrast turn-around is thus

$$\rho_{ta}/\rho_{univ} = [a_{univ}/a_{ta}]^3 = 9\pi^2/16.$$

After turn-around the density contrast between the top hat region and the surrounding universe increased as the overdensity collapsed. During the collapse the total energy was conserved: gravitational energy was converted to kinetic energy. A final equilibrium state was attained when the kinetic motion resisted further contraction. The *virial theorem*, described in Appendix G, relates the mean kinetic energy of the matter particles to their mean potential energy at equilibrium:

$$\langle \mathrm{PE}_{\mathrm{vir}} \rangle + 2\langle \mathrm{KE}_{\mathrm{vir}} \rangle = 0. \tag{15.29}$$

The gravitational energy of a sphere of uniform density ρ_s and radius R, taking $M(<r)$ to be the mass within radius $r < R$ from the center, is

$$\begin{aligned} \langle \mathrm{PE} \rangle &= -\int_0^R \frac{GM(<r)}{r}(4\pi r^2 \rho_s \, \mathrm{d}r) \\ &= -\int_0^R \frac{16}{3} G\pi^2 \rho_s^2 r^4 \mathrm{d}r = -\frac{3GM^2}{5R}, \end{aligned} \tag{15.30}$$

where M is the mass of the whole sphere. For a uniform sphere the virial equation gives

$$\langle \mathrm{KE}_{\mathrm{vir}} \rangle = 3GM^2/[10R_{\mathrm{vir}}], \tag{15.31}$$

where M is again the mass of the sphere, essentially that of the fluctuation, and R_{vir} its radius. The total energies at turn-around and at virialization are equal. Ignoring the small kinetic energy at turn-around, energy conservation gives

$$\langle \mathrm{PE} \rangle_{\mathrm{ta}} = \langle \mathrm{PE} \rangle_{\mathrm{vir}} + \langle \mathrm{KE} \rangle_{\mathrm{vir}} = (1/2)\langle \mathrm{PE} \rangle_{\mathrm{vir}}, \tag{15.32}$$

where the second equality comes from applying the virial theorem, Equation (15.29). From this we learn that $R_{\mathrm{vir}} = R_{\mathrm{ta}}/2$, and that the density of the fluctuation increased by a factor 8. In parallel the density of the universe fell while the fluctuation virialized. Virialization occurs when $\phi = 2\pi$ so that $t_{\mathrm{vir}} = 2t_{\mathrm{ta}}$. Thus the universe expanded by a factor $2^{2/3}$ and its density fell by a factor 4. Overall the contrast between the density of the fluctuation and that of the surrounding universe grew by a factor

$$\mathrm{contrast} = \frac{9\pi^2}{16} \times 8 \times 4 \approx 200.$$

This is very similar to the contrast between the galaxy cluster density, in baryonic and dark matter, and the corresponding mean density as observed in the universe today. Further contraction to reach the overdensities of galaxies and stars depended on baryons being able to shed energy in electromagnetic radiation, in photons that escape the contracting mass—something dark matter cannot do.

The time it takes for a structure to collapse can be estimated using the free fall or dynamic time $\sqrt{1/G\rho}$. This neglects the dissipative processes involved, but it serves to show that the galaxies we see, whose overall matter content is denser, collapse sooner than the galaxy clusters. For a galaxy cluster with a density ~200 times the density of the universe the estimate is ~3 Gyr. Thus galaxy clusters are still forming. In the case of the Milky Way, taking a mass of $1.5 \times 10^{12}\ M_\odot$, the virial radius comes out at 18 kpc. This compares with the 8 kpc that the Sun lies from the center of the Galaxy. The Milky Way density is now higher, $\sim 10^5$ times the density of the universe thanks to the contraction after virialization made possible by the radiative cooling of the baryons. This contraction was happening in parallel and faster than the virialization of larger structures.

We can also make a crude estimate of the redshift z, at the time when virialization occurred. At virialization the density contrast between matter in the Milky Way Galaxy and in the universe as a whole was about 200. Currently the contrast is 10^5. Now the density of matter has varied like $(1 + z)^3$ for most of the time involved. Hence, virialization would have taken place when

$$(1 + z)^3 = 10^5/200,$$

making z around 6.9.

15.5 Baryonic Gas Cloud

The virial theorem can be applied to the gas cloud within a dark matter halo. In this case Equation (15.31) becomes

$$\langle KE \rangle_b = 3GMM_b/[10R_{vir}], \tag{15.33}$$

where M is the total mass and M_b the baryonic mass within the radius R_{vir}. The kinetic energy of the baryons

$$\langle KE \rangle_b = [3/2]\frac{k_B T_{vir} M_b}{\mu m_H},$$

where, as before, μ is the molecular weight and m_H the mass of the hydrogen atom. Then inserting this expression in Equation (15.33) gives

$$M_b = f_b M = \frac{5f_b R_{vir} k_B T_{vir}}{G\mu m_H}, \tag{15.34}$$

where f_b is the fraction of baryonic matter. Now

$$M_b = (4/3)\pi R_{vir}^3 \rho_b, \tag{15.35}$$

where ρ_b is the baryon density in the cloud. Eliminating R_{vir} using the last two equations gives

$$M_b = \left[\frac{5f_b k_B T_{vir}}{G\mu m_H}\right]^{3/2} \left[\frac{3}{4\pi\rho_b}\right]^{1/2}. \tag{15.36}$$

Note that the baryonic virial mass is very similar to the Jeans mass of Equation (15.4). A virialized baryonic cloud can collapse by radiating the energy released; but a virialized dark matter halo cannot contract further. Now $\rho_b = n\mu m_H$ where n is the number density of baryons. Then inserting the constants, the virial baryon mass is

$$M_b = 10^5 T_{vir}^{3/2} f_b^{3/2} n^{-1/2} M_\odot. \tag{15.37}$$

15.6 Growth of Structures

We now look at how large-scale structures developed in the later universe from the density fluctuations revealed by the CMB. Two key features that should carry through in the baryon structures are the acoustic oscillations, and the near scale invariant (Harrison–Zeldovich) power spectrum. Then the basic question is this: does the ΛCDM model predict correctly the evolution of such features to the low redshift era, ~13.8 Gyr later? In order to get answers the structures observed at low redshift have been analyzed in a way similar to the method applied to the CMB. An essential step is to Fourier analyze the angular distribution of matter across the sky: this gives the angular frequency distribution. This is the same process as projecting a train of sound waves into its frequency spectrum. If there are structures of size r, however distributed across the sky, the Fourier component of angular frequency $k = 2\pi/r$ will be large. A crucial property of the Fourier components of the perturbations is that while a component remains in the linear regime it evolves independently of the others. That means that when the small-scale overdensities collapse the larger-scale overdensities continue to evolve linearly. Otherwise the large-scale structures at low redshift would not preserve their features post decoupling. Explicitly for perturbations $\delta(\mathbf{r})$:

$$\delta_{\mathbf{k}} = \int \delta(\mathbf{r}) \exp[i\mathbf{k} \cdot \mathbf{r}] \, \mathrm{d}\mathbf{r}, \tag{15.38}$$

where $\mathbf{k}$ is the wavenumber with a corresponding wavelength $2\pi/k$. The scale is chosen so that the large volume integrated over is unit volume. The reverse transformation is

$$\delta(\mathbf{r}) = \frac{1}{(2\pi)^3} \int \delta_{\mathbf{k}} \exp[-i\mathbf{k} \cdot \mathbf{r}] \, \mathrm{d}\mathbf{k}. \tag{15.39}$$

The distribution of fluctuations in the CMB is independent of direction, so that it and its Fourier transform are isotropic. It follows that variation with k rather than with $\mathbf{k}$ is adequate to describe the distribution fully. The quantum fluctuations during inflation had a Gaussian distribution. As discussed in Section 12.4 this Gaussianity was inherited by the fluctuations in the CMB, and hence by the baryonic density fluctuations, in both space and in the Fourier transform. The practical measure used to quantify the distribution in k is the *power spectrum* averaged over all orientations of $\mathbf{k}$:

$$P(k) = \langle |\delta_{\mathbf{k}}|^2 \rangle. \tag{15.40}$$

$P(k)$ is parametrized as a power of k

$$P(k) \propto k^n, \tag{15.41}$$

and in the case of fluctuations that are scale invariant the power n would be exactly unity.

Mirroring the analysis applied to the CMB, the correlation between the over-densities at points $\mathbf{x}$ and $\mathbf{x} + \mathbf{r}$ is averaged over a normalizing volume that for convenience we again set to unity:

$$\xi(\mathbf{r}) = \langle \delta(\mathbf{x})\, \delta(\mathbf{x} + \mathbf{r}) \rangle = \int \delta(\mathbf{x})\, \delta(\mathbf{x} + \mathbf{r})\, \mathrm{d}\mathbf{x}. \tag{15.42}$$

According to the Wiener–Khinchin theorem this auto-correlation function and the power spectrum of fluctuations are Fourier transforms of one another

$$\xi(\mathbf{r}) = \frac{1}{(2\pi)^3} \int P(k) \exp[-i\mathbf{k} \cdot \mathbf{r}]\, \mathrm{d}\mathbf{k}. \tag{15.43}$$

Performing the angle integrals gives

$$\xi(\mathbf{r}) = \frac{1}{(2\pi)^3} \int P(k) \frac{\sin kr}{kr} 4\pi k^2 \mathrm{d}k. \tag{15.44}$$

This equation is used as a tool for extracting the power spectrum of spatial correlations in matter in the universe at low redshift. Such correlations have evolved from those in matter at decoupling. In principle the current correlations should be predictable by using the laws of physics to emulate the evolution. Using such techniques, evidence is presented below that the baryonic acoustic oscillations persist in the matter content of the universe at low redshift.

The observed power spectrum of the CMB fluctuations has n slightly smaller than unity. This near scale invariance was interpreted in Chapter 13 as a consequence of slow roll inflation. It is to be expected that the power spectrum of fluctuations in the distribution of matter in the universe should inherit the same functional form. A widely used way to quantify perturbations in the recent universe starts from the fractional excess in the mass in a volume, paralleling the overdensity definition:

$$\delta M = (M - \langle M \rangle)/\langle M \rangle,$$

where $\langle M \rangle$ is the mass calculated taking the mean density of the universe. Then the parameter used is the root mean square deviation of the fractional excess enclosed within a spherical volume V of radius R:

$$\sigma_R^2 = \frac{V}{(2\pi)^3} \int W^2(kR) P(k)\, 4\pi\, k^2\, \mathrm{d}k, \tag{15.45}$$

where $W(kR)$ is the Fourier transform of the spherical top hat distribution with uniform density $3/4\pi R^3$ out to a radius R:

$$W(x) = \frac{3j_1(x)}{x} = \frac{3[\sin x - x \cos x]}{x^3},$$

$j_1(x)$ being a spherical Bessel function. Taking a power spectrum $P(k) \propto k^n$ and using $x = kR$ again gives

$$\sigma_R^2 \propto \frac{9V}{2\pi^2 R^{3+n}} \int_0^\infty x^n [j_1(x)]^2\, \mathrm{d}x \propto R^{-[3+n]}. \tag{15.46}$$

One standard choice for the parameter to quantify the clumpiness of matter is σ_8, the variance of fluctuations of mass within volumes of a size, 8 Mpc, comparable to that of a typical galaxy cluster. A value for σ_8 of ~0.8 has been determined from overall fits in the ΛCDM model, compatible with the CMB, the BAO, and also the SN Ia data discussed in Chapter 17.

With the Harrison–Zeldovich spectrum $P(k) \propto k$ we have $\sigma_R \propto R^{-2} \propto M^{-2/3}$. The small-scale fluctuations have larger deviations from the average mass than do large-scale fluctuations. Thus the probability of a fluctuation becoming non-linear and collapsing is greatest among the small-scale fluctuations. Denser mass fluctuations have additionally shorter free fall times. These two features assist in giving the observed hierarchical clustering in the universe. The smaller scale dark matter accumulations containing the future stars collapsed before the larger-scale accumulations that gave rise to galaxies, and galaxy clusters continue to form into the present era.

In Section 15.2 it was shown that dark matter streams out from fluctuations whose masses are smaller than the Jeans mass. The effect is to wash out such low mass fluctuations. In turn this depresses the power spectrum below the Harrison–Zeldovich prediction at the corresponding large wave numbers. If the particles constituting dark matter had been light, that is *hot dark matter (HDM)* as distinct from CDM then their relativistic streaming at early times in the life of the universe would have dispersed the smaller dark matter accumulations. We now explore the effect of having hot dark matter, choosing a mass $m_{\rm dm} = 2.5\,{\rm eV}\ c^{-2}$ for the individual dark matter particles. As a simple approximation we assume the dark matter particles streamed at the speed of light while the temperature T was high enough that their kinetic energy exceeded their rest energy $m_{\rm dm}c^2$. Below this temperature we also assume that dark matter stopped streaming. Then

$$3k_{\rm B}T \approx m_{\rm dm}c^2, \tag{15.47}$$

which yields a temperature of 9650 K. The universe was then still radiation dominated, so that the relationship between the time t, the scale factor a, and the temperature was

$$t \propto a^2 \propto T^{-2}.$$

As a reference point we take matter/radiation equality, for which Section 12.2 gives $T = 9384$, $a = 1/3444$, and $t = 50,000$ yrs. Thus dark matter would make the transition to non-relativistic motion when

$$t = 50,000\,(9384/9650)^2 = 47,200\ \text{yr}.$$

At that time the particle horizon, the limit of causal contact since the universe began, was $ct = 47,200$ light years or 14.48 kpc. Also

$$a = [9384/9650]/3444 = 1/3540,$$

so that the comoving size of that horizon, that is its size now, would be 51 Mpc. Under our approximation of streaming at velocity c structures of any smaller size would be dispersed. The corresponding mass of dark matter would have been

$$M = (4\pi/3)\rho_{\rm m0}[51\ \text{Mpc}]^3 = 4.43 \times 10^{45}\ \text{kg} = 2.23 \times 10^{16}\ \text{M}_\odot.$$

Thus if the dark matter particles had mass 2.5 eV c^{-2}, then any dark matter fluctuations lighter than 2.23×10^{16} $M_{\odot}$ would be dispersed by the streaming of the dark matter while relativistic. This would eliminate structures as large as the Local Group of galaxies, the one including the Milky Way and M31. HDM would require that the first structures were the largest: smaller structures could only result from the fragmentation of the larger structures, and appear later. This is the reverse of what is observed: galaxies formed at high redshifts while clusters are still forming at present. Particle physics experiments put an upper limit of ~1 eV c^{-2} on the neutrino masses, hence neutrinos are highly relativistic and can be ruled out as the dark matter particles. Figure 15.3 shows the fluctuation power spectra as a function of the fluctuation mass. It shows the initial Harrison–Zeldovich spectrum, and the mass spectra of the surviving fluctuations for CDM and HDM. The mass assumed for the dark matter particles is 2.5 eV c^{-2}. Figure 15.4 shows the same power spectra, now as a function of the wavenumber. Both figures demonstrate the severe and unrealistic loss of galaxy sized accumulations of matter with HDM.

A compilation of measurements of the matter power spectrum from a wide range of cosmological probes is shown in Figure 15.5 (Planck Collaboration et al., 2020). The solid line is an overall fit using the ΛCDM model. The goodness of the fit is strong evidence in particular for CDM and the near scale invariance of perturbations during inflation. As we shall see next, data incorporated in Figure 15.5 provide another, crucial test of how well the ΛCDM model describes the development of the universe.

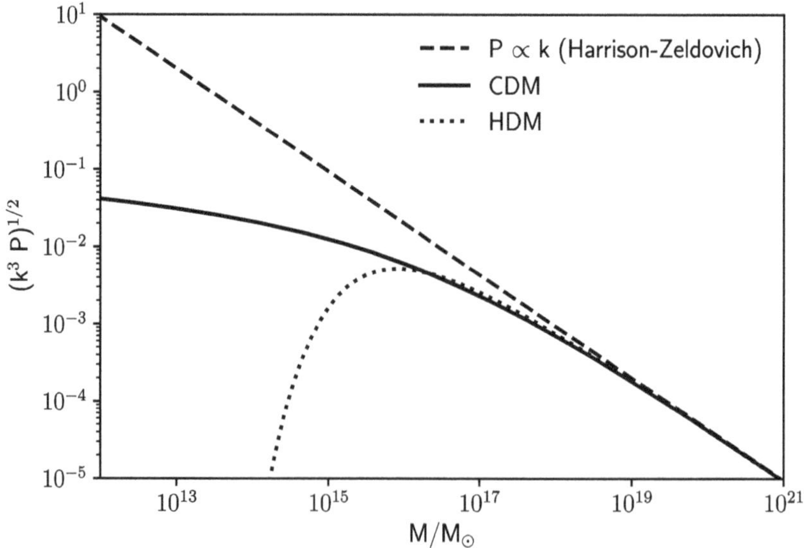

Figure 15.3. The matter power spectrum (in arbitrary units) as a function of mass at matter–radiation equality. The input Harrison–Zeldovich spectrum is shown as well as the spectrum expected after damping by HDM, and after damping by CDM. Courtesy Dr Sean McGee, Institute for Gravitational Wave Astronomy and School of Physics and Astronomy, Birmingham University.

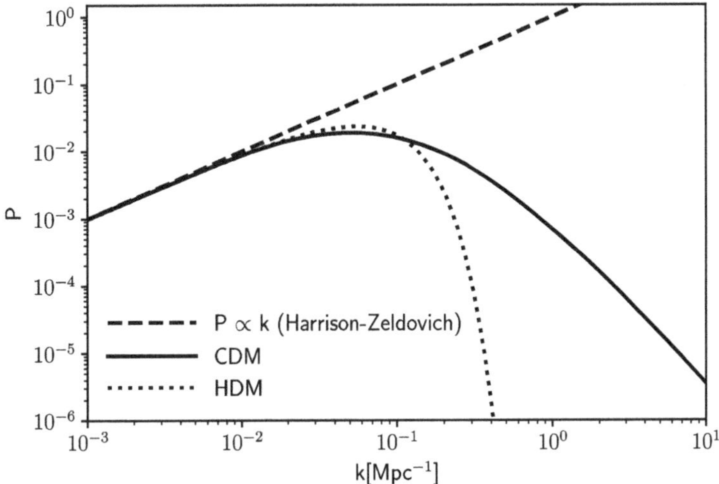

Figure 15.4. Matter power spectrum (in arbitrary units) as a function of wavenumber at radiation-matter equality. The input Harrison–Zeldovich spectrum is shown as well as the spectrum expected after damping by HDM and after damping by CDM. Courtesy Dr Sean McGee, Institute for Gravitational Wave Astronomy and School of Physics and Astronomy, Birmingham University.

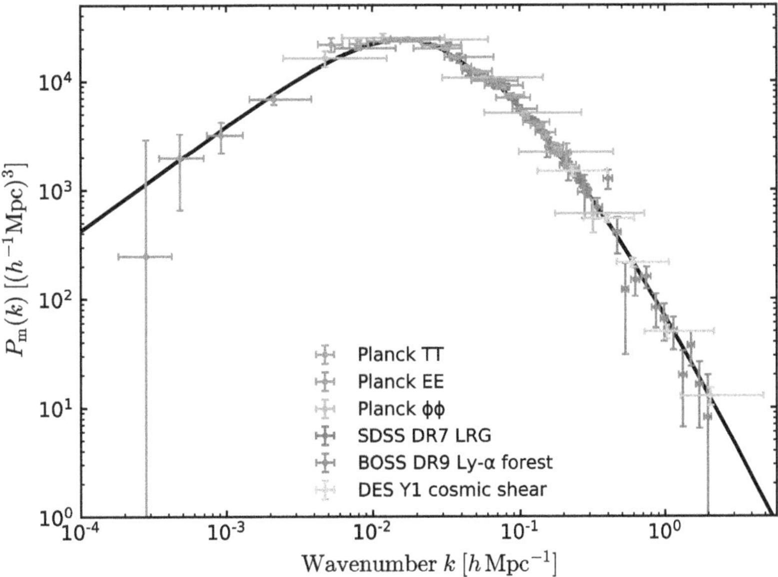

Figure 15.5. Matter power spectrum inferred from different cosmological probes. The curve is from an overall fit to this and other data using the ΛCDM model. SDSS LRG refers to luminous red galaxies from the Sloan Digital Sky Survey. h is the ratio of the observed Hubble constant to 100 km s^{-1} Mpc^{-1}. Figure 19 from Planck Collaboration et al., (2020). Reproduced with permission © ESO.

15.7 Baryon Acoustic Oscillations

According to the interpretation of the CMB given in Chapter 12 oscillations in the baryon plus photon plasma cause the peaks and troughs in the temperature and polarization correlations in Figures 12.3 and 12.6. These are called the baryon acoustic oscillations. The dominant peak was due to those plasma oscillations that had traveled exactly one half wavelength at the time of decoupling. Hence the angle of maximal correlation observed on the Earth is that subtended by this *sound horizon*. As well as being preserved in the CMB this correlation should be preserved equally in the distribution of baryonic matter, and should be detectable in the correlations between overdensities in the distribution of galaxies. One important point to note is that the plasma waves had all orientations in space; consequently correlations in galactic density are expected not only in angular separation but also in redshift along the line of sight.

The sound horizon distance at decoupling, 0.146 Mpc taken from Equation (12.17), has grown with the expansion of the universe. Its present and comoving size is thus

$$r_s \approx 1090 \times 0.150 = 163 \text{ Mpc}. \tag{15.48}$$

The baryons were dragged for some time after decoupling by the more numerous photons, being released at a redshift around 1020. When a careful determination is made using the ΛCDM model the comoving horizon distance falls to 153.19 Mpc. It is clear that in order to detect correlations on this scale a galaxy survey over a volume of Gpc^3 is required. One such survey is the SDSS BOSS survey made with the 2.5 m diameter objective Sloan telescope at Apache Point NM. The survey covered a volume of 13 Gpc^3, spread across 8500 square degrees over the sky, and covering redshifts from 0.2 to 0.7. This survey collected the positions and redshifts of 700,000 galaxy. The WiggleZ and 6dFGS[4] collaborations had comparable sized data sets. The required search for correlations can be three dimensional: in the two orthogonal transverse angles and in redshift separation. The correlation is determined by making a comparison between the observed galaxy distribution and a random galaxy distribution. The latter is simply a randomly redistributed version of the actual sample. The correlation coefficient for galaxy densities at a given separation is then

$$\xi = (DD - 2DR + RR)/RR, \tag{15.49}$$

where DD is the number of data/data coincidences, DR the data/random coincidences and RR the random/random coincidences.

Correlations are measured as a function of the separation $\Delta z/z$ in redshift, and in the orthogonal angles $\Delta\Theta_x$ and $\Delta\Theta_y$. From the observed correlations the size of the sound horizon is determined and this size is compared to that expected if the

[4] The WiggleZ Dark Energy Survey was a redshift survey of 240,000 galaxies using the 3.9 m AAT telescope at Siding Spring Observatory NSW; the 6dF Galaxy Survey observed with redshifts, 136,000 galaxies with the 1.2 m UK Schmidt telescope at the same location.

evolution of the universe follows the prediction of the ΛCDM model. Two conversion factors are required. First, the separation between galaxies at measured redshifts z and $z + \Delta z$, expressed as a comoving distance between the galaxies, is:

$$s_{\Delta z} = \frac{\Delta z}{z}\left[\frac{cz}{H(z)}\right]. \tag{15.50}$$

Second, the comoving distance between galaxies separated by an angle $\Delta\Theta$, expressed in terms of the angular diameter distance d_{A}, is:

$$s_{\Delta\Theta} = \Delta\Theta[(1 + z)d_{\mathrm{A}}]. \tag{15.51}$$

We can also convert volumes by making use of Equations (15.50) and (15.51): the comoving volume enclosed by a difference in redshift of $\Delta z/z$ and the two orthogonal angle separations, $\Delta\Theta_x$ and $\Delta\Theta_y$, is

$$\Delta V = D_V^3(z)[\Delta z/z]\Delta\Theta_x\Delta\Theta_y, \tag{15.52}$$

where the three dimensionally-averaged conversion factor is

$$D_V(z) = [cz(1 + z)^2 d_{\mathrm{A}}^2 H^{-1}(z)]^{1/3}. \tag{15.53}$$

The correlations observed by the SDSS collaboration are shown in Figure 15.6 and reveal clear baryon acoustic oscillations. On the left the correlations are plotted against the comoving distance. The quantity h appearing on the axes is H_0 divided by a nominal value of 100 km s^{-1} Mpc^{-1}. On the right is the Fourier transform, that is the power spectrum plotted against the wavenumber. This latter corresponds to a section of Figure 15.5 at high magnification. The two measurements of the sound horizon from the CMB and from these galaxy overdensity correlations at low redshift are in excellent agreement. The fit shown in Figure 15.6 corresponds to a comoving sound horizon distance r_s, at the time baryons are released from the radiation drag, of 153.19 Mpc.

Figure 12.4 shows the restrictions placed on the acceptable ranges of values of Ω_{m} and Ω_Λ found by comparing the size of the BAO at the time of the CMB and at the era of low z. You can see that because the expansion of the universe since decoupling has been mainly controlled by matter the constraint from BAO is a good guide to the value of Ω_{m}. That there should be a common intersection point in Figure 12.4 is a necessary test of the ΛCDM model, and the intersection fixes the values for Ω_{m} and Ω_Λ. Beyond that, because the intersection point lies on the line $\Omega_\Lambda + \Omega_{\mathrm{m}} = 1.0$, then the universe must be flat.

The successful extrapolation made using the ΛCDM model from the CMB oscillations to galaxy clustering shows that this model is a signal success. One view is to accept that this model provides an accurate description of the development of the universe and make inferences on this basis. The sound horizon at any redshift then becomes a standard ruler supplementing the standard candles as a tool for measuring cosmological distance. One application is to use the measurement of $\Delta z/z$ where Δz spans the sound horizon to determine the value of the Hubble

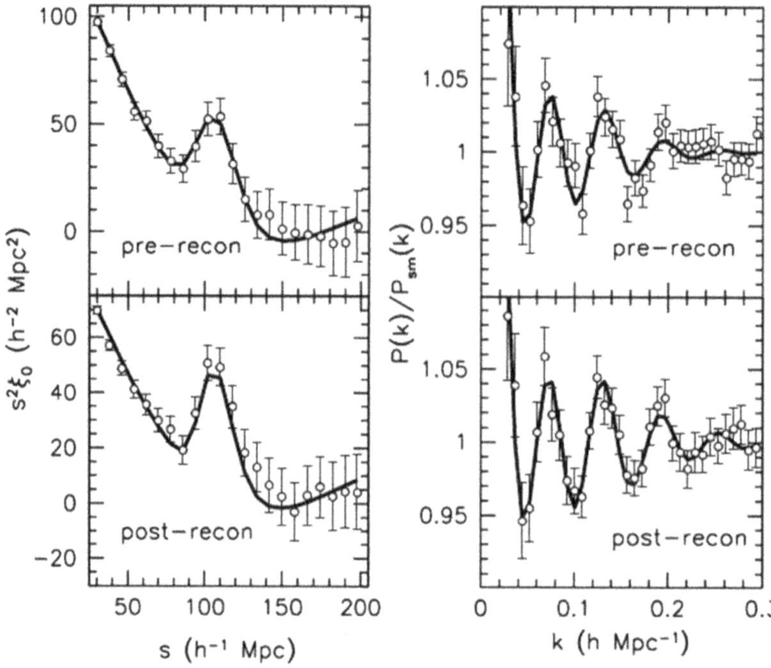

Figure 15.6. Correlations in the galaxy distribution density measured in the Sloan Digital Sky Survey: Baryon Oscillation Spectroscopic Survey. On the left, the correlation coefficient plotted against the comoving separation. On the right, the power spectrum plotted against the wavenumber. Reconstruction removes distortions from bulk flows estimated using a three-dimensional map of the galaxy positions to infer their peculiar velocities. Figure 11 from Anderson et al. (2014). Courtesy: Royal Astronomical Society, London and Oxford Univ. Press, Oxford.

constant at an earlier era, $H(z)$. A second application is to use the measured angular size of the sound horizon to get a precise value of the angular diameter distance. In this way the authors find $d_A(z = 0.57)$ is 1386 ± 26 Mpc and $H(z = 0.57)$ is 94.1 ± 4.7 km s^{-1} Mpc^{-1}. These measurements of distance at relatively large redshifts complement those made with supernovae to establish the current accelerating expansion under the influence of the cosmological constant. This topic is returned to in Chapter 17.

15.8 Exercises

1. Given that HDM was made up of particles of mass 10 eV calculate the following: the time at which the motion becomes non-relativistic, the dark matter density at that time, the mass of dark matter then in causal contact.
2. What percentage error is made in calculating the speed of sound by setting it to $c/\sqrt{3}$ at matter–radiation equality?
3. In Figure 15.1 explain why the spacing of the radiation oscillations tightens up and by how much.

4. The rotation velocity of matter at the outer edge of a spiral galaxy of radius R is v. Show that the rotation period at this radius is very similar to the time it took the galaxy to collapse from a density perturbation. Take the dark matter halo to be spherical.

Further Reading

Rich J 2009 *Fundamentals of Cosmology* (2nd ed.; Berlin: Springer). This text gives much useful coverage.

References

Anderson, L., Aubourg, É., Bailey, S., et al. 2014, MNRAS, 441, 24

Birkhoff, G. D., & Langer, R. E. 1923, Relativity and Modern Physics (Cambridge, MA: Harvard Univ. Press)

Planck Collaboration, Aghanim, N., Akrami, Y., et al. 2020, A&A, 641, A1

Ian R Kenyon

Chapter 16

Baryonic Structures

16.1 Introduction

Immediately following decoupling, the structure of the universe was relatively simple. Overdensities of dark matter originating from the quantum fluctuations during inflation were steadily growing; baryonic matter followed the dark matter. The only radiation was the infrared CMB, steadily cooling as its wavelength expanded with the universe. The universe entered the *dark ages*. In time the overdense concentrations of dark matter virialized, so reaching an equilibrium. Thereafter the accompanying baryonic overdensities could radiate energy away and collapse further. Baryonic matter accumulations reached densities at which nuclear burning is ignited and they became visible stars. The dark ages ended when the first stars began to emit energy as radiation, a period known as *first light*. Ultraviolet radiation, principally from the stars, *reionized* the hydrogen gas that still made up the most part of baryonic matter. What we observe now, as in Figure 1.3, are the structures of visible matter. One early research paper likened the appearance of the universe to foam on the top of washing up. Matter is distributed in filaments that intersect in nodes, and between the filaments are vast, nearly empty voids. Galaxies densely populate the threads and strings and form clusters and superclusters at the nodes.

The processes of structure formation involve physical effects on scales remote from processes observed on Earth. Nonetheless cosmologists have deduced the broad principles involved. Making the assumption that the laws of physics remain valid at all eras and on all scales has led to a consistent description of the processes at work.

Simulations of the evolution of the universe have been helpful in supplementing and interpreting the astronomical observations. One large-scale program of 2001 was carried out at the Max Planck Institute for Astrophysics in Munich. This

doi:10.1088/2514-3433/acc3ffch16

Millennium simulation followed the evolution in the expanding universe, since the decoupling era, of 10^{10} collisionless (dark matter) particles under their mutual gravitational attraction alone: each of mass 10^9 $M_\odot$ so that 100 would equal a dwarf galaxy in mass. The volume was a periodic cubic box of side length 500 Mpc/h, where $h = 100/H_0$, and the initial distribution of perturbations was made scale invariant, taking the amplitude from that seen in the CMB. The results reproduced with surprising fidelity the organization and evolution of the universe as seen at low redshifts. Figure 16.1 shows the simulation of the evolved *cosmic web* at present, the width of the image corresponding to 500 Mpc/h. The brightness indicates the overdensity and hence the regions populated by baryonic matter that we see as stars, galaxies, and clusters. For comparison with the actual universe at low redshift see Figure 1.3.[1] The filaments and nodes are plainly visible, with overdensities comparable to those in nature. The intervening voids are also as large and as sparsely populated as in the real universe. Such good agreement with the observed structure is another indicator of the validity of our model of the universe.

In this chapter the formation of stars, galaxies and clusters of galaxies in their dark matter halos is followed roughly chronologically. Cooling mechanisms and reionization are described first. Then the way that stars form within large molecular

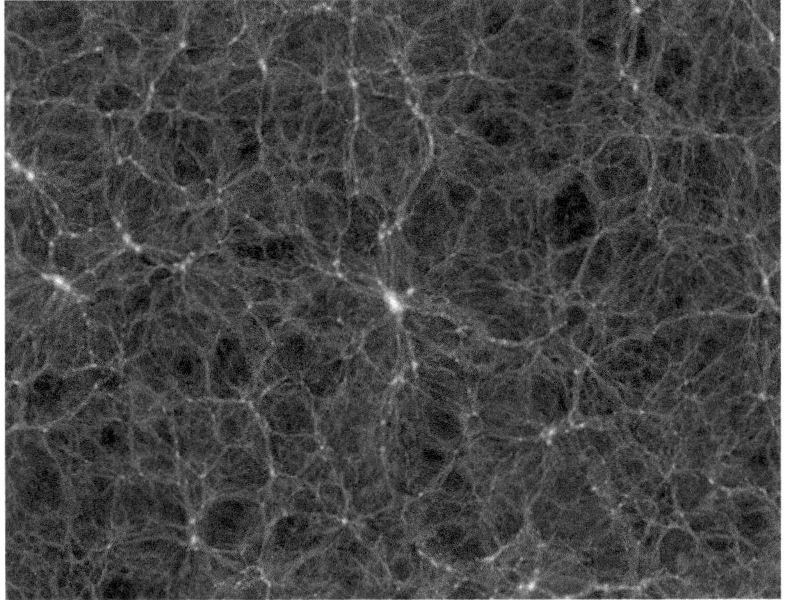

Figure 16.1. Millennium simulation of the evolution of the universe: carried out at the Max-Planck Institute for Astrophysics (Springel et al. 2005). The width of the image is 500/hMpc. h is the usual factor $H_0/100$ km s^{-1} Mpc^{-1}. Courtesy Professor Springel.

[1] More recently the simulation has been extended to include baryons, a project called MilleniumTNG: https://arxiv.org/abs/2210.10060.

clouds is outlined. Next the properties of galaxies, galaxy clusters, and superclusters are discussed. Reference will be made to Chapter 8 in which we got to know the monster black holes powering the AGNs of galaxies, and the mechanism by which stars collapse to compact objects. The Milky Way is looked at in more detail, and the nearby Coma cluster and supercluster. The last, but not least, topic concerns the intergalactic matter which, unexpectedly, holds much more of the baryonic matter than do the stars.

16.2 The Cooling of Baryonic Matter

Unlike dark matter, baryonic matter can radiate photons and thus lose energy. Hence baryonic matter can cool, and contraction can continue after virialization. During this contraction both the density and the temperature rise, and these affect the Jeans mass: using Equations (15.4) and (15.11) we find that

$$M_J \propto T^{3/2}\rho^{-1/2}. \tag{16.1}$$

Whether contraction does happen or not is therefore dependent on how the temperature depends on the density. In the case of the primordial baryonic gas of monoatomic hydrogen atoms: $T \propto \rho^{\gamma-1}$, where γ is the ratio of heat capacities. In adiabatic conditions, i.e., no heat transfer, $\gamma = 5/3$, so that the Jeans mass would increase as the density increased, making for stability. However baryons can cool by radiative processes, keeping the temperature constant, so that $\gamma = 1$. In this latter case the Jeans mass falls, and further collapse becomes possible. As the density increases the Jeans mass falls progressively: galaxy-sized accumulations of matter fragment and it is from such fragments that stars will form.

Several processes are of importance in cooling baryonic matter fluctuations. One requirement is that the cooling time t_{cool} should be shorter than the free fall time t_{free} in order to be effective. A second requirement is that both times must be shorter than the Hubble time, in order that the structures we see today could have had time to form. The first of the important interactions contributing to cooling is *Bremsstrahlung*

$$e + X^+ \rightarrow e + X^+ + \gamma, \tag{16.2}$$

in which the path of a high energy electron swerves in the intense electric field of a proton or other nucleus and the 4-momentum it loses is carried off by the emitted photon. The photons escape from the matter accumulation, so cooling it. The bremsstrahlung cross-section rises with energy making bremsstrahlung the dominant cooling process at temperatures above 10^6 K, at which temperature the photons are predominantly X-rays. Expressing the electron energy as $(3/2)k_BT$ we see that a temperature of 8.2×10^6 K is required to power 1 keV X-ray emission. If n_e is the electron density and the temperature is T, then the rate at which energy is radiated in Bremsstrahlung is

$$L_X = 1.4 \times 10^{-40} n_e^2 \sqrt{T}\ \mathrm{W\,m^{-3}}. \tag{16.3}$$

The thermal energy of the baryonic matter is

$$E = (3/2)n\, k_B T \text{ J m}^{-3}, \tag{16.4}$$

where n is the number density of electrons plus protons plus helium nuclei. For simplicity this is taken to be $2n_e$. Then the time taken to cool by bremsstrahlung is

$$t_{\text{cool}} = E/L_X = 2.96 \times 10^{17} \frac{\sqrt{T}}{n_e} \text{ s} = 9.38 \frac{\sqrt{T}}{n_e} \text{ Gyr}. \tag{16.5}$$

For example, take a virialized halo at a redshift of 4, about 11.8 Gyr ago, at a temperature 10^6 K. The mean electron and proton density at redshift 4 would have been larger than the mean density today by a factor 5^3 due to the expansion of the universe. A virialized fluctuation would be 200 times denser; in all by a factor 25,000 denser. This gives a cooling time of 1.5 Gyr, showing that there was adequate time for this phase of cooling.

At temperatures lower than 10^6 K cooling is principally via collisions between electrons and atoms, or between atoms and atoms. In these processes an atom is excited from its ground state by absorbing kinetic energy: this is followed by its de-excitation with the emitted photon carrying off energy from the matter fluctuation. Symbolically the generic process is

$$\text{Y} + \text{X} \to \text{Y} + \text{X}^*, \quad \text{then } \text{X}^* \to \text{X} + \gamma, \tag{16.6}$$

where X^* is an excited state of the X atom and Y another atom or an electron. This is called cooling by line emission. In the primordial gas the spectral lines are just those of helium and hydrogen. Immediately after decoupling, and until stars form, the atoms are numerically 90% hydrogen atoms and 10% helium atoms. The dominant hydrogen transition is the Lyα transition between the 2p first excited state and the 1s ground state, with the emitted photon having 10.2 eV energy (121.6 nm wavelength). Already we know from the analysis of decoupling that the hydrogen clouds became neutral once their temperature fell below 4000 K, terminating the processes described above. Further cooling would be made possible by the collisional excitation and decay of either fine and hyperfine levels in heavy metals. However these processes *only become available after metals are produced* by nuclear processing in stars or supernovae.

Figure 16.2 illustrates the effects of cooling for baryonic matter with baryon number density plotted against temperature. The diagonal lines mark the virial fluctuations of fixed total mass, taken from Equation (15.37) with the baryon mass fraction $f = 0.14$. The right-hand gray area marks the region in which the cooling time by these radiative mechanisms is less than the free fall time in primordial matter with the same metallicity as the Sun. This includes the effect of bremsstrahlung and a downward peak between 4000 K and 5000 K due to transitions in atomic hydrogen and helium. The region populated by the observed galaxies, with bound masses of roughly 10^8–10^{13} $M_\odot$, lies neatly within the region of effective cooling. On the other hand, clusters, which typically have total masses of order 10^{14} $M_\odot$, lie outside, and could only condense after fragmentation.

Some further processes are required to cool baryonic matter so that density fluctuations can contract into stars. Looking at our own fairly typical galaxy is

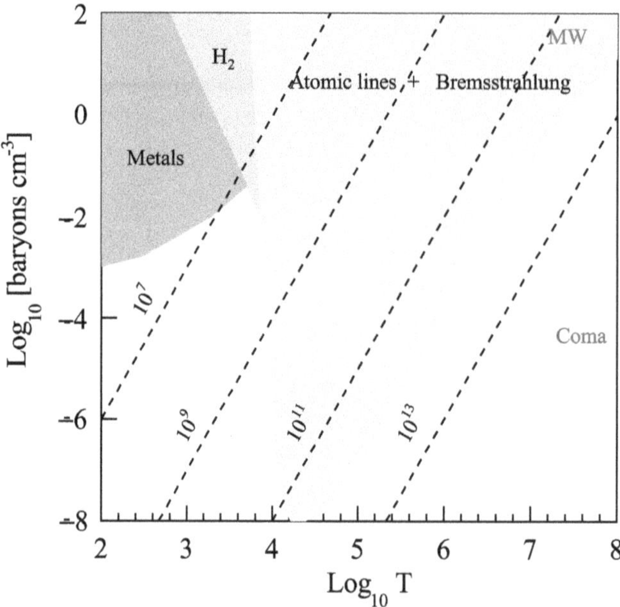

Figure 16.2. A sketch of baryon density versus temperature for cosmic structures. The diagonal lines are at fixed total mass. In the differently shaded regions different cooling mechanisms operate as described in the text. The locations of the Milky Way Galaxy and the Coma cluster are indicated. Heavily adapted from Figure 3 in Blumenthal et al. (1984). Courtesy Springer.

useful. The *interstellar medium* (ISM) is a flattened disk of cold gas, mainly hydrogen, of which 20% is in molecular form and 1% is dust. The dust grains are typically 0.35–1.0 nm across so they radiate and absorb well at comparable wavelengths, that means in the near-infrared. Molecular hydrogen is produced by the processes in which free electrons act as catalysts:

$$e^- + H \to H^- + \gamma\text{: followed by } H^- + H \to H_2 + e^-,$$
$$\text{or } H + H^+ \to H_2^+ + \gamma\text{: followed by } H_2^+ + H \to H_2 + H^+$$

The molecular hydrogen is distributed in *giant molecular clouds* (GMCs) of masses $\sim 10^4$–$10^6\ M_\odot$ and average densities $\sim 10^{-18}$ kg m^{-3}. It is in these clouds that star production is seen to be concentrated in the Milky Way. Prolific star production is seen too in starburst galaxies, again dominantly in GMCs. The gray region in Figure 16.2 marked "H_2" is the region where transitions between molecular rotational and vibrational states of hydrogen are effective in cooling the molecular clouds. These transitions involve meV energies rather than eV energies, and carry the cooling correspondingly further. Cooling proceeds in two steps: kinetic energy is transferred to molecular excitation in collisions, then photons carry off the excitation energy and escape from the cloud of hydrogen. One representative transition is the forbidden dipole transition at 43.9 meV, corresponding to 510 K thermal excitation. This molecular cooling takes the hydrogen molecules to temperatures of order 200 K. The molecular clouds in our Galaxy contain metals and dust whose transitions in the infrared are important for cooling in the area marked "metals" in Figure 16.2. The end result of this sequence of cooling processes is that the GMCs reach temperatures of around 10 K. The mass distribution of young stars that are observed today is discussed in Section 16.5.1, masses are of order $M_{\odot}$ being typical.

16.3 First Light and Reionization

The earliest stars were formed from the elements available after BBN, so that the potential for cooling by metals and dust was absent. These stars' masses have been estimated by simulations of the early universe. A rough estimate can be made by taking the density in GMCs to be the same then as currently. The GMC temperature would have been that attainable by cooling with molecular hydrogen and having no contribution from metals or dust; namely about 200–300 K compared to 10 K attained in current GMCs. Then using Equation (16.1) gives a Jeans mass of order 100 $M_{\odot}$. Such stars would have surface temperatures of 10^5 K, with luminosity 10^6 $L_{\odot}$, and with peak emission at wavelengths well into the ultraviolet. Lacking metals these monster stars are expected to burn hotter and exhaust their fuel more quickly. To date, none of these early population-III stars (that is, with null metallicity) have been detected. The JWST, by probing to higher redshifts than before, will give direct information on this era. The ultraviolet radiation from the earliest stars and AGNs would have begun ionizing the surrounding neutral hydrogen gas. Eventually, as the number of sources steadily increased, the regions of ionization overlapped, ending the dark ages of the universe. An estimate of when reionzation occurred is described next. It relies on measurements made by the Planck Collaboration of the polarization correlations of the CMB.

Once the hydrogen gas was ionized the CMB radiation could Thomson scatter from the newly freed electrons. This renewed Thomson scattering increases the polarization of the CMB beyond that caused by the last scattering: additionally it washes out the finer angular correlations, so that the high ℓ oscillations are reduced to the level shown in Figure 12.6. Now the probability of scattering over a path element $c\mathrm{d}t$ traversed in a short time $\mathrm{d}t$ is $n_e\sigma_T c\,\mathrm{d}t$, where σ_T is the Thomson scattering cross-section, and n_e is the electron number density. Hence the corresponding change $\mathrm{d}I$ in the CMB intensity I is

$$\mathrm{d}I = In_e\sigma_T c\,\mathrm{d}t. \tag{16.7}$$

Thus for the full path

$$I = I_{\mathrm{dec}}\exp(-\tau),$$

where I_{dec} is the CMB intensity at decoupling, I is the received intensity, and

$$\tau = \int_{\mathrm{dec}}^{t_0} n_e\sigma_T c\mathrm{d}t. \tag{16.8}$$

τ is called the *optical depth*. The effect on the observed polarization of the CMB was isolated in their analysis by the Planck Collaboration yielding a value of 0.056 for τ.

This value for τ is what we use to estimate when the universe reionized. Here the simplifying assumption is made that the universe was neutral up to some redshift z_{re}, and fully ionized from then on. At times later than this redshift boundary the electron density expressed in terms of the current density, $n_{\mathrm{e}0}$, is $n_{\mathrm{e}} = n_{\mathrm{e}0}/a^3$. Then

$$\tau = n_{\mathrm{e}0}\sigma_{\mathrm{T}} c \int_{a_{\mathrm{re}}}^{1} \mathrm{d}t/a^3. \tag{16.9}$$

Now

$$\mathrm{d}t = \mathrm{d}a/\dot{a} = \mathrm{d}a/[aH]$$

so that

$$\tau = n_{\mathrm{e}0}\sigma_{\mathrm{T}} c \int_{a_{\mathrm{re}}}^{1} \mathrm{d}a/[Ha^4].$$

The limits of integration lie in a period that is matter dominated. Therefore we can approximate Equation (10.32) adequately by

$$H = H_0\sqrt{\Omega_{\mathrm{m}0}}\, a^{-3/2}. \tag{16.10}$$

Performing the integration for τ, with H substituted in this way, gives

$$\tau = \frac{2}{3}\frac{n_{\mathrm{e}0}\sigma_{\mathrm{T}}\, c}{\sqrt{\Omega_{m0}}\, H_0}[a_{\mathrm{re}}^{-3/2} - 1], \tag{16.11}$$

which when rearranged becomes

$$a_{\mathrm{re}} = \left[1 + \frac{3\sqrt{\Omega_{m0}}\, H_0\tau}{2n_{\mathrm{e}0}\sigma_{\mathrm{T}}\, c}\right]^{-2/3}.$$

Using $\tau = 0.056$ as measured by the Planck Collaboration, and inserting 1.32×10^{26} m for c/H_0, 0.30 for Ω_{m0}, 6.65×10^{-29} m^2 for σ_{T} and finally 0.25 m^{-3} for $n_{\mathrm{e}0}$ gives

$$a_{\mathrm{re}} \approx 0.126 \text{ and } z_{\mathrm{re}} \approx 6.9,$$

making the universe about 760 Myr old. Next we investigate whether, as should be the case, the ultraviolet radiation from the earliest stars was adequate to reionize the universe by this time.

Only the hottest stars, O-type stars, are hot enough to radiate substantial numbers of UV photons capable of ionizing hydrogen, that is with energy above 13.6 eV. These are called Lyman continuum *LyC* photons. It follows that if we estimate the total flux of ionizing photons emitted by a typical O-type star and multiply this by density of these stars, then the product will give the density of the hydrogen atoms ionized in the lifetime of such stars. This takes a few steps. First a typical O-type star, of mass 50 $M_{\odot}$ and surface temperature 50,000 K emits a flux of LyC ionizing photons $\sim 2 \times 10^{49}$ s^{-1}. Second, Figure 16.3 shows the measured star formation rates per comoving volume against redshift (Madau & Dickinson 2014): at redshift ~7 a total stellar mass of 10^7 $M_{\odot}$ was created per year per Gpc^{-3}. Now O-type stars make

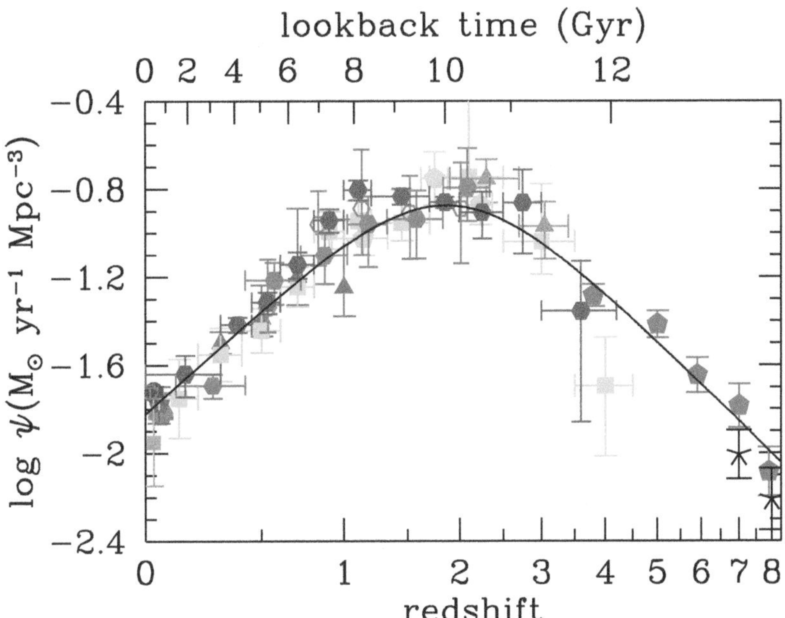

Figure 16.3. Star formation rates versus redshift. Far ultraviolet and infrared rest-frame measurements. The far-UV is obscured by intervening dust, while the latter IR penetrates it. The reference volume is the comoving volume. Adapted from Figure 9 in Madau & Dickinson (2014). Courtesy Professor Piero Madau and Annual Reviews Inc.

up some 10% of the stellar material. Whence 2×10^4 O-type stars of mass 50 $M_\odot$ were born every year per Gpc^{-3}. These stars live for around 4 Myr. Hence at any one moment there were $2 \times 10^4 \times 4 \times 10^6$ *live* O-type stars each emitting 2×10^{49} ionizing photons per second, in all 1.6×10^{60} per second per Gpc^3. This converts to $5.0 \times 10^{64}\ \mathrm{Myr}^{-1}\ \mathrm{Mpc}^{-3}$. Of these most do not escape from the gas cloud where the star is born. Taking the escape fraction to be 10%, the flux of ionizing escaping photons

$$\dot{n} = 5.0 \times 10^{63}\ \mathrm{Myr}^{-1}\ \mathrm{Mpc}^{-3}.$$

There are presently about 0.25 protons per (comoving) meter cubed or 7.4×10^{66} per comoving Mpc^3. The time needed to accumulate enough photons to ionize that number of H atoms in one comoving Mpc^3 is $7.4 \times 10^{66}/5.0 \times 10^{63}$ Myr. That is of the order of a billion years. At the level of precision of our calculation, the flux of LyC photons from O-type stars was adequate to bring about a timely reionization of the universe. The relatively rarer quasars present in the early universe would have been another important source of ionizing photons.

16.4 The Lyα Forest

Figure 16.4 shows the spectra of three quasars at redshifts 1.3, 2.9, and 5.8. The wavelengths have been translated to the rest frame of the quasar in each case (McQuinn 2016). Spectra like these provide information about the *intergalactic*

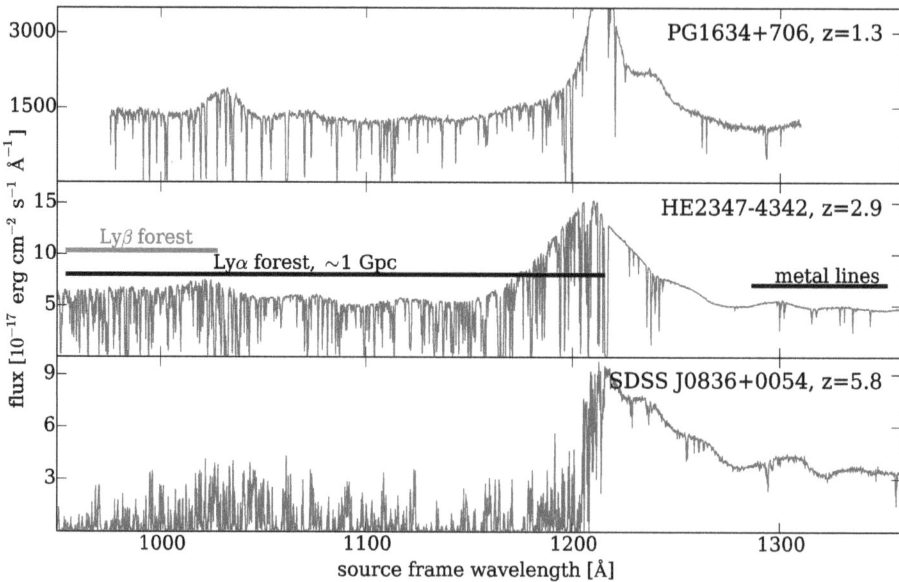

Figure 16.4. The optical spectra of quasars at redshift 1.3, 2.9, and 5.8 plotted against wavelength in the source's rest frame. This shows the Lyα absorption by neutral hydrogen between the quasar sources and the Earth. Figure 1 from McQuinn (2016). Courtesy Professor Matthew McQuinn and Annual Reviews Inc.

medium (IGM) from 1 Gyr after the Big Bang, and in particular on the progress of reionization. The broad dominating emission peak in each spectrum is the Lyα line emission from hydrogen atoms around the quasar; in the Lyα transition the electron drops from the excited state (2p) to the ground state (1s). The transition energy is 10.2 eV, and the wavelength, well into the ultraviolet, is 121.6 nm. The other component of the quasar radiation visible in the figure is a smooth continuous spectrum, punctured by numerous absorption lines. The absorptions involve the same Lyα transition: photons in the continuum resonantly scattering off neutral hydrogen atoms in the gas clouds between the quasar and the Earth. Any scattered photons are deflected and lost from the flux traveling toward the Earth. In the case of a cloud of neutral (atomic) hydrogen, in a cloud at redshift z it will scatter out quasar radiation of wavelength $121.6 \times (1 + z)$ nm: that is blueshifted (leftward) from the quasar's Lyα line at $121.6 \times (1 + z_{\text{quasar}})$. The many absorption lines seen in the figure are known as the *Lyα forest.* They are of great interest to cosmologists. Firstly because the pattern of absorption lines provides a map in redshift of where clouds of neutral hydrogen gas were located along the line of sight, and secondly because the depth of an absorption line reveals the integrated density of neutral hydrogen gas along the line of sight through that cloud. The quasar Lyα emission line is far broader than the absorption lines because emission is from hydrogen gas in violent motion close to the AGN. Many millions of quasar spectra have been recorded; and from the Lyα forest in each information has been obtained about redshift distribution of the neutral component of hydrogen clouds. In addition the lateral extent of these clouds has been extracted from the absorption spectra of quasars at small angular separations. Comparing the three spectra in the figure shows that the Lyα forest absorption increases from redshift 1.3 to 2.9 to 5.8. Thus, as expected, the neutral component of the intergalactic gas increases with redshift toward the dark ages when there were no stars. We can calculate the optical depth of these clouds for photons arriving with the Lyα wavelength. We now show that only a tiny proportion of neutral atoms in the hydrogen clouds is enough to decimate any photon cohort arriving with the Lyα wavelength.[2]

Take the Lyα transition frequency to be ν_0 and consider the scattering from a neutral cloud when the scale factor was a. The cross-section for scattering photons that would reach Earth with a frequency ν is:

$$\sigma(\nu/a - \nu_0) = \frac{e^2}{4\varepsilon_0 m_e c} f\phi(\nu/a - \nu_0) = 1.14 \times 10^{-6}\, \phi(\nu/a - \nu_0)\ \mathrm{m}^2,$$

where m_e is the electron mass and $f = 0.416$ is the oscillator strength of the Lyα transition. ϕ is the line shape of the Lyα transition peaking at the frequency ν_0, and normalized so that $\int \phi(\nu - \nu_0)\mathrm{d}\nu = 1$ s. The line shape is sharper than the Doppler shift width of the absorption features so that it may be treated here as a Dirac-δ function. This gives the convenient result that for any generic function $F(a)$

$$\int \phi(\nu/a - \nu_0)F(a)\mathrm{d}a = F\left(\frac{\nu}{\nu_0}\right)\frac{\nu}{\nu_0^2},$$

where the range of integration includes the value $a = \nu/\nu_0$ for which the δ-function diverges. First we assume all the hydrogen atoms remain neutral. Then using Equation (16.8) again the optical depth for photons arriving at the Earth with frequency ν is

$$\tau(\nu) = \int_{aq}^{1} \sigma(\nu/a - \nu_0)n_b(a)\mathrm{d}t = \int_{aq}^{1} \sigma(\nu/a - \nu_0)n_b(a)\frac{c\,\mathrm{d}a}{aH(a)},$$

where $n_b(a)$ is the baryon density at a scale factor a. The lower limit of integration is the scale factor at the quasar aq. Evidently the δ-function will pick out the clouds around the scale factor $a = \nu/\nu_0$. $n_b(a) = n_0\,[a]^{-3}$, with n_0 taken to be the current baryon number density. We make the substitution for $H(a)$ given in Equation (16.10), appropriate when matter dominates. This gives

$$\tau(\nu) = \int_{aq}^{1} \sigma(\nu/a - \nu_0)\frac{n_0 c}{a^{5/2}\sqrt{\Omega_{m0}}H_0}\mathrm{d}a.$$

After integration

$$\tau(\nu) \approx \frac{\sigma_0 n_0 c}{H_0\sqrt{\Omega_{m0}}}[\nu_0^{1/2}/\nu^{3/2}].$$

[2] This analysis is based on notes and comments by Professor Tom Theuns: http://icc.dur.ac.uk/tt/IGM.pdf. Professor Matthew McQuinn gave useful advice on this section.

If the cloud lies at redshift z, then $\nu = \nu_0/(1 + z)$, and the optical depth is

$$\tau(\nu) \approx \frac{\sigma_0 n_0 c}{H_0\sqrt{\Omega_{\rm m0}}} \frac{(1+z)^{3/2}}{\nu_0} \approx 3\ 10^4\,(1+z)^{3/2},$$

where we have been assuming that all the hydrogen atoms remain neutral. Absorption saturates when τ reaches unity; hence it only requires a tiny fraction of the hydrogen to remain neutral at redshift 5, of order 10^{-5}, to produce complete absorption. Consequently, in Figure 16.4, the incomplete absorption seen in the spectrum of the quasar at redshift 5.8 demonstrates that reionization is very close to being total at this redshift. The current view based on detailed analysis is that reionization took place between redshifts 12 and 6 (360 Myr and 920 Myr, respectively). The observed quasars get rarer as the redshift grows. Despite that it may be possible to gather further precise information using the recently launched JWST with its larger mirrors, and with its detectors having better infrared sensitivity than Hubble.

16.5 Formation of Stars

Giant clouds of mainly molecular hydrogen gas, GMCs, within galaxies are the primary sites of star formation. They are always associated with star clusters younger than $\sim 10^7$ yr old, indicating that they are shorter lived than galaxies. In our Galaxy the sites are mainly concentrated in the disk of the Galaxy. The nearest giant molecular cloud, 150 pc distant, is the Taurus GMC; it is highly structured and studded with tens of young stars. Giant molecular clouds are typically tens of parsecs across and have masses of 10^4–10^6 $M_\odot$: the mean gas number densities are $\sim 10^9$ m^{-3}, with cores up to 1000 times denser. The clouds maintain uniform temperatures around 10K despite a wide range of densities. It follows that collisions between hydrogen molecules are not energetic enough to excite transitions from the ground state of a hydrogen molecule. However in current galaxies the reprocessing of primordial matter in earlier stars followed by the dispersion of material in supernovae has laced the clouds with many other molecules such as CO, H_2O, and HCN at densities $\sim 10^{-4}$ that of hydrogen. These heavier molecules have lower energy molecular rotational and vibrational transitions which facilitate cooling.

CO has, for example, a prominent transition at 2.6 mm (115 GHz and 0.47 meV). This only requires excitation by a collision imparting 0.47 meV, corresponding to 5 K thermal excitation. Hence the transition is excited copiously at the temperature of the molecular clouds. These organic molecules are cooled by radiating their excitation energy, and they in turn cool the hydrogen molecules in collisions. Radiation from organic molecules not only provides the cooling mechanism for the GMCs, but also its detection reveals the extent and density of GMCs.[3] Despite the intense ultraviolet radiation from the hottest stars in the same galaxy the giant molecular clouds remain cold. This is because the dust and metals incorporated in

[3] Many telescopes in use are sensitive to the infrared radiation from such organic molecular transitions: for example the 50 m diameter Large Millimeter Telescope at 4640 m altitude in Sierra Negra, Mexico, sensitive to 0.85–4 mm wavelength radiation.

these molecular clouds absorb the ultraviolet radiation and re-emit the incoming energy, principally as infrared radiation. This cocoons the cold core of each cloud. Eventually the radiation from stars formed inside a GMC will disperse its gas and halt further star formation.

The linewidths of emission from gas molecules in GMCs indicate that the molecules have a velocity dispersion of ~10 km s^{-1}. This far exceeds the dispersion expected in a gas at 10 K, of only 0.2 km s^{-1}. The reason for the elevated velocity dispersion is thought to be supersonic turbulent flow in the GMCs. Likely sources of turbulence are outflows from supernovae, winds from massive stars, radiation pressure and cosmic ray streaming. The outcome is that GMCs show a filamentary structure with nodes on all scales. GMCs are not themselves bound by self-gravity: to see this we can use a virial parameter comparing the kinetic and gravitational energies of a molecule at the cloud's surface

$$\alpha_{\rm vir} = \langle {\rm KE} \rangle / \langle {\rm GE} \rangle \approx \frac{\sigma^2 R}{GM}$$

where M is the cloud mass, σ the line-of-sight velocity dispersion and R the cloud radius. In the case of clouds in our Galaxy $\alpha_{\rm vir}$ ranges up to 100: they are not going to collapse as single items. Analogous to the three-dimensional Jeans mass there is a critical mass per unit length above which a filament in a GMC will collapse. Two radial forces are at work (Inutsuka & Miyama 1997): the first is from the outward pressure

$$F_{\rm p} \sim c_{\rm s}^2 / R, \tag{16.12}$$

where R is the outer radius of the filament and $c_{\rm s}$ is the velocity of sound in the filament. The other is the inward gravitational force

$$F_{\rm g} \sim -\frac{GM_{\rm line}}{R}, \tag{16.13}$$

where $M_{\rm line}$ is the mass per unit length. Then the critical mass per unit length at which these forces balance is

$$M_{\rm crit} \sim 2c_{\rm s}^2 / G. \tag{16.14}$$

The velocity of sound $c_{\rm s}$ in a GMC at temperature T is

$$c_{\rm s} = \left[\frac{k_{\rm B}T}{m_m}\right]^{1/2} = 203\left[\frac{T}{10{\rm K}}\right]^{1/2} {\rm m\ s^{-1}}, \tag{16.15}$$

where m_m is the mass of a hydrogen molecule. Then the critical mass per unit length is

$$M_{\rm crit} = 1.24 \times 10^{15}\left[\frac{T}{10\ {\rm K}}\right]{\rm kg\ m^{-1}} = 19.1\left[\frac{T}{10\ {\rm K}}\right] M_{\odot}\ {\rm pc^{-1}}.$$

Figure 16.5 shows the column density in the far-infrared (70–500 μm) observed with the Herschel Space Telescope (André et al. 2010) over the molecular cloud in the

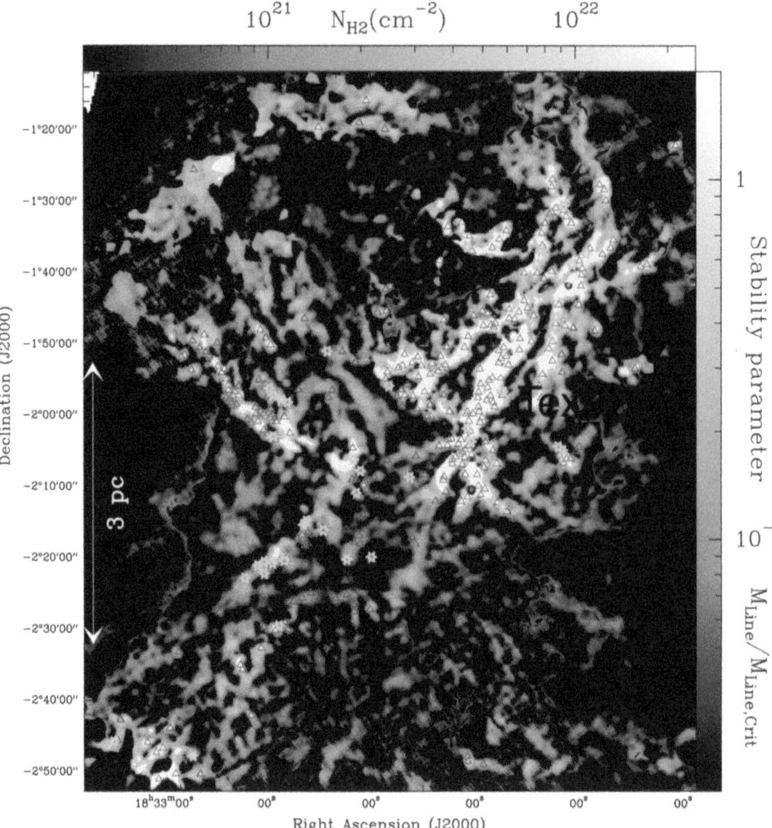

Figure 16.5. Column density in the far-infrared across a molecular cloud in the Aquila Rift. This was observed with the Herschel Space Telescope in the range 70–500 μ m. The shading indicates the fraction of the critical mass per unit length along the filaments. White indicates where this exceeds 0.5. The green and blue points mark the O-type protostars and the bound pre-stars, respectively. Figure 1(a) from André et al. (2010). Reproduced with permission © ESO.

Aquila rift.[4] The filaments are typically 0.05 pc across. In the figure the identified bound prestellar cores and Class-O protostars are shown to be concentrated along those filaments where the mass per length is above half the critical value.

Once collapse commences, flow and accretion from the environment will result in a near spherical *core*. We can calculate the Jeans mass once a spherical core has formed. The local molecular number density n_{core} is $\sim 10^9$–10^{12} m^{-3} and the mass density ρ_{core} is $\sim 10^{-18}$–10^{-15} kg m^{-3}. Using Equation (15.1) the free fall time for the core is

$$t_{\text{free}} = [G\rho_{\text{core}}]^{-1/2} = 0.21 \text{ Myr} \left[\frac{n_{\text{core}}}{10^{11}\,\text{m}^{-3}}\right]^{-1/2},$$

[4] This molecular cloud blocks visible radiation coming from stars in the galactic plane.

so that the formation time is of order Myr. The Jeans mass given by Equation (15.4) is

$$M_J = (\pi/6)c_s^3\left[\frac{\pi^3}{G^3\rho_{\text{core}}}\right]^{1/2}.$$

Referencing this to typical values of temperature and molecular number density gives

$$M_J = 1.22\,M_\odot\left[\frac{T}{10\text{K}}\right]^{3/2}\left[\frac{n_{\text{core}}}{10^{11}\text{m}^{-3}}\right]^{-1/2}. \tag{16.16}$$

Molecular clouds have uniform temperatures thanks to the turbulence and hence the Jeans mass falls as the density increases. As a result fragmentation into smaller collapsing bodies happens. This can continue while the fragments can radiate their binding energy efficiently. A limit is imposed by self-absorption of radiation on the dust and metals present in the collapsing gas. The limit is reached when the density reaches $\sim 10^{-10}$ kg m^{-3} or 10^{17} molecules per cubic meter. Inserting this value in Equation (16.16) gives a minimum mass for stars of around 0.001 M$_\odot$, that is around the mass of the planet Jupiter.

The star formation rate in our Galaxy is equivalent to 1 $M_\odot$ per year. Now the mass of gas in the Milky Way is some 2×10^9 $M_\odot$, so that this rate of star formation would have exhausted the supply in 2 billion years, yet the Galaxy has lasted for around 12 billion years. This discrepancy is resolved if there has been repeated recycling of the baryons in stars, which is necessary anyway to explain the metals and dust now present. Another source of additional gas is through accretion from the wider environment, which can continue over a long period. Star formation takes place in the giant molecular clouds that are overtaken by the turbulence of the rotating spiral arms of a galaxy. In such cold regions the accumulations inferred from Equation (16.16) to collapse are of order $M_\odot$. Once star formation occurs the radiation from the new stars disrupts the parent cloud. As a result little of the baryonic matter ends up in stars: it is buffeted about by supernovae, winds from massive stars, outflows from AGNs, and on the larger scale by galactic collisions, and other magneto-hydrodynamic processes. Below we shall review the evidence that the biggest portion of the baryons, 86%, is in the form of warm and hot diffuse ionized clouds pervading and enveloping galaxies and clusters of galaxies.

16.5.1 The Initial Mass Function

The accessible property of stars is their luminosity across the electromagnetic spectrum. This varies with time in a well-understood way from the moment a star enters the main sequence. Hence the distribution of the initial masses of the stars can be inferred, with a compensation made for stars no longer likely to remain visible. Taking only the stars in a galaxy, this distribution is known as its *initial mass function* (IMF). The IMF has been determined for many galaxies at low redshift, and these have been found to be broadly similar.

Uniformity of these IMF is consistent with the idea that stars are born in similar environments, namely giant molecular gas clouds, with similar metallicities. This was not the case in the early universe, when dust and metals had yet to be produced. Then the energy radiated during the collapse of a protostar escaped directly rather than being re-absorbed by dust and metals. Consequently it required higher temperatures and pressures to ignite hydrogen fusion, and hence the stars would have been more massive. Nuclear burning in these massive stars would have been correspondingly more rapid, so that none could remain today.

The template for the IMFs at low redshifts is shown in Figure 16.6. Stars with the highest luminosity, the hottest, having masses above 15 $M_\odot$ are called O-type. Stellar luminosity is roughly proportional to the mass cubed in main sequence stars. A 50 $M_\odot$ star has a luminosity $L = 20,\,000\ L_\odot$ in the B-band where $L_\odot$, where the Sun's B-band luminosity is 1.54×10^{26} W. This makes the mass to light ratio for such 50 $M_\odot$ stars to be $50/20{,}000 = 0.0025[M/L]_\odot$. At the other end of the stellar mass scale are red dwarf or M-type stars with masses up to 0.5 $M_\odot$ these account for three quarters of the visible stars in the Milky Way; and beyond them lie the brown dwarfs with masses less than 0.08 $M_\odot$, making them too light to ignite hydrogen fusion in their cores.

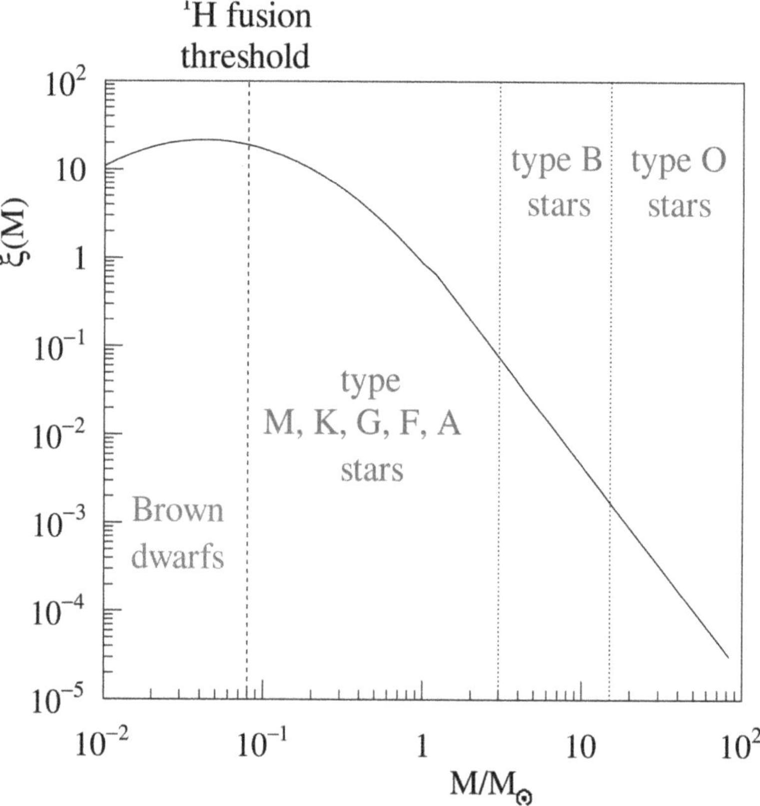

Figure 16.6. Stellar initial mass function. The low mass stars contribute most of the galactic mass; the heaviest stars contribute most of the galactic radiation.

Overall the mass in a galaxy is concentrated in the overwhelmingly numerous low luminosity stars, while the heaviest stars, the O-type stars (>15 $M_\odot$) and B-type stars (3–15 $M_\odot$), provide most of the radiation. In some galaxies there is copious star formation while others are less active. Taking all stars from local galaxies and averaging gives a mass to light ratio about 4 $[M/L]_\odot$. Our Sun is fairly average in this as in many other ways. The IMF shown in Figure 16.6 [5] has, for masses above $M_\odot$, a power law dependence, the *Salpeter IMF*:

$$\xi(M) = \alpha M^{-\beta}, \tag{16.17}$$

where $\xi(M)\mathrm{d}M$ is the number of stars with masses in the interval M to $M + \mathrm{d}M$. The power β is typically 2.35, and α is a constant. At lower masses the dependence is normal in log M, the *Chabrier IMF*:

$$\xi(M) = \frac{1}{M}\exp\left[-(\log_{10}M - \log_{10}M_m)^2/[2\sigma^2]\right], \tag{16.18}$$

centered on the mass $M_m = 0.2\ M_\odot$ and with variance $\sigma = 0.54$. This region includes the brown dwarfs. The formulae given for the two mass regions join smoothly if α is set to 0.5.

The Salpeter and Chabrier IMFs are simply empirical fits to the observed IMF; and the interpretation of the IMF is still under discussion. A few remarks can be made. As noted earlier in Equation (16.1) the Jeans mass

$$M_\mathrm{J} \propto T^{3/2}\rho^{-1/2},$$

so that when a fluctuation of mass $M \geqslant M_\mathrm{J}$ collapses isothermally the Jeans mass falls. Thus within a GMC the collapsing masses are expected to undergo a cascade of fragmentations. At some point cooling by radiation will no longer be large enough to maintain the existing temperature and the core begins to heat up: if massive enough fusion of hydrogen into helium will commence.

In addition the gas is subject to the turbulence and shocks in the cloud so that after multiple steps of fragmentation the final stellar mass is a product of many random effects. In that case the mass distribution would be expected to have the observed log normal distribution. The power law tail of the IMF at large masses can be seen as the outcome of the scale invariance in the early steps of fragmentation. Scale invariance here means that a mass M splitting into fragments of masses $aM, bM,\ldots$ is just as likely as a mass m later in the cascade splitting into fragments with masses $am, bm,\ldots$.

16.6 Galaxies

The giant galaxies mostly have total baryonic plus dark matter masses in the range 10^7–10^{13} $M_\odot$. There are three broad classes of giant galaxies: around three quarters are spirals, the rest elliptical or the rarer irregularly shaped galaxies. Figure 16.7 shows a barred spiral galaxy, UGC12158 in Pegasus, similar to our own. These are

[5] The stars were ordered alphabetically according to the intensity of the hydrogen spectral lines by Williamina Fleming in 1880. Although the ordering is now by temperature, these alphabetic labels were retained.

Figure 16.7. The barred spiral galaxy UGC12158 similar to the Milky Way. The bright regions are lit by stars formed within giant molecular clouds. Credit: ESA/Hubble and NASA.

galaxies in which intense star formation is in progress at the bright points along the spirals. A plausible view is that the mass concentrations from which spiral galaxies developed would have lacked spherical symmetry. Gravitational collapse would have been most rapid along the minor axis, leading to a pancake final shape. During the collapse of this disk toward its core (harboring a black hole) angular momentum would be conserved: then, just as in the case of a skater pulling her arms into her body while rotating, this spun up the rotation of the disk. Our Sun, for example, takes 240 Myr to complete its journey round the Galaxy's axis. The spiral arms are located where shock waves rotating around the core produce high pressure and turbulence, and hence a burst of star production. These shock waves travel maintaining constant angular velocity; by contrast, as we have seen, the gas in the Galaxy rotates with constant linear velocity. As a result, the path of the shock waves trace out spirals in the gas. Along each spiral arm the UV radiation from the young massive O-type and B-type stars ionizes the gas around them giving the beads of light seen in Figure 16.7. The radiation from these massive stars gives spiral galaxies their characteristic blue/white coloration, and the lifetime of such stars determines the width of the spiral arms.

Figure 16.8 is a sketch of a section made through the rotation axis of our Galaxy. The stars form a disk which extends out to around 17 kpc; 95% of them are concentrated in a thickness ~0.3 kpc. The bulge is the section taken across the bar, itself a few kpc in

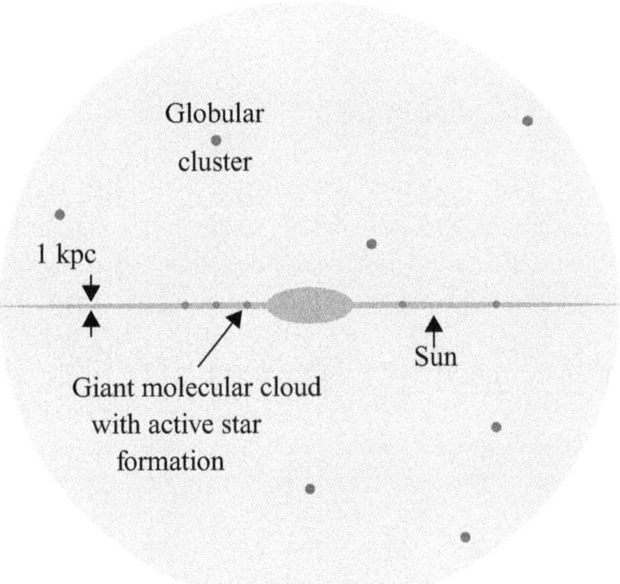

Figure 16.8. A sketch of a section through the Milky Way Galaxy. At the center of the bulge lies a black hole of mass $\sim 4 \times 10^6\ M_\odot$. The disk's radius is around 17 kpc. The globular clusters are dispersed in the baryonic halo, while the molecular clouds are largely confined to the disk. Gas clouds in the halo are likely to be atomic rather than molecular hydrogen. The dark matter halo is roughly spherical and extends well beyond the sketch to a radius of order 100 kpc.

radius with mass $\sim 10^{10} M_\odot$. At the center lies the black hole near Sgr A* with mass $\sim 4 \times 10^6\ M_\odot$. Dark matter forms a halo that, judging from Figure 1.7, extends beyond 50 kpc. The total mass of the stars is $\sim 7 \times 10^{10}\ M_\odot$ and their collective luminosity is $\sim 2 \times 10^{10}\ L_\odot$. Isolated stars, gas clouds, and around 150 globular clusters form a spherical halo, contributing something like a percent of the total stellar mass. Globular clusters are collections of 10^5–10^6 densely packed stars, in volumes $\sim 1\ \text{pc}^3$. One possible view is that in the hierarchical ordering of baryonic structures, with low mass structures developing first, globular clusters would have formed earlier than galaxies. Some globular clusters do have very low metallicity, consistent with this interpretation. However the origin of globular clusters is still under study and not yet settled.

It was explained earlier, in Chapter 1, that a dark matter halo is essential to explain why the rotational velocity of the visible matter v remains constant out to radii at the limit of detectability. The total mass, baryonic plus dark matter, within radius R of the central black hole, $M(<R)$, is given by the radial equation of motion of a star of mass m at that radius

$$GM(<R)m/R^2 = mv^2/R,$$

so that

$$M(<R) = v^2 R/G. \tag{16.19}$$

Inserting the rotational velocity of the Sun, 230 km/s, at 8 kpc from the central black hole gives

$$M(<R) = 1.22 \times 10^{12} \left[\frac{R}{100 \text{ kpc}} \right] M_{\odot}. \tag{16.20}$$

Recently the motion of stars traveling at high velocity in orbits far out of the galactic plane have been analyzed and this gave an independent estimate of 1.2–1.9 × 10^{12} $M_{\odot}$ for the total mass of our Galaxy. Using Equation (15.1) the free fall time of a galaxy radius R and mass M is

$$t_{\text{free}} = \left[\frac{4\pi R^3/3}{GM} \right]^{1/2} = 10^9 \left[\frac{R}{100\text{kpc}} \right]^{3/2} \left[\frac{10^{12}\text{M}_{\odot}}{M} \right]^{1/2} \text{yr}.$$

Thus our Galaxy took ~1 Gyr to form. The Milky Way is typical in having many gravitationally bound satellite galaxies. The satellite galaxies range from dwarfs formed from thousands of stars to the Large Magellanic Cloud with mass around 10^{11} $M_{\odot}$. Many tens of the attendant low luminosity dwarf galaxies were only detected in recent decades. Given a strictly hierarchical evolution of the universe, globular clusters and dwarf galaxies would be the first stellar structures to form.

In contrast to spiral galaxies, the elliptical galaxies contain predominantly older stars and are therefore reddish in color. They lack the gas and dust, whose presence was seen to be important in stellar nurseries. Stars in elliptical galaxies have randomly oriented elliptical orbits around the center of mass, very different from the dominant uniform rotation in spiral galaxies. Both the coloration and random motion are consistent with the elliptical galaxies being the outcome of galaxy–galaxy interactions or mergers. In such events the impact of gravitational interaction would disrupt existing coherent rotation in the participant galaxies. Figure 16.9 illustrates the process: images of observed pairs of galaxies in various stages of merging were selected and ordered by the authors. In the collision the stars, being relatively widely spaced, would not collide. However the gas clouds would collide, heat up and escape the gravitational pull of the resultant elliptical galaxy. There would be one burst of star formation and when this subsided there would remain the collection of yellow and red stars that we now observe.

Spiral galaxies are found in groups such as the Local Group containing the Milky Way. This group extends over a volume of dimension ~3 Mpc, and comprises around thirty much smaller galaxies and one other equally large galaxy M31. By contrast elliptical galaxies are usually found in clusters of hundreds of large galaxies, where galaxy–galaxy collisions are likely to be more frequent.

16.6.1 Low Redshift Galaxies' Stellar Mass Distributions

Large surveys of the luminosity and spectra of galaxies at low redshifts have been carried out. The GAMA Collaboration compiled around 250,000 galaxy spectra covering ~250 deg^2 of the sky (Driver et al. 2022). The stellar mass content is extracted from the observed luminosities of the galaxies, using models that include all the stages of stellar evolution, and which take account of redshifts. A *galactic*

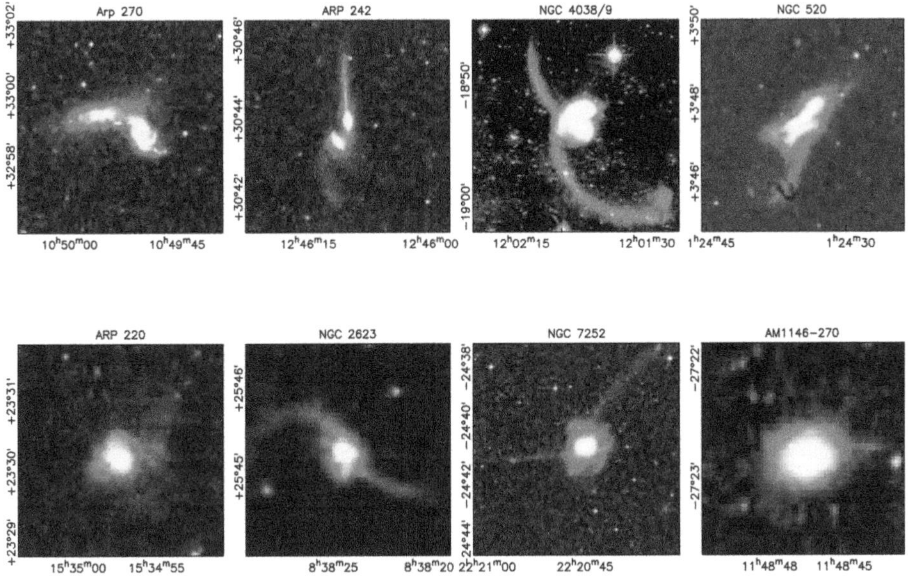

Figure 16.9. A representative sequence of galaxy mergers presented in what would be chronological order: reading left to right, and top to bottom. Figure 1 in Read & Ponman (1998). Courtesy Professors Read and Ponman, the Royal Astronomical Society, London and Oxford Univ. Press, Oxford.

stellar mass function (GSMF) $\phi(M)\,\mathrm{d}M$ is defined to be the number of galaxies per Mpc^3 each containing a total stellar mass in the interval M to $M + \mathrm{d}M$. This distribution is parametrized using a *Schechter function*:

$$\phi(M)\mathrm{d}M = \phi^*\left[\frac{M}{M^*}\right]^{\alpha} \exp[-M/M^*]\,\frac{\mathrm{d}M}{M^*}. \tag{16.21}$$

Here ϕ^* is the normalizing galaxy number density per Mpc^3, and M^* a characteristic galactic stellar mass. Integration over all galactic stellar masses gives the total stellar mass density

$$J = \int M\phi(M)\,\mathrm{d}M = \phi^*\,M^*\,\Gamma(2+\alpha),$$

where Γ is the gamma function, and for the natural numbers (1, 2, 3,...) $\Gamma(n) = (n-1)!$ holds. With the Schechter parametrization the total mass density is equal to, within a small numerical factor, the product of the reference mass density and reference number density. The corresponding *galaxy stellar mass density distribution* is defined by $\phi(M)M\mathrm{d}M$. Figure 16.10 shows compilations of both the GSMF and the galaxy stellar mass density distribution made by the GAMA Collaboration (Driver et al. 2022) at redshifts $z < 0.1$. For the Hubble constant $70\ \mathrm{km\ s^{-1}\ Mpc^{-1}}$ was used.[6] The solid curve is the fit made to the data using a double

[6] If the Hubble constant is larger than $70\ \mathrm{km\ s^{-1}\ Mpc^{-1}}$ by a factor h_{70} then the entry should be multiplied by this factor. dex^{-1} indicates the entry is that for a decade of stellar mass centered on the stellar mass entry.

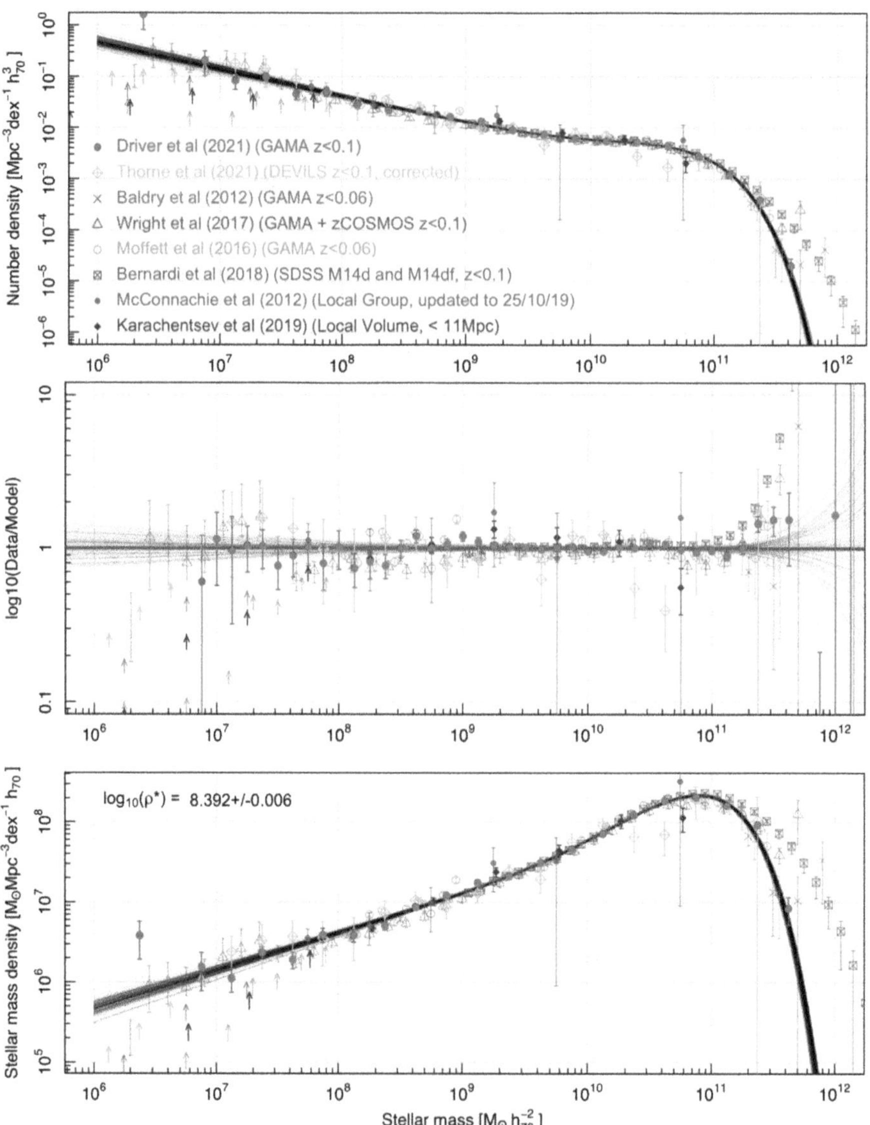

Figure 16.10. Measurements of low $z < 0.1$ galaxy stellar mass distributions compiled by the GAMA Collaboration, comprising several hundred thousand galaxy spectra. The solid line shows their fit to the data. The upper panel shows the GSMF. The middle panel shows the ratio of the data to the fit. The gray lines show examples of how randomly displacing the data points by their errors would affect the fit. The lower panel shows the distribution of the stellar mass density. Figure 12 from Driver et al. (2022). Courtesy of the Royal Astronomical Society, London and Oxford Univ. Press, Oxford.

Schechter function. This fit yields the following parameters: $M^* = 5.63\ 10^{10} M_\odot$, $\log_{10}\phi_1^* = -2.44$, $\log_{10}\phi_2^* = -3.20$, $\alpha_1 = -0.466$ and $\alpha_2 = -1.53$. There is a sharp cutoff in the stellar mass distribution at the high mass end: while at the lower end the lowest accessible mass is less than a factor one hundred larger than the Jeans mass.

Taking these two features together the authors argue that their parametrization offers a plausible extrapolation to cover the full mass distribution. On this basis the authors infer that at $z < 0.1$ the stellar mass density and the fraction this contributes to the critical density are, respectively:

$$\rho_* = 2.97 \times 10^8\ M_\odot\ \text{Mpc}^{-3} \text{ and } \Omega_* = 2.2 \times 10^{-3}. \tag{16.22}$$

This means that only 5% of the baryons have been captured in stars up to our era.

In addition to the elliptical and spiral galaxies there are irregularly shaped galaxies. These are generally smaller galaxies and in number form about a quarter of all galaxies. Some owe their shape to collisions and interactions, others are in the process of consolidation.

16.7 Clusters and Superclusters of Galaxies

Our Galaxy has a companion, the Andromeda galaxy (M31), at a distance of roughly 0.8 Mpc and of similar size to our Galaxy. Together they dominate a local cluster of about 30 galaxies, which is roughly 3 Mpc across. The Virgo cluster is a nearby assembly of ~1500 galaxies spread across ~4 Mpc about 16 Mpc from Earth. These two clusters, in turn, are members of the Virgo supercluster containing ~100 clusters, about 33 Mpc across and with mass $\sim 10^{15}\ M_\odot$. This supercluster is part of a larger filamentary structure. About 10^7 such superclusters lie within the visible part of the universe. The largest supercluster of galaxies is the *Hercules–Corona Borealis Great Wall.* It has a volume 3 Gpc by 2.5 Gpc by 300 Mpc and lies at distance from Earth of 3 Gpc.

How much further is this sequence of growing structures likely to extend? We can try simplistic reasoning. The separation of the stars in globular clusters is $\sim 10^{-2}$ pc. Suppose we take this as a measure of the size of the average overdense region from which these stars condensed. Then by comparison the largest structures, the superclusters of dimensions 100 Mpc are 10^{10} times larger. This can be compared to the expected range in dimensions expected for the overdense quantum fluctuations generated during inflation. Equal periods of e-fold growth would have produced equal numbers of fluctuations. Hence the range in physical dimensions (not overdensity) would have been at least 10^{30}. Ten is comfortably less than thirty, and we can infer that structures considerably larger, but also rarer, than the superclusters may exist. However, from the time that dark energy took control no further growth of perturbations could occur. What remains of these larger-volume, slightly overdense regions are tenuous clouds of matter that can never contract.

Using Equation (15.1) the free fall time for cosmological structures is

$$t_{\text{free}} = 3.2 \times 10^{10} \left[\frac{R}{10\ \text{Mpc}}\right]^{3/2} \left[\frac{10^{15} M_\odot}{M}\right]^{1/2} \text{yr},$$

where M is the structure's mass and R its radius. An overdensity with supercluster size or larger would not have had time to collapse and would have fragmented. Instead the more compact clusters would have collapsed and virialized. The gravitational attraction between the structures that did form continues to pull them together. For example, the Milky Way and M31 are approaching at about

120 km s^{-1} indicating the continued effect of the dark and baryon matter gravitational well they share.

The best studied cluster, the Coma cluster, is a large, densely populated cluster containing over 10^3 galaxies. It lies close in the sky to the constellation Coma Berenices: the viewing direction is perpendicular to our galactic plane, and hence clear of most of our Galaxy's dust. The mean velocity of recession of the galaxies making up the Coma cluster, v is 6933km s^{-1}. Hence the mean redshift is 0.0231 and the corresponding distance from us $v/H_0 = 102$ Mpc. This cluster is small enough to have had time to collapse and virialize. Zwicky applied the virial theorem to the motion of galaxies in this cluster and deduced the presence of the unseen dark matter, with many times the mass of the visible matter. We can update Zwicky's analysis using recent data. The velocity dispersion of the galaxies along the line of sight σ, is ~1000km s^{-1}. We calculate the mass M within the virial radius R from the center of the cluster, where the density falls to 200 times the critical density, that is within 3 Mpc. Then the virial theorem gives

$$M = 3\sigma^2 R/G, \tag{16.23}$$

where the factor 3 converts the line-of-sight dispersion to the three-dimensional dispersion. The mass inferred is $2 \times 10^{15} M_\odot$. With luminosity in the visible spectrum $10^{13} L_\odot$ the mass to light ratio is 200 times that of the Sun, and 50 times that for our Galaxy. The inference from updating Zwicky's analysis is that dark matter dominates over baryonic matter.

In Figure 16.11 we see what Zwicky could not see, a composite view, 800 kpc wide, covering the Coma cluster. The color shows where there is X-ray emission

Figure 16.11. Composite image of the Coma cluster. The pink and blue color indicates the X-ray emission from gas at millions of degrees kelvin, and the white indicates the optical emission from galaxies. The X-ray image was taken by the *Chandra* satellite (https://chandra.harvard.edu/) and the optical image by the Sloan Digital Sky Survey (http://www.sdss.org). Courtesy *Chandra* and SDSS Collaborations.

from gas at ~8 keV; superimposed in white is the optical image of the constituent galaxies in the same region. X-rays are strongly absorbed by the Earth's atmosphere, so that it was only when satellites were launched in the 1970s, equipped with X-ray detectors, that X-ray emission could be observed and studied. The X-ray emission seen in Figure 16.11 provides direct evidence that the greater part of baryonic matter in the universe lies in diffuse gas clouds enveloping the galaxy clusters. Including these diffuse gas clouds builds up the baryonic contribution to 4.5% of the critical density at the present time, the value used throughout this text.

16.8 Intergalactic Baryonic Matter

Figure 14.2 shows that the baryon to photon ratio predicted in two ways, from the observed ratios of the light elements and from the CMB correlations, are mutually compatible. Fits to both data sets require that baryonic matter contributes 0.045 of the critical density: the rest is dark matter and dark energy with a touch of radiation. Equation (16.22) reveals that the stars themselves make up only 5% of this 0.045 baryon fraction: that translates to *a mere 0.0027* of the energy of the universe. Gas between stars, between galaxies and between clusters of galaxies only accounts for a further 8% of the expected baryonic matter. The missing 87% of the baryonic mass is gas held in the dark matter potential wells that confine clusters of galaxies.

There is X-ray emission from the whole volume of clusters of galaxies and this emission is highest from the center of each cluster. Continuum spectra are observed, which match those expected from baryon plasmas in thermal equilibrium. Taking this interpretation to be correct, in the case of the Coma cluster the equilibrium temperature of the plasma is 9×10^7 K (8 keV). There are spectral lines on top of the X-ray continuum, lines corresponding to emission from heavily ionized iron atoms. The intensities of the spectral lines are consistent with the interpretation that the source is a thermal plasma at the temperature inferred from the continuum. The metal content of the plasma inferred from the X-ray spectra is $\sim 0.5\, Z_\odot$, where $Z_\odot$ is the Sun's metallicity. Thus the X-rays are emitted by gas that has already been processed within stars. Unlike galaxies that are open systems, the clusters retain the enriched material created in their comoving volume. The violent motion during the collapse that accompanies virialization stirs up and ionizes the gas in the cluster. As a result the velocity of the gas is expected to be comparable to the velocity dispersion of the galaxies in the cluster. If the velocity dispersion is σ, the kinetic energy per baryon would be $\sim \mu\sigma^2$ where μ is the mean mass of the baryons in this overall *neutral* plasma. Now the primordial mix contains nine hydrogen ions for every helium ion and 11 electrons, hence

$$\mu \approx (9m_\mathrm{p} + 4m_\mathrm{p} + 11m_\mathrm{e})/21 \approx 0.62m_\mathrm{p},$$

where m_p is the proton mass and m_e that of the electron. On that basis we should find that the kinetic energy per particle equals the thermal energy of the X-ray source per particle. The measured ratio between these quantities

$$\beta = \frac{\mu\sigma^2}{k_\mathrm{B}T}$$

has a distribution around unity for the clusters studied, which supports the plasma interpretation.

Assuming the plasma observed in a cluster has reached equilibrium after virialization, then the cluster mass including its dark matter content can be deduced. For simplicity the gas cloud is taken to be spherically symmetric. In hydrostatic equilibrium the radial pressure gradient in the gas at radius r balances the inward gravitational force

$$\mathrm{d}P_\mathrm{g}/\mathrm{d}r = GM(<r)\rho_\mathrm{g}/r^2, \tag{16.24}$$

where P_g is the gas pressure, ρ_g the gas density at radius r, and $M(<r)$ is the total mass inside the sphere of radius r. Using the perfect gas law at radius r

$$P_\mathrm{g} = \rho_\mathrm{g} k_\mathrm{B} T_\mathrm{g}/\mu. \tag{16.25}$$

Differentiating this with respect to the radius gives

$$\frac{\mathrm{d}P_\mathrm{g}}{\mathrm{d}r} = [k_\mathrm{B}/\mu]\left[T_\mathrm{g}\frac{\mathrm{d}\rho_\mathrm{g}}{\mathrm{d}r} + \rho_\mathrm{g}\frac{\mathrm{d}T_\mathrm{g}}{\mathrm{d}r}\right]. \tag{16.26}$$

Combining Equations (16.24) and (16.26) removes the unmeasurable pressure gradient, and finally gives an expression for the enclosed total mass in terms of physical parameters at that radius

$$M(<r) = \frac{k_\mathrm{B} T_\mathrm{g} r}{G\mu}\left[-\frac{\mathrm{d}\ln\rho_\mathrm{g}}{\mathrm{d}\ln r} - \frac{\mathrm{d}\ln T_\mathrm{g}}{\mathrm{d}\ln r}\right]. \tag{16.27}$$

This estimate of the total mass in galaxy clusters is independent of the estimate made using the virial theorem. The two methods give consistent results: in the case of the Coma cluster $\sim 2 \times 10^{15}\ M_\odot$.

The bulk of the baryons, 87% by mass, lies in tenuous clouds of ionized gas in the intergalactic regions, and radiate negligibly in the visible spectrum. There are two distinct components. Around 40% is diffuse ionized intergalactic gas with temperature less than 10^5 K that is spread in the vast regions that started underdense in the early universe and remained underdense. The last 47% of baryonic matter is the ionized *warm-hot intergalactic matter* (WHIM) at temperatures 10^5–10^8 K (0.01–10 keV) that stretches in threads and clumps matching the structures expected for the dark matter distribution. Its density is only 1–1000 baryons per m^3. Detectors aboard satellites recorded the thermal radiation emitted by the WHIM: XMM-Newton (X-ray Multimirror Mission) detecting X-rays from 0.1 to 12 keV and *Chandra* detecting 0.1–10 keV X-rays (0.12–12 nm). The commonest element in the universe after hydrogen and helium is oxygen which exists in highly ionized states in the WHIM: for example, OVI the quintuply ionized state with three retained electrons has transitions at 103 nm and 22 nm. Such lines are detected and from their intensity it is deduced that the metallicity of the WHIM is about $0.5Z_\odot$. The WHIM is evidently reprocessed material. Nuclear burning in stars, the explosions that terminate the life of massive stars, and AGNs all generate massive outflows of matter and radiation. These mechanisms ionize and inject metals into the intergalactic clouds.

16.9 Exercises

1. A galaxy has a luminosity 10^8 $L_\odot$, half coming from within a radius of 4 kpc. The observed line-of-sight velocity dispersion of its stars is 0.1 km s^{-1}. Calculate the galaxy's mass, its mass to light ratio and its overdensity.
2. Can you explain why accretion onto an AGN brings an electron with each proton? What is the momentum flux, that is momentum per unit time, carried by radiation from the AGN with luminosity L? For this part refer to Section 1.14. What is the momentum flux across unit area of the accreting surface, taking this to be spherical and radius r? Writing σ_T as the Thomson scattering cross-section, how much momentum does an electron receive in Thomson scattering from the outgoing radiation per unit time? What is the inward gravitational force on the electron and its proton partner? From these results show that there is a limiting luminosity that prevents more rapid accretion: the Eddington limit. What is this limit for an AGN of mass 10^6 $M_\odot$?
3. A group of interacting galaxies has a velocity dispersion of 350 km s^{-1}. What temperature would the intergalactic gas reach? At what wavelength would this gas radiate most strongly?
4. In a galaxy group the gas is in hydrostatic equilibrium, so that the radial pressure gradient balances the gravitational attraction. Suppose the gas has a uniform temperature of 1.5×10^7 K and the density varies as 1/r out to a radius r of 200 kpc. Calculate the total mass enclosed within this radius.
5. The gas in the group of the previous exercise has mass 8×10^{11} $M_\odot$ and the light from the galaxies is 10^{10} $L_\odot$. Assuming a mass to light ratio of the stars is four times that of the Sun determine the gas/star mass ratio, and the ratio of the total mass to the baryonic mass.
6. Using the initial mass function of Equation (16.17) above a mass cutoff of 0.25 $M_\odot$, calculate the fraction of stars with masses above 5 $M_\odot$ and the fraction of the total mass that they contribute.

Further Reading

Sparke L S and Gallagher J S 2007 *Galaxies in the Universe: An Introduction* (Cambridge: Cambridge Univ. Press). This presents a clear account of the astrophysics of galaxies since the early universe at roughly the same level as the text here.

Mo H, van den Bosch F and White S 2010 *Galaxy Formation and Evolution* (Cambridge: Cambridge Univ. Press). This is an advanced text with thorough coverage at graduate level. Not for the faint-hearted.

References

André, P., Men'shchikov, A., Bontemps, S., et al. 2010, A&A, 518, L102
Blumenthal, G. R., Faber, S. M., Primack, J. R., & Rees, M. J. 1984, Natur, 311, 517
Driver, S. P., Bellstedt, S., Robotham, A. S. G., et al. 2022, MNRAS, 513, 439
Inutsuka, S., & Miyama, S. M. 1997, ApJ, 480, 681
Madau, P., & Dickinson, M. 2014, ARA&A, 52, 415
McQuinn, M. 2016, ARA&A, 54, 313
Read, A. M., & Ponman, T. J. 1998, MNRAS, 297, 143
Springel, V., White, S. D. M., Jenkins, A., et al. 2005, Natur, 435, 629

AAS | IOP Astronomy

Introduction to General Relativity and Cosmology (Second Edition)

Ian R Kenyon

Chapter 17

The Dark Sector

17.1 Introduction

The existence and dominance of dark matter over baryonic matter has been integral to the account of the evolution of the universe given in earlier chapters. In this chapter the evidence for dark matter is reviewed, and considered further. That other mysterious entity, dark energy, has been introduced and the ΛCDM model has been seen to provide a plausible interpretation of the evolution of the universe. Crucial evidence for the existence of dark energy came late last century; it was discovered that the expansion of the universe is accelerating now and has been accelerating for several gigayears. This evidence will be reviewed in the latter part of the chapter.

17.2 Dark Matter

A key observation made by Zwicky in the 1930s is that when the virial theorem is applied to the motion of galaxies in the Coma cluster, the inferred gravitational mass attracting the galaxies is larger than the mass of the visible matter in stars by a factor of order several hundred. This calculation was updated in Section 16.7. Additionally, as shown in Figure 1.7, Vera Rubin and colleagues found that the rotational velocity of the gas in each galaxy observed remains constant well beyond the volume of the stellar population. As explained in Section 1.8 this observation requires the presence of invisible dark matter in galaxies.

One possible interpretation of dark matter is that it is in the form of elementary particles. Taking this interpretation of dark matter, the development of structures in the universe would depend on whether the dark matter particles were relativistic or non-relativistic. If relativistic, the rapid streaming of dark matter would have smoothed out the smallest dark matter structures in the universe. As explained in Section 15.6 this is contrary to what actually happened. The development was hierarchical; stars and galaxies formed first, only later collecting in clusters. Dark

doi:10.1088/2514-3433/acc3ffch17

matter is therefore non-relativistic: cold dark matter. Finally the success, discussed in Section 15.7, of the prediction for how the baryon acoustic oscillations developed in the matter era, requires dark matter at a density consistent with other measurements.

These dark matter particles could be supersymmetric particles. Supersymmetry is the remaining unexploited symmetry consistent with the special theory of relativity. Supposing supersymmetry to have been unbroken at the extreme energy density in the universe shortly after inflation, then fermions (with half-integral spins in units $\hbar$) and bosons (with integral spins) would have been interconvertible via supersymmetric particles, whereas now when this symmetry has been broken fermions and bosons are quite distinct. Any number of one species of boson, for example photons, can exist in the same quantum state. In the case of a species of fermion, for example electrons, a quantum state can only ever contain a single fermion. Once the symmetry was broken the supersymmetric particles would have decayed to the lightest neutral supersymmetric particle; making the latter the dark matter particles in today's universe. Pursuing this interpretation, the supersymmetry breaking transition must have occurred before the electroweak symmetry breaking transition. Otherwise the supersymmetric particles would be as light as the Higgs boson ($\sim 10^3$ GeV c^{-2}) and would have been produced and detected from the 13 TeV collisions between protons at the Large Hadron Collider at CERN. Supersymmetric particles more massive than the Higgs would become non-relativistic, and hence CDM, long before the decoupling of matter from radiation. Many experiments have been carried out to detect the huge numbers of such supersymmetric dark matter particles that would currently penetrate all space (including the Earth). Searches are made for examples of nuclei recoiling from collisions with these invisible particles. None has been found. The null results impose very severe limits on the interaction strength of dark matter supersymmetric particles with baryons, orders of magnitude weaker than the weak nuclear force. A second candidate dark matter particle is the *axion*. Its existence would solve a fundamental conceptual difficulty with the standard model of elementary particle physics, namely the absence of any charge-parity violation in the strong force. Axions would have minuscule masses in the range under 10^{-3} eV c^{-2}. An axion would convert to a pair of photons in a strong magnetic field. Attempts to detect axions have so far also failed. Axions, despite their low mass, would have been non-relativistic from early in the life of the universe. Thus, either axions or supersymmetric particles could fulfill the role of dark matter.

From early in the life of the universe dark matter only interacted gravitationally with baryonic matter and electromagnetic radiation. Dark matter could therefore collapse directly into the gravitational potential wells resulting from quantum fluctuations during inflation. At the same time the baryons and photons formed a plasma that oscillated within these wells. In Chapter 15 we saw how dark matter accumulations became the scaffolding of today's galaxies and clusters of galaxies. To summarize:

- There is about six times more dark matter than baryonic matter.
- Dark matter interacts gravitationally with itself and with baryonic matter.
- Dark matter has no other detectable interaction.

- Dark matter is most likely made up of non-relativistic particles, having, as yet, undetermined masses, but certainly less massive than planets, and more likely on the scale of elementary particles.

Dark matter is an essential component of the ΛCDM model that provides a consistent model of the evolution of the universe.

17.3 Gravitational Lensing

Dark matter was discovered through its gravitational interaction with stars and galaxies. It bends spacetime. Consequently concentrations of dark matter deflect electromagnetic radiation, of whatever wavelength, whether γ-rays, X-rays, light, or radio waves. Figure 1.1 shows a special case: one galaxy is aligned with a farther galaxy so that its gravitational attraction focuses light from the far galaxy into an Einstein ring. The deflection produced is determined by the total mass in the nearer galaxy within the ring, whether dark or baryonic. Equation (7.15) gives the angular deflection for light passing at a distance b from the center of a spherically symmetric mass M

$$\alpha = \frac{4GM}{bc^2}, \tag{17.1}$$

valid for the cosmological cases of interest, where the deflection is small. This example indicates how cosmologists use the deflection produced by a galaxy cluster to determine its total baryonic and dark matter mass.

More often the lensing and lensed structures are not so well aligned. This can make the image an arc rather than the full circle; alternatively multiple images are seen of an individual source. Figure 7.7 shows an early observation of two images being identified as those from the same source, the quasar UM673: their spectra and redshifts are identical. Note that the surface brightness of the smeared-out image produced by gravitational lensing is identical to the surface brightness the image would have in the absence of lensing. It is simply that a bigger area of the surface of the source is now visible. As a consequence the flux of light collected by a telescope is increased. This effect has made it possible to detect sources at higher redshifts than would otherwise be possible. The cases described above are all examples of *strong gravitational lensing*.

Figure 17.1 illustrates such lensing. A source S is imaged at I by a lens formed by a nearer galaxy cluster. Given the large separations between the source, the lensing cluster and the observer, it is a fair approximation to suppose the mass of the cluster is concentrated in a thin lens surface LL, and for simplicity we go further and take it to be point-like. All the angles are small so we have the image height

$$\theta_s D_s + \alpha D_{\ell s} = \theta_i D_s,$$

giving a deflection in the observer's plane of

$$\theta_i - \theta_s = \alpha D_{\ell s}/D_s. \tag{17.2}$$

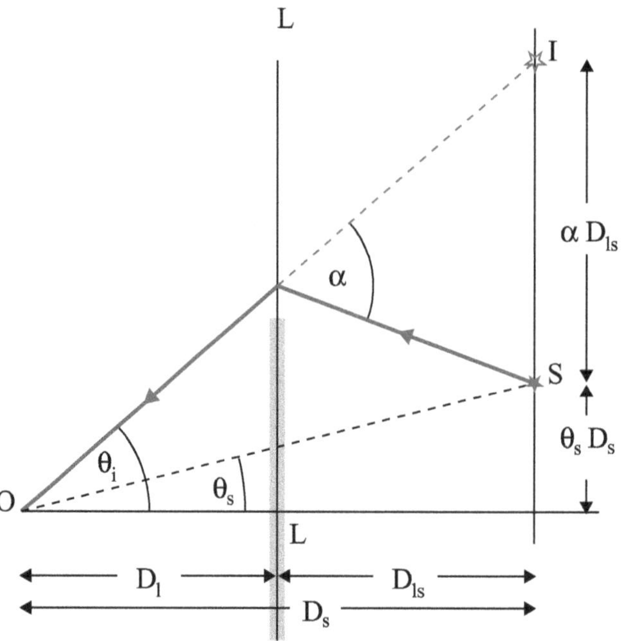

Figure 17.1. Gravitational lensing. A source at S is imaged at I by a galaxy cluster in the plane LL. The lateral dimension shown here is hugely magnified; in reality all the angles are very small.

The distance of closest approach of the ray to the lensing cluster, given by Equation (17.1), is

$$b = \theta_i D_\ell. \tag{17.3}$$

If the source is directly behind the center of the cluster $\theta_s = 0$ and an Einstein's ring would result with $\theta_i = \theta_E$. Using Equations (17.1), (17.2), and (17.3) gives

$$\theta_E = \frac{4GM}{c^2 b}\frac{D_{\ell s}}{D_s} = \frac{4GMD_{\ell s}}{c^2\theta_E D_\ell D_s}, \tag{17.4}$$

so that the Einstein ring has angular radius

$$\theta_E = \sqrt{\left[\frac{4GM}{c^2}\frac{D_{\ell s}}{D_\ell D_s}\right]}. \tag{17.5}$$

Taking a source at 1 Gpc and a lensing cluster of mass $10^{15}\ M_\odot$ at half that distance, the Einstein ring would have an angular radius of 95 arcsec. Keeping the lens halfway to the source the angular radius scales as $\sqrt{M/D_s}$.

Rewriting Equation (17.2), using Equation (17.1) to replace α and Equation (17.3) to replace b, gives

$$\theta_i - \theta_s = \frac{4GM}{c^2 b}\frac{D_{\ell s}}{D_s} = \frac{4GMD_{\ell s}}{c^2\theta_i D_\ell D_s} = \theta_E^2/\theta_i, \tag{17.6}$$

and rearranging this we have

$$\theta_i^2 - \theta_s\theta_i - \theta_E^2 = 0. \tag{17.7}$$

In this case when the alignment is not exact there are two images with

$$\theta_i = \frac{1}{2}\left[\theta_s \pm \sqrt{\theta_s^2 + 4\theta_E^2}\right]. \tag{17.8}$$

Taking the positive sign gives $\theta_i > \theta_E$, and the image lies outside the Einstein ring. Taking the negative sign makes θ_i negative and $|\theta_i| < \theta_E$, so the image lies inside the ring, below the axis (the horizontal line) in the figure. This second image is inverted top to bottom. Lensing galaxies and clusters are extended objects, not necessarily axially symmetric, so that images also depend on the geometry of the lens.

A less restrictive example is that of an extended lensing galaxy axially symmetric around a line drawn from the observer. Thanks to the symmetry the deflection produced is the same as that of a point object on axis; the mass of this equivalent object is equal to the integrated mass of the actual lensing object within radius b. A further useful simplification is to take the galaxy's mass to have uniform *surface mass density* Σ across LL in Figure 17.1. Then the mass doing the lensing is

$$M = \pi b^2\Sigma. \tag{17.9}$$

The gravitational attraction on the ray passing through the lensing surface off axis by b due to all the mass outside the circle of radius b cancels exactly. This result parallels the result that a body at radius r within a spherically symmetric mass distribution feels no effect from matter outside radius r. A critical surface mass density is defined as that just sufficient to produce an Einstein ring. Now in the case of an Einstein ring Equations (17.5) and Equation (17.3) give

$$\theta_E^2 = \left[\frac{b}{D_\ell}\right]^2 = \frac{4GM}{c^2}\frac{D_{\ell s}}{D_\ell D_s}. \tag{17.10}$$

Rearranging this equation gives the critical mass required to produce an Einstein ring,

$$M = \frac{b^2c^2D_s}{4GD_\ell D_{\ell s}}, \tag{17.11}$$

whence the *critical surface mass density* is

$$\Sigma_{cr} = \frac{M}{\pi b^2} = \frac{c^2}{4\pi G}\frac{D_s}{D_\ell D_{\ell s}}. \tag{17.12}$$

Notice that the result is independent of b. Whatever the area of the lensing body there will be an Einstein ring at its edge if the surface density has the critical value Σ_{cr}. The *convergence* is defined as the ratio of the surface mass density to the critical surface mass density

$$\kappa = \Sigma/\Sigma_{cr}. \tag{17.13}$$

If the convergence is larger than unity there will be multiple images. Requiring the convergence κ to be greater than unity provides a convenient definition of the strong gravitational lensing regime. Where the angle between the lines of sight through the source and through the lens is large, κ will be much less than unity. In such cases the effect of lensing will only amount to a distortion of the image. This is the *weak gravitational lensing* regime.

17.4 MACHOs

Dark matter could plausibly exist as dark stars lighter than the brown dwarfs. These objects are called massive astrophysical compact halo objects or *MACHOs*. Extensive searches have been made by the OGLE Collaboration (Wyrzykowski et al. 2011) for MACHOs in the Large and Small Magellanic Clouds, satellite galaxies of our own galaxy. Thanks to the relative motion of all three, the Earth, a suitable located MACHO and a background star, all could at some moment line up. Then the MACHO would briefly pass in front of the star. The consequent gravitational lensing would cause the distant star to brighten just as briefly. In order to exclude cases where the distant star was itself changing state, it was required that the temporary brightening was the same all across the distant star's spectrum. The OGLE Collaboration found only three possible examples in 13 years. This result imposed an *upper limit* on the contribution of MACHOs to dark matter of order 1%.

17.5 Cosmic Shearing

In weak gravitational lensing the lensed source is well outside the Einstein ring of the lens and the distortion of the image is small. Weak lensing changes the apparent shape of a galaxy. If an elliptic galaxy is being imaged, then in general, the ratio of the major to the minor axis, its ellipticity, is altered. This effect is known as shearing, and the shearing produced by the totality of matter between a distant galaxy and the observer is called *cosmic shearing*.

If cosmic shearing can be measured then in principle the density of matter in the universe could be determined, using our knowledge of how gravitational fields deflect the path of light. The ellipticity of the image of a galaxy will be the sum of its intrinsic ellipticity and gravitationally induced shear. Of course we know neither the actual ellipticity nor the orientation of a lensed galaxy. What makes it possible to unscramble the shear is that the ellipticities of galaxies with small angular separation in the sky will be correlated because the light rays from both galaxies have undergone very similar gravitational lensing. Measuring the correlation between the ellipticities of large numbers of pairs of galaxies with similar redshifts is the key to eliciting the shear and hence in determining the integrated density of the intervening matter. The parameter to be measured is the correlation of the shear between pairs of galaxies. Shears are all of order 1% so that to an excellent approximation the shear is linearly dependent on the difference between the gravitational fields over the paths from the two sources to the observer. This gives shear a formal structure in common with gravitational waves. Gravitational waves are shown in Chapter 9 to have the quadrupole modes expressed in Equation (9.7).

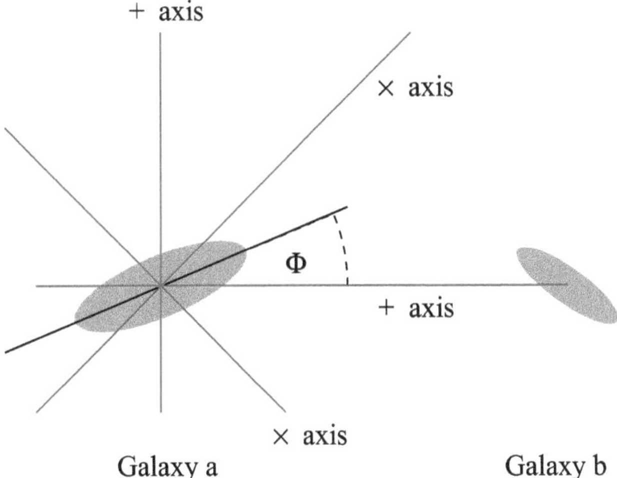

Figure 17.2. Axes are shown for the shear modes of a galaxy referenced to a second galaxy. The +-axes define one mode of shear, with shrinkage along one axis and expansion along the other. Similarly the ×-axes rotated by 45° apply for the orthogonal shear mode.

The corresponding quantities here are simply the shear components in the image plane. Figure 17.2 shows the reference axes for the two independent shear modes drawn at the left-hand galaxy. The axes of the two modes, labeled + and ×, make an angle of 45°. The difference in distortion between the imaging of nearby regions depends on the differential:

$$\frac{(\partial\theta_{\mathrm{s}})_k}{(\partial\theta_{\mathrm{i}})_j} = A_{kj} \tag{17.14}$$

where k and j are labels for axes at right angles in the image plane. As before, i and s label image and source. Guided by Equation (9.7), A is written

$$A = (1 - \kappa)\begin{bmatrix} 1 + \gamma_+ & \gamma_\times \\ \gamma_\times & 1 - \gamma_+ \end{bmatrix}, \tag{17.15}$$

where κ is the convergence, while γ_+ and $\gamma_\times$ are the projections of the shear $\boldsymbol{\gamma}$ along the respective axes. Thus in terms of its two components

$$\boldsymbol{\gamma} = (\gamma_+,\ \gamma_\times), \tag{17.16}$$

where

$$\gamma_+ = \gamma\cos 2\phi \ \text{ and } \ \gamma_\times = \gamma\sin 2\phi, \tag{17.17}$$

with ϕ being the angle shown in Figure 17.2. Again, as for gravitational waves, the doubled angle 2ϕ appears here because of the way quadrupoles (spin-2 tensors) transform under rotations. We can separate out the magnification

$$\mu = 1/(\text{determinant } A) = \frac{1}{(1-\kappa)^2 + \gamma^2} \approx 1 + 2\kappa. \tag{17.18}$$

Now the measurable quantity, from which the ellipticity of a galaxy is determined, is the quadrupole moment. Continuing with the same perpendicular axes in the image plane, the quadrupole moments are defined by

$$Q_{kj} = \frac{\int I(\boldsymbol{\theta})\theta_k\theta_j \mathrm{d}\boldsymbol{\theta}}{\int I(\boldsymbol{\theta})\mathrm{d}\boldsymbol{\theta}}, \tag{17.19}$$

where $I(\boldsymbol{\theta})$ is the brightness at each two-dimensional point $\boldsymbol{\theta}$ in the image plane. Then the + and × ellipticity components are

$$\boldsymbol{\varepsilon} = \frac{Q_{xx} - Q_{yy}}{Q_{xx} + Q_{yy}}, \frac{2Q_{xy}}{Q_{xx} + Q_{yy}}, \tag{17.20}$$

irrespective of the orientation of the perpendicular x- and y-axes in the image plane. As noted a few lines above, the observed ellipticity of a galaxy is the sum of the intrinsic ellipticity and the shear:

$$\boldsymbol{\varepsilon} = \boldsymbol{\varepsilon}_{\text{int}} + \boldsymbol{\gamma}.$$

The range of ellipticity of galaxies is around 0.3, while the shear is only around 0.01. Therefore any attempt to determine the shear has to be statistical and requires a data set of many millions of galaxies. The strength of the correlation between the ellipticities of pairs of galaxies falls as their separation increases. Hence the galaxy pairs must be grouped according to their angular separation in the sky. For each range of angular separation the mean correlations of interest are:

$$\xi_\pm = \langle(\varepsilon_+)_a(\varepsilon_+)_b \pm (\varepsilon_\times)_a(\varepsilon_\times)_b\rangle \tag{17.21}$$

where a and b label the two galaxies in a pair and the average is taken over all pairs in the selected range of angular separation. Correlations like $\langle(\varepsilon_+)_a(\varepsilon_\times)_b\rangle$ between the orthogonal modes should vanish, which provides an experimental test for the validity of the interpretation given here for the shear observed. Another test involves randomly interchanging the galaxy data set and checking that correlations disappear. The DES Collaboration surveyed 26 million galaxies over an area of 1321 square degrees with redshifts measured to lie in the range 0.2–1.3. In Figure 17.3 the mean correlations are plotted against the angular separation of the galaxy pair (Troxel et al. 2018). The curves superposed in the figure are the prediction for the shear made using the ΛCDM model parameters. This prediction depends on the mean density of dark matter, expressed by the parameter Ω_{m}, and on the clumpiness of matter expressed by the parameter σ_8 introduced in Section 15.6. Measurements of cosmic shearing determine the product $S_8 = \sigma_8\sqrt{\Omega_{\text{m0}}/0.3}$. More recently the DES Collaboration (DES Collaboration et al. 2022) combining data from shearing, galaxy–galaxy lensing and galaxy clustering using 10^8 galaxies find

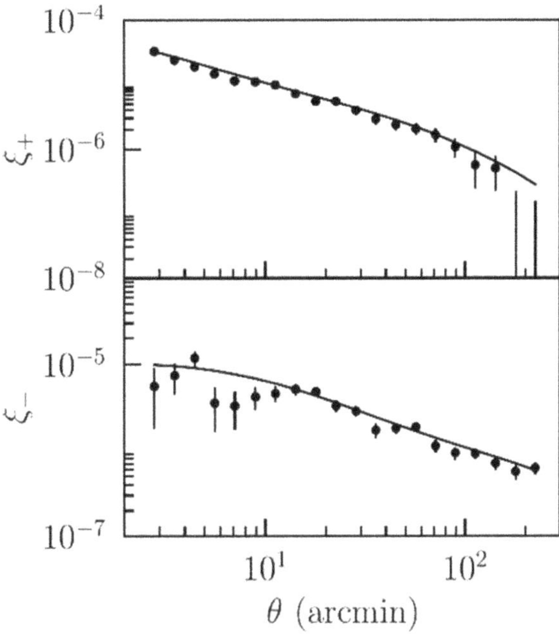

Figure 17.3. Lensing correlations measured by the DES Collaboration. The curves are predictions made with the ΛCDM model. Adapted from Figure 3 from Troxel et al. (2018). Courtesy American Physical Society.

$S_8 = 0.816 \pm 0.008$ and $\Omega_{\mathrm{m}0} = 0.306 \pm 0.006$ in the ΛCDM model. (This is little different from the value 0.30 for $\Omega_{\mathrm{m}0}$ used in calculations in this book.)

It is worth emphasizing that determinations of the distribution of the total mass and its distribution from cosmic shearing and galaxy–galaxy lensing are critical because they only rely on the presumption that the deviation of light rays due to matter is that given by general relativity.

17.6 The Bullet Cluster

Weak gravitational lensing has been used to determine the mass distribution in the Bullet cluster of about 40 galaxies at redshift 0.296 and hence 1.14 Gpc from the Earth. The mass distribution obtained is shown in both parts of Figure 17.4 by the green contour lines: the outer line corresponds to a convergence of 0.16, and successive lines mark increases of 0.07 in convergence. The mass is concentrated in two components whose centers of mass are 0.66 Mpc apart. In the left-hand plot the mass distribution is superposed on the optical image of the Bullet cluster; while in the right-hand plot the X-ray intensity in the interval 0.5 to 5 keV is superposed. The heating of the gas, evidenced by the X-ray flux, indicates that what we see is the result of an on-going collision between two clusters with the lighter partner traveling to the right. Now the stars in galaxies are so widely separated from one another that the stars in the two galaxy clusters pass one another without colliding. By contrast the gas in the two galaxies does collide, gets heated and radiates the X-rays seen in the right-hand image. The obvious bullet seen in that image is gas in the smaller

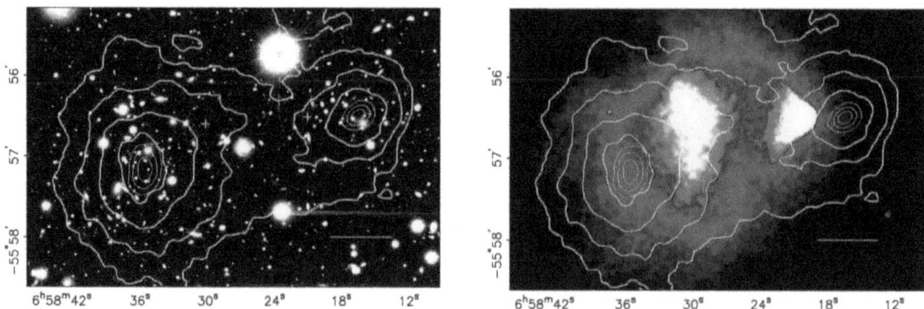

Figure 17.4. The Bullet cluster. In both plots the green lines are the matter density contours determined from weak lensing measurements. On the left they are superposed on the visible galaxies; on the right superposed on the X-ray intensity in the energy range 0.5 to 5.0 keV. At the outer contour the convergence $\kappa = 0.16$, and going inward the contours mark successive increases of 0.07. Figure 1 from Clowe et al. (2006). Courtesy Professor Gonzalez for the authors.

cluster which is being stripped by ram pressure exerted by the intracluster medium. Forward of the visible bullet there is a shock wave traveling at 4700 km s^{-1}; and in the region between the bullet and the surface of the shock wave the temperature of the emitted X-rays climbs to 28 keV. By contrast the temperature both in the bullet and beyond the shock wave is much lower. (Expressing electron energy as $(3/2)k_BT$ gives a temperature of 8.2×10^7 K to power 10 keV X-ray emission.)

A significant feature seen in Figure 17.4 is the precise overlap between the total mass distribution detected by weak gravitational lensing and the distribution of the stars. Evidently dark matter in the galaxy clusters keeps pace with the stars, rather than falling behind with the X-ray emitting baryonic gas. The passage of the dark matter must therefore also be collisionless. These observations confirm the existence of dark matter and show *directly* that its only interaction with baryonic matter is gravitational.

17.7 Dark Energy

Dark energy, currently the dominant contributor to the energy content of the universe only interacts gravitationally. Unlike dark matter it has uniform density, constant in time, and its gravitational effect is repulsive. First we bring together the steps that were followed in the reasoning for the existence of dark energy, and then describe the use of data from observations of SNe Ia (supernovae type Ia), that sealed the argument for the existence of dark energy.

The starting point in the reasoning is this: if the universe obeys the Friedmann equation then a critical energy density, $3H^2c^2/[8\pi G]$, makes the universe precisely flat. Dark and baryonic matter provide only 31% of this critical density. However the location of the angular peaks of the CMB perturbations are where they would be in a flat universe. We showed in Chapter 13 that extrapolating back to a fraction of a second from the origin of the universe, the departure of the universe from flatness would have been vanishingly small. Such a degree of fine-tuning seems unlikely;

more plausibly the energy density was exactly the critical density and would then remain so to the present. Inflation in the first $\sim 10^{-36}$ s was invoked to ensure that, whatever the starting conditions, the universe became as flat as required, and crucially made the universe as homogeneous as it is today. The conclusion then drawn is that some undetected material, *dark energy*, makes up the deficit needed to reach the critical density. It contributes the bulk of the energy, around 70% at the current epoch.

Significantly, the properties of this dark energy are, so far as we can tell, consistent with the cosmological constant in Einstein's equation. If this association is correct, then as discussed in Chapter 6, dark energy exerts a repulsive gravitational force, the opposite from matter. It appears to fill space uniformly; increasing with the volume of space, and expanding space indefinitely. This makes for a grim future, though many gigayears away.

17.8 SNe Ia and the Distance Scale

The direct evidence for dark energy, and with it the resurrection of Einstein's cosmological constant, emerged late last Century. Supernovae were proposed as standard candles to extend the cosmic distance scale. The objective had been to detect the expected deceleration of the expansion of the universe under the universal gravitational attraction of matter. In fact the reverse was observed, the expansion of the universe is accelerating and has been doing so for around 5 Gyr.

One class of supernovae, SNe Ia, was found to be suitable standard candles and measurements of the distance of these supernovae now extend to $z \sim 2$. This covers a time period going back ~10 Gyr. The contrasting interpretation of the redshift of a SN Ia, with and without dark energy, is illustrated in Figure 17.5 by the curves marked "Λ" and "No Λ", respectively. The supernova is marked by a star in the figure. If an observer, who presumes that $\Lambda = 0$, measures the redshift, then he/she would infer that the supernova occurred more recently than it actually did. The conclusion would be that the supernova was unexpectedly dimmer than it should have been at that redshift. This unexplained dimming observed for SNe Ia (Perlmutter et al. 1997; Riess et al. 1998) convinced cosmologists that the expansion of the universe had begun accelerating, following the red path in the figure. It is the repulsive gravitational interaction of dark energy that is responsible for this acceleration.

An SN Ia is a supernova produced from the white dwarf in a binary star pair, the other star being a giant star from which matter funnels onto the white dwarf. At some point the mass of the white dwarf reaches the Chandrasekhar limit of 1.4 $M_\odot$ so that its temperature and pressure are enough to ignite thermonuclear fusion. Around 0.5 $M_\odot$ is converted into ^{56}Ni, which then decays with a half life of 6.1 days to ^{56}Co, and this in turn decays with a half life of 77.2 days to ^{56}Fe. ^{56}Fe has the largest binding energy per nucleon of all nuclear species so burning terminates. The total energy released is around 10^{44} J. In the initial burning phase the radiated power peaks at around 10^{36} W so that the SN Ia outshines its parent galaxy. Si, Ni, Co, and Fe spectral lines are seen but no hydrogen lines. The SN Ia cools over weeks

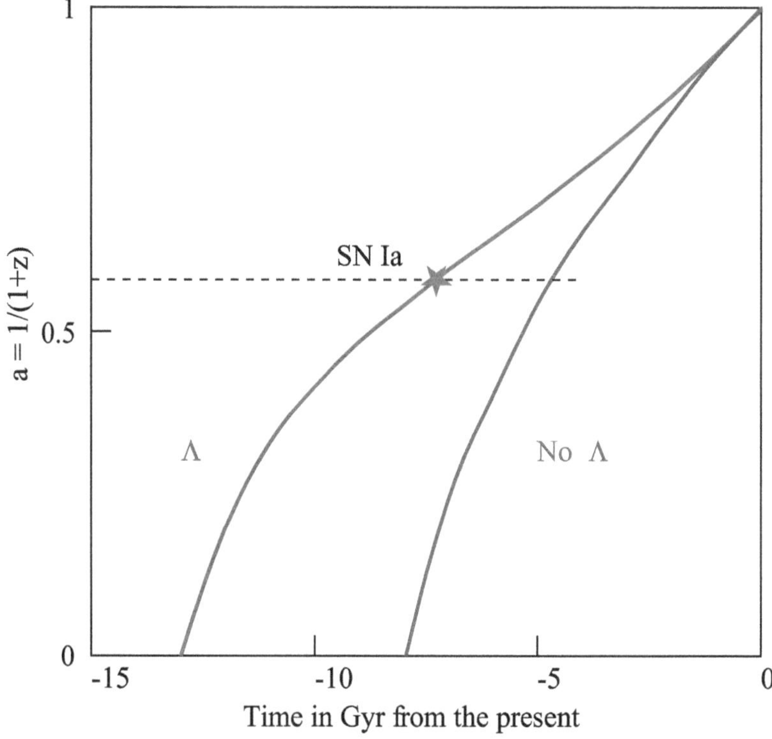

Figure 17.5. Evolution of the scale parameter and redshift with time for universes assuming there is dark energy, and assuming there is no dark energy. Note that where the curve is concave down there is deceleration, and where concave up, acceleration.

radiating the energy generated by the radioactive decays.[1] Riess, Schmidt, Perlmutter and colleagues followed the development of the SNe Ia spectra from the initial rapid rise in intensity through the slow fall, the *light curve*. They demonstrated that the peak brightness and the decay rate of SNe Ia varied little and that variations of these observables were tightly correlated one with another. This made it possible to use SNe Ia as standard candles. An alternative interpretation to account for the systematic dimming was that it was due to dust around the SNe Ia. However such dimming would be accompanied by reddening of spectra. Spectroscopy was used to study this possibility: the effect was found to be small and could be compensated reliably.

Three different types of observation are used to measure distance as a function of redshift, commonly called the distance ladder; the ranges overlap enough to cross-calibrate their results. At short distances within the Galaxy parallax measurements have been made using earlier the satellite *Hipparcos* and later *Gaia*. From within the Galaxy and out to neighboring galaxies Cepheid variable stars are standard candles: the frequency of their luminosity variation is constant and correlates in a

[1] SNe Ia also occur when two white dwarfs merge.

predictable manner with their luminosity. More recently Cepheid variables have been supplemented by the TRGB method discussed in Section 1.4. The use of SNe Ia as standard candles has carried the scale to distant galaxies with redshifts currently above 2.0. Because the absolute peak brightness of SNe Ia is not well determined from first principles, the overlap with shorter range standard candles is essential in calibrating the supernovae Ia distance scale.[2]

The observed redshift and luminosity distance of a standard candle can be connected using in turn Equations (11.10) and (11.1):

$$d_{\mathrm{L}} = d_{\mathrm{P}}(1 + z) = c(1 + z)\int_{t}^{t_0} \frac{\mathrm{d}t'}{a(t')}, \tag{17.22}$$

where z is the redshift at time t, and now is the time t_0. Then Equation (10.30) can be used to give an integral in a with explicit dependence on the proportions of matter and dark energy in the universe:

$$d_{\mathrm{L}} = \frac{c(1 + z)}{H_0}\int_{a(z)}^{1}\left[\frac{\mathrm{d}a}{(\Omega_{\mathrm{m}0}a + \Omega_{\mathrm{r}0} + \Omega_{\Lambda 0}a^4)^{1/2}}\right]. \tag{17.23}$$

Inserting values of $\Omega_{\mathrm{m}0}$, etc. from a model of the universe in Equation (17.23) yields one estimate of the luminosity distance. The luminosity distance to a SNe Ia can also be determined directly from the difference between the predicted absolute magnitude M of the SNe Ia and its measured apparent magnitude m: using Equation (1.19) we get

$$M - m = -5\log_{10}[d_{\mathrm{L}}/10\ \mathrm{pc}]. \tag{17.24}$$

If the model of the universe describes its evolution correctly, including any period of accelerated expansion, then the two estimates of the luminosity distance should coincide.

In practice the SNe Ia data is presented as a plot of the *distance modulus* $\mu = M - m$ against the redshift, with a prediction made using the model of the universe superposed. The intention had been to track the slowing down of the expansion with time, as was expected in a matter-dominated universe. Perlmutter, Riess, Schmidt, and their colleagues found the reverse: the expansion for the last 5 Gyr has been speeding up thanks to the repulsive gravitational force of dark energy. Figure 17.6 shows the distance moduli versus redshift from a sample of 1550 SNe Ia measured by the Pantheon+ Collaboration (Brout et al. 2022). Corrections have been applied to compensate for the effect of the difference in luminosities, depending on the type of host galaxies, and for other small biases. The solid line is a fit to the data made with the ΛCDM model. There is a point of inflection at around redshift 0.5 (5 Gyrs ago) when dark energy began to dominate and the expansion of the universe accelerated. The broken gray line shows the expected bias due to very low redshift effects arising from peculiar velocities. In the lower section the

[2] My thanks to Professor Brout for clarifying several features of SN Ia observations.

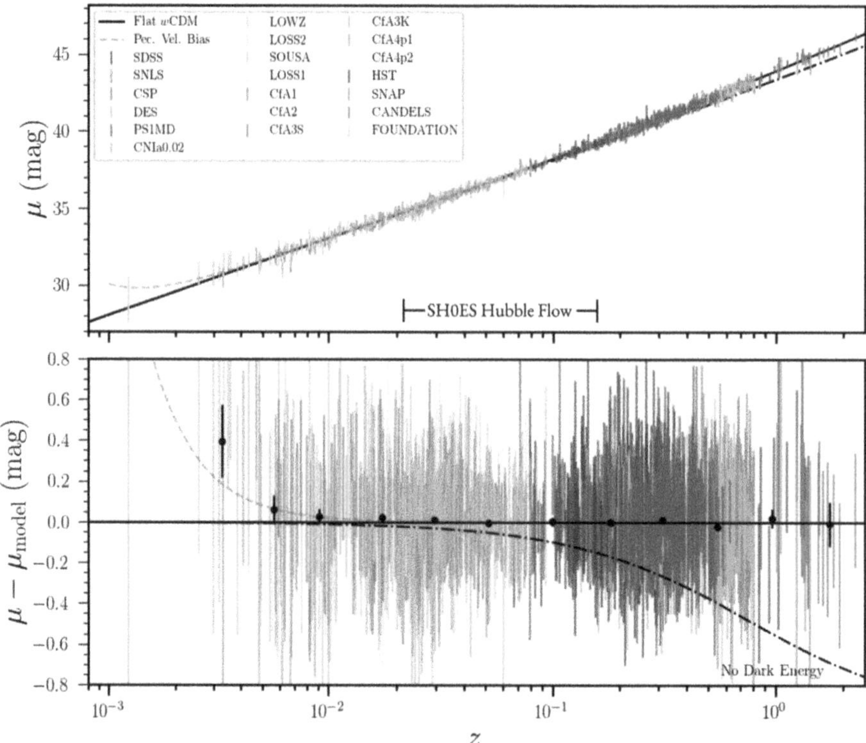

Figure 17.6. The distance modulus versus redshift survey from 1701 SNe Ia light curves compiled by the Pantheon+ Collaboration with redshifts from 0.001 to 2.26: Figure 4 from Brout et al. (2022). Courtesy Professor Brout for the Pantheon+ Collaboration: ArXiv:2202.04077. The solid lines show the predictions using the ΛCDM model. The dot-dash lines show the predictions for no dark energy (courtesy Professor Brout).

differences between the data points and the fit are shown: the black dots with error bars show the data averaged over intervals in redshift. A prediction made with all matter and no dark energy is indicated by the dashed–dotted lines: these diverge from data with increasing redshift. They lie over 0.6 units in magnitude below the data points at redshift 1.0; meaning that if there were no dark energy the supernovae at redshift 1.0 would be a factor of 2 brighter. This difference is many times the statistical variation indicated by the black error bars, which is convincing evidence for dark energy. Currently the universe is in a transitional state: until recently matter dominated, but from hereon it will be dark energy that does so. Riess, Schmidt, and Perlmutter shared the Nobel Prize in Physics for their work in 2011.

The surprising result that the expansion of the universe is now accelerating can be checked by using the ruler provided by the baryon acoustic oscillations introduced in Section 15.7. The ruler's length is the scale factor $D_V(z)$ given by Equation (15.52), which is obtained from isotropic acoustic scale fits to galaxy surveys. Figure 17.7 shows the variation of $D_V(z)$ with redshift during the period of acceleration at low

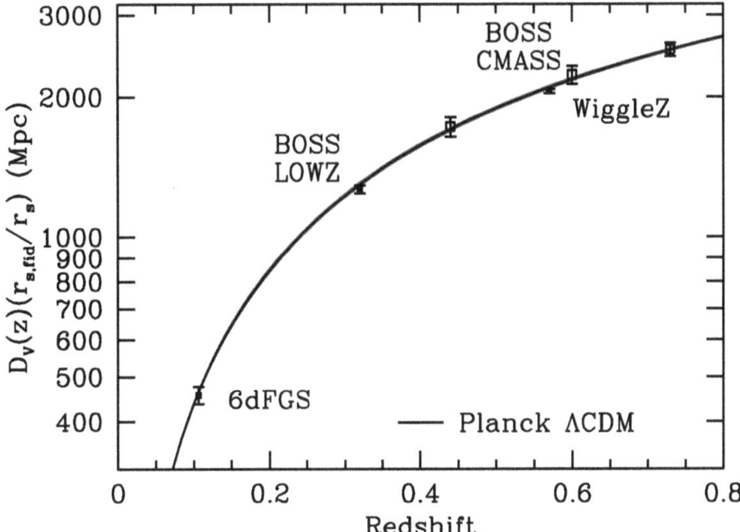

Figure 17.7. The scale factor $D_V(z)$ measured from the baryon acoustic oscillations plotted against redshift. The curve is a prediction made using the ΛCDM model of the universe. Figure 21 from Anderson et al. (2014), Courtesy Royal Astronomical Society, London and Oxford Univ. Press, Oxford.

redshift.[3] The superimposed curve is the prediction of the ΛCDM model, which matches the observed change in the ruler's length with redshift. This confirms the evidence from the observations of SNe Ia that currently the dark energy has taken over from matter as the dominant energy component in the universe.

Referring back, Figure 12.4 shows the constraints imposed in $\Omega_{\Lambda 0}$ Ω_{m0} space by the separate CMB, BAO, and SNe Ia data sets in the ΛCDM model. The three acceptable regions overlap in the neighborhood of the parameter choice used throughout this book: that is where $\Omega_{\Lambda 0}$ is 0.70 and Ω_{m0} is 0.30. This shows that a simultaneous fit to the data from all eras exists, giving confidence that the model describes the evolution of the universe adequately. We can look at each of the three constraints in turn.

First the observed angular spacing of the thermal perturbations of the CMB requires the universe to be close to having perfect flatness; the constraint is that $\Omega_{\Lambda 0} + \Omega_{m0}$ is close to unity. Next the spacing of the BAO grew after decoupling during a matter dominated era and so depends primarily on the fraction of dark plus baryonic matter in the universe: this gives a constraint at roughly constant Ω_{m0}. Finally the acceleration of the expansion of the universe evidenced by SNe Ia data is determined by the difference between the gravitational repulsion due to dark energy ($\Omega_{\Lambda 0}$) and the gravitational attraction due to matter (Ω_{m0}): this gives a band in Figure 12.4 at right angles to the first, CMB defined band. It evidently helps in showing the consistency of the data sets, and in determining the values of $\Omega_{\Lambda 0}$ and

[3] The factor multiplying $D_V(z)$ compensates for the small shift in the parameters of the ΛCDM model used as input into the analysis and their values emerging from the fit to the data.

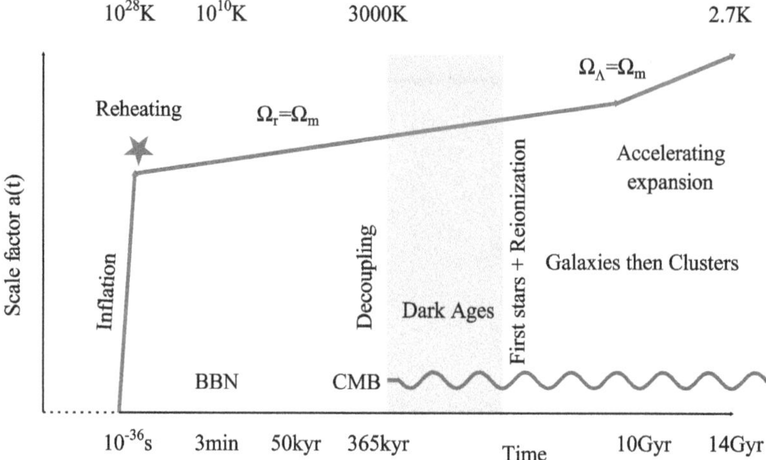

Figure 17.8. Outline of the evolution of the universe as interpreted in the text using the ΛCDM model.

Ω_{m0}, that the three data sets chosen give bands in the figure that intersect at large angles. For reference, Figure 10.3 shows how the energy content of the universe changed with time according to the ΛCDM model.

Figure 17.8 summarizes the life history of the universe.

17.9 What Is Dark Energy?

The choice for the equation of state for dark energy has been $p = w\varepsilon$, with $w = -1$. This is what it would be if dark energy is simply Einstein's cosmological constant. Fits to cosmological data using the ΛCDM model are made with a variable w: these fits require w to lie slightly more positive than, and within errors, consistent with -1.

An interpretation broadly equivalent to the cosmological constant is that dark energy is a scalar field like the Higgs field filling all space, and also like the field responsible for inflation in the early universe. Being a scalar field the associated particles would be spinless, like the Higgs particle.

In quantum field theory the vacuum possesses energy, which should be equivalent to, or contribute to dark energy. This vacuum energy arises from the spontaneous creation and decay of particle–antiparticle pairs in the vacuum. The process must be consistent with the uncertainty principle: if a pair is created with energy ε, it can only exist for a time $\hbar/\varepsilon$. Calculations of atomic energy levels and basic properties like the magnetic moments of electrons and muons must take account of these virtual particle states. Such calculations give agreement at the limit accessible with current experimental precision. Leaving out virtual particle states would both be inconsistent mathematically and inconsistent with the data. One celebrated example is the Lamb shift of 1000 MHz in frequency in the 2p → 1s transition in hydrogen. However, despite expectation, a simple calculation of vacuum energy density due to virtual particle excitation shows how wildly this energy density differs from the dark energy density. For reference the dark energy density is

$$\xi_{\rm cosmo} = \rho_c \Omega_{\Lambda 0} c^2 = 5.65 \times 10^{-10}\ \mathrm{J\ m^{-3}}. \tag{17.25}$$

We start by considering the volume in momentum space between momenta of magnitude p and $p + \mathrm{d}p$, which is simply $\mathrm{d}V = 4\pi p^2 \mathrm{d}p$. Then the corresponding volume in the six dimensional momentum × physical space taking unit volume in physical space is also $\mathrm{d}V$. The number of quantum states in this product volume is determined by applying the uncertainty principle: states can only be packed so close that their spacing does not violate the rule that $\Delta p \Delta r \geqslant \hbar$ in any momentum × position coordinate pair product. Thus the number of distinct quantum states in $\mathrm{d}V$ is $\mathrm{d}V/\hbar^3$. For simplicity the particles in the quantum states are assumed to be bosons, like photons, so that any number can in principle occupy one quantum state. The energy of one quantum is $\varepsilon = \sqrt{p^2c^2 + m^2c^4}$ if the particles have mass m. Now the upper limit on the possible momentum fluctuations is much larger than masses of particles we know about: hence we can take $\varepsilon = pc$ within most of this available momentum space. The zero-point fluctuation in energy due to the creation of virtual particle–antiparticle pairs is one half quantum per quantum state, that is $pc/2$. The possible range of momenta of excitation is from zero to some maximum. This maximum is considered to lie at the Planck scale, discussed in Section 8.4. There the limiting Planck momentum $P_{\rm P}$ is given by Equation (8.26). Consequently the total zero-point energy in quantum fluctuations in a unit volume of physical space is

$$\xi_{\rm part} \sim \int_0^{P_{\rm P}} \left[\frac{4\pi p^2}{\hbar^3}\frac{pc}{2}\mathrm{d}p\right] \sim \left[\frac{P_{\rm P}}{\hbar}\right]^3 P_{\rm P}c \sim 10^{113}\ \mathrm{J\ m^{-3}}. \tag{17.26}$$

The two estimates of the vacuum energy density, one cosmological in Equation (17.25), and the other from elementary particle physics in Equation (17.26), are totally incompatible. The estimates are based on our current otherwise successful understanding of cosmology and particle physics. How this disagreement will be resolved is unclear: it is perhaps in the class of the black body radiation problem that triggered Planck's discovery of quantization.

17.10 Exercises

1. What is the angle deflection of a lightray just grazing a neutron star of mass 1.4 $M_\odot$ and radius 10 km?
2. Show that the expansion rate of a universe changes from decelerating to accelerating when the matter density is twice the dark energy density.
3. Calculate the Einstein ring radius for a lensing mass of 10^3 $M_\odot$ halfway between the observer and a source at 2 Gpc distance. In the same configuration, what is the critical surface mass density?
4. Starting from Equation (17.23) make an estimate of the ratio between the apparent luminosity of a source at redshift 2 in a ΛCDM universe and one in a universe with $\Omega_{\rm m0} = 1.0$.
5. Suppose that during the thermonuclear burning phase of an SN Ia a mass 0.5 $M_\odot$ is converted to ^{56}Ni. Estimate the total energy released. Make use of the binding energies of nuclear species. The nucleus ^{56}Ni decays first by electron

capture to ^{56}Co. This is followed by the emission of a gamma-ray with energy 1.75 MeV; and a decay constant 1.31×10^{-6} s^{-1}, corresponding to a half life of 6.1 days. The gamma-ray energy is converted in the star to radiation across the spectrum. Make an estimate of the initial luminosity. Show that the thermonuclear energy release is enough to disrupt the star. Take the SN Ia progenitor to have mass 1.0 $M_{\odot}$ and radius 7000km.

6. Suppose that black holes of mass 10^{-6} $M_{\odot}$ make up the dark halo of our Galaxy. What would their spacing be? What would be the chance for finding one within one AU from the Sun?

Further Reading

The 2011 Nobel lectures given by Saul Perlmutter and Adam Riess concerning the discovery of the acceleration of the expansion of the universe are well worth reading/viewing. They can be found at Nobel.org.

References

Anderson, L., Aubourg, É., Bailey, S., et al. 2014, MNRAS, 441, 24

Brout, D., Scolnic, D., Popovic, B., et al. 2022, ApJ, 938, 110

Clowe, D., Bradač, M., Gonzalez, A. H., et al. 2006, ApJ, 648, L109

DES CollaborationAbbott, T., Aguena, M. C., Alarcon, A., et al. 2022, PhRvD, 105, 023520

Perlmutter, S., Gabi, S., Goldhaber, G., et al. 1997, ApJ, 483, 565

Riess, A. G., Filippenko, A. V., Challis, P., et al. 1998, ApJ, 116, 1009

Troxel, M. A., MacCrann, N., Zuntz, J., et al. 2018, PhRvD, 98, 043528

Wyrzykowski, L., Skowron, J., Kozłowski, S., et al. 2011, MNRAS, 416, 2949

AAS | IOP Astronomy

Ian R Kenyon

Appendix A

The Particles and Forces

Around 5% of the energy in the universe is carried by baryonic matter; the rest is carried by dark matter and dark energy, neither of which is directly detectable. Baryonic matter includes free protons and neutrons as well as nuclei into which they are bound by their mutual strong interaction. This is nature's strongest interaction, about 10^2 times stronger than the next strongest, the electromagnetic interaction. Nuclei, including protons and neutrons, that all feel the strong force are called *hadrons*. Electrons do not feel the strong force, but are nonetheless classed as baryonic matter by convention in the context of cosmology. Atoms consisting of an electron cloud around a nucleus are equally baryonic matter. Most of what has been learnt about the universe has come from observing electromagnetic radiation from baryonic matter. Neutrons and protons are massive compared to electrons, 938.3 MeV c^{-2} and 939.6 MeV c^{-2}, respectively, compared to 0.511 MeV c^{-2}. The masses are such that neutrons can decay into protons, or given enough energy, a proton can convert to a neutron: thus

$$\begin{aligned} \mathrm{n} &\rightarrow \mathrm{p} + \mathrm{e}^{-} + \bar{\nu}_{\mathrm{e}}, \\ \bar{\nu}_{\mathrm{e}} + \mathrm{p} &\rightarrow \mathrm{e}^{+} + \mathrm{n}. \end{aligned} \tag{A.1}$$

The two processes introduced here occur through the third force, the weak interaction, orders of magnitude weaker than the electromagnetic interaction. The processes indicated involve the electron neutrino ν_{e} and its antiparticle $\bar{\nu}_{\mathrm{e}}$. Neutrinos have minuscule masses, less than 1.0 eV c^{-2}, and feel neither the strong nor the electromagnetic force. Everyone feels the weak force: protons, neutrons, electrons, and neutrinos. The e^{+}, appearing in the second equation, signifies the positron, the antiparticle of the electron. Like other antiparticles it has the same mass but the opposite charge to its particle partner. An antiparticle interacts through the same forces as its particle partner. Although weak, the weak force is what powers the Sun and other astrophysical processes. It requires the high pressure and temperature that develop in the interior of stars when they collapse under their gravitational

doi:10.1088/2514-3433/acc3ffch18

self-attraction. The underlying process going on in the core of the Sun at a temperature of 1.5×10^7 K is

$$4\,{}^1\mathrm{H} \rightarrow {}^4\mathrm{He} + 2e^+ + 2\nu_e, \tag{A.2}$$

where, using nuclear notation, ^{1}H is a proton and ^{4}He a helium nucleus containing four nucleons (two protons and two neutrons). The strong force has a range of 10^{-15} m, which determines the size of nuclei. The weak force is also short range; the electromagnetic force has infinite range, and propagates at the speed of light, c.

Lastly all matter and radiation feels the gravitational force: like the electromagnetic force, it too has infinite range and propagates with the speed of light. Gravitation is by far the weakest force in absolute terms, 10^{36} times weaker than the electromagnetic force between a pair of protons. The gravitational force is unique in its operation: the presence of all matter/radiation distorts spacetime and all matter/radiation follows the wrinkles produced in spacetime. The other three interactions act in spacetime, not on it. Galaxies and their component stars must evidently be overall electrically neutral to high precision (one part in 10^{20}) because, wherever tested, their motion is that determined by gravitation rather than by the more powerful electromagnetic force (Table A.1).

Electromagnetic radiation comes in quanta with energy $E = hf$, where f is the frequency of the radiation and h is Planck's constant. These massless quanta are called photons. In visible radiation of wavelength 500 nm (green) the individual photon's energy is 2 eV. The cosmic microwave radiation that bathes the universe has a black body spectrum at a temperature of 2.7 K, with a corresponding energy of $k_B T = 230\,\mu$eV. In the current era the density of these microwave photons is 4×10^8 m^{-3}, of cosmic background neutrinos, at 1 K, 10^8 m^{-3}: the mean number density of baryons in the universe is 10^9 times less than that of the photons. Earlier, the densities were greater but the ratios were the same, since the formation of the baryons.

For systems in thermal equilibrium the ambient temperature determines whether particle behavior is relativistic or non-relativistic. If the temperature is high enough that the mean particle energy, $E \sim k_B T$, is much greater than the rest energy mc^2 then Equation (1.25) reduces to

$$E = pc, \tag{A.3}$$

Table A.1. Particles and Their Interactions

Particle	Strong	EM	Weak	Charge	Mass
n	Yes	(Yes)	Yes	0	939.6 MeV c^{-2}
p	Yes	Yes	Yes	+e	938.3 MeV c^{-2}
e	No	Yes	Yes	−e	0.511 MeV c^{-2}
ν_e	No	No	Yes	0	~0.5 eV c^{-2}
γ	No	Yes	No	0	0

The neutron electromagnetic interaction is solely through its magnetic moment. All feel the gravitational force.

which is relativistic, and, of course, holds at all temperatures met for photons and, similarly in practice, for neutrinos. At the other extreme for baryons at low enough temperatures Equation (1.25) reduces to

$$E = mc^2 + p^2/(2m), \tag{A.4}$$

the non-relativistic behavior. Roughly speaking the changeover happens for nucleons around GeV energies (10^{13} K) and at MeV energies (10^{10} K) for electrons. Within instants after the Big Bang energies dropped well below 1 GeV so that nucleons have behaved non-relativistically for the greater part of the life of the universe. We shall see that whether particles are non-relativistic or relativistic has far-reaching consequences. The history of the universe depends strongly on whether the particles that make up the mysterious dark matter would be relativistic (light and hot) or non-relativistic (heavy and cold). The latter case is strongly favored by all current observations.

The protons and neutrons are themselves bound states of yet more fundamental particles, known as quarks. There are three quarks in each proton and in each neutron, bound together by the strong force. The force acts through the exchange of gluons, which are the equivalent of the photons that bind electrons to nuclei to form atoms. The parallel goes further. There is a residual attenuated electric dipole force between atoms, the Van der Waals force: correspondingly the residual attenuated strong force between protons and neutrons is what binds them into nuclei.

Introduction to General Relativity and Cosmology
(Second Edition)

Ian R Kenyon

Appendix B

Variational Methods

The general problem is to calculate for what path between A and B in spacetime the path integral

$$I = \int_A^B L \, \mathrm{d}\tau$$

has a stationary (maximum or minimum) value. L is a function of the coordinates x^μ and their derivatives $q^\mu = \mathrm{d}x^\mu/\mathrm{d}\tau$ appearing as the integrand in Equation (5.19). τ is the proper time or an equivalent scalar used to parametrize the path. The end points are fixed so that only variations on the route between A and B are to be considered. The requirement that the path is stationary can be written symbolically as

$$\delta I = \delta \int_A^B L \, \mathrm{d}\tau = 0.$$

Suppose that the stationary path is known and that x^μ and q^μ are the coordinates and derivatives along the path. Next consider the small excursion from this path shown in Figure B.1 over which these quantities become $x^\mu + \delta x^\mu$ and $q^\mu + \delta q^\mu$, respectively. Then the change in the path integral is

$$\delta I = \int_A^B \left(\frac{\partial L}{\partial x} \delta x + \frac{\partial L}{\partial q} \delta q \right) \mathrm{d}\tau.$$

where for clarity the superscript μ has been suppressed. Integrating the second term by parts gives

$$\left[\left(\frac{\partial L}{\partial q} \right) \delta x \right]_A^B - \int_A^B \delta x \left[\frac{\mathrm{d}(\partial L/\partial q)}{\mathrm{d}\tau} \right] \mathrm{d}\tau.$$

doi:10.1088/2514-3433/acc3ffch19

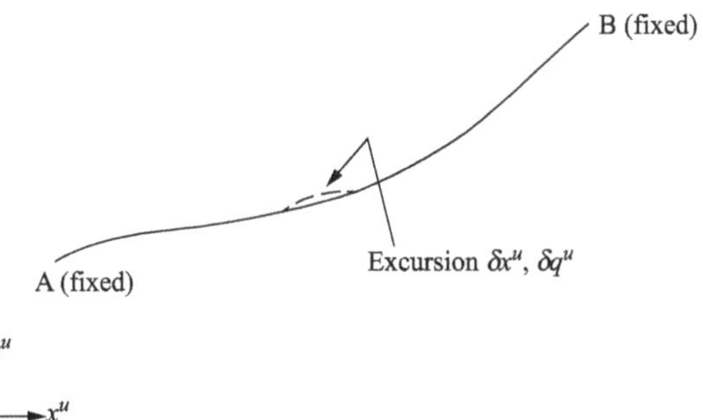

Figure B.1. Schematic diagram of a small excursion from a stationary path.

Of these terms, the first vanishes because the end points of the path are fixed so that $\delta x = 0$ at A and B. Thus

$$\delta I = \int_{\mathrm{A}}^{B} \left[\frac{\partial L}{\partial x} - \frac{\mathrm{d}(\partial L/\partial q)}{\mathrm{d}\tau} \right] \delta x \, \mathrm{d}\tau.$$

This variation must be zero for any choice of δx, which means that the term in square brackets must be identically zero along the whole path. Thus the stationary path has the differential equation (with superscript μ restored)

$$\frac{\partial L}{\partial x^{\mu}} - \frac{\mathrm{d}(\partial L/\partial q^{\mu})}{\mathrm{d}\tau} = 0. \tag{B.1}$$

If L is replaced by $\sqrt{L}$, the result is unchanged. In order to obtain the geodesic equation from Equation (B.1) we make the substitution

$$L = g_{\alpha\beta} q^{\alpha} q^{\beta}.$$

Differentiating, and remembering that the coordinates and velocities are independent quantities, we obtain

$$\begin{aligned}
\frac{\partial L}{\partial x^{\mu}} &= g_{\alpha\beta,\mu} q^{\alpha} q^{\beta} \\
\frac{\partial L}{\partial q^{\mu}} &= g_{\alpha\mu} q^{\alpha} + g_{\mu\beta} q^{\beta} \\
\frac{\mathrm{d}(\partial L/\partial q^{\mu})}{\mathrm{d}\tau} &= g_{\alpha\mu,\sigma} q^{\alpha} q^{\sigma} + g_{\mu\beta,\sigma} q^{\beta} q^{\sigma} + 2 g_{\mu\alpha} \frac{\mathrm{d}q^{\alpha}}{\mathrm{d}\tau}.
\end{aligned}$$

Substituting these values into Equation (B.1) gives

$$g_{\alpha\beta,\mu} q^{\alpha} q^{\beta} - g_{\alpha\mu,\sigma} q^{\alpha} q^{\sigma} - g_{\mu\beta,\sigma} q^{\beta} q^{\sigma} - 2 g_{\mu\alpha} \frac{\mathrm{d}q^{\alpha}}{\mathrm{d}\tau} = 0.$$

The first three terms simplify using Equation (5.17)

$$-2\Gamma_{\mu\beta\alpha} q^{\alpha} q^{\beta} - 2 g_{\alpha\mu} \frac{\mathrm{d}q^{\alpha}}{\mathrm{d}\tau} = 0,$$

that is

$$g_{\mu\nu}\Gamma^{\nu}{}_{\beta\alpha}q^{\alpha}q^{\beta} + g_{\mu\nu}\frac{\mathrm{d}q^{\nu}}{\mathrm{d}\tau} = 0,$$

where in the second term the equality of $g_{\mu\nu}$ and $g_{\nu\mu}$ has been used, and the repeated index has been changed from α to ν. This reduces to

$$\Gamma^{\nu}{}_{\beta\alpha}q^{\alpha}q^{\beta} + \frac{\mathrm{d}q^{\nu}}{\mathrm{d}\tau} = 0.$$

Finally, expressing q^{μ} in full, we have

$$\Gamma^{\nu}{}_{\beta\alpha}\frac{\mathrm{d}x^{\alpha}}{\mathrm{d}\tau}\frac{\mathrm{d}x^{\beta}}{\mathrm{d}\tau} + \frac{\mathrm{d}^2x^{\nu}}{\mathrm{d}\tau^2} = 0, \tag{B.2}$$

which is the geodesic Equation (5.15).

If L is independent of one of the coordinates x^{ν}, then the corresponding component of Equation (B.1) becomes much simpler

$$\frac{\mathrm{d}(\partial L/\partial q^{\nu})}{\mathrm{d}\tau} = 0, \tag{B.3}$$

while the remaining three component equations with $\mu \neq \nu$ retain the more general form of Equation (B.1). In this case $\partial L/\partial q^{\nu}$ is a constant of the motion, a result which is frequently used in Chapter 7. The reader may recognize that the variational methods used here are closely related to the Lagrangian analysis of classical mechanics. In that case L is the Lagrangian $T - V$, where T is the kinetic energy and V is the potential energy. In the cases of interests T is purely a quadratic function of the momentum so that, if V is independent of some coordinate x^{μ}, the result Equation (B.3) follows. This is the basis of the conservation laws of classical mechanics. If V, and hence L, is independent of the position coordinates, as is usually the case, then

$$\frac{\partial L}{\partial q^{\mu}} = \frac{\partial T}{\partial q^{\mu}} = \frac{\partial(q^2/2m)}{\partial q^{\mu}} = \frac{q_{\mu}}{m}$$

is a conserved quantity. In other words, the linear momentum is conserved. Similar calculations lead to the conservation of energy and angular momentum. The corresponding conserved quantities for Schwarzschild spacetime are discussed in Chapter 7 during the study of planetary orbits round the Sun.

Appendix C

The Schwarzschild Metric

The spherically symmetric metric satisfying Einstein's equation for empty space is the Schwarzschild metric. This is straightforward to demonstrate using the additional assumption that the metric is static. The most general symmetric metric is

$$ds^2 = a\ c^2\ dt^2 - b\ dr^2 - 2e\ dt\ dr - r^2\ d\Omega^2,$$

where a, b and c are functions of r and t only. A new time coordinate t' can be chosen for which

$$c\ dt' = \left(a^{1/2}\ c\ dt - \frac{e}{a^{1/2}\ c}\ dr\right)f(t,\ r),$$

where f is the integrating factor needed to make the right-hand side a perfect differential. Then

$$ds^2 = Ac^2\ dt'^2 - B\ dr^2 - r^2\ d\Omega^2.$$

If the metric is static then A and B depend on r only. Dropping the prime gives

$$ds^2 = Ac^2\ dt^2 - B\ dr^2 - r^2\ d\Omega^2 \tag{C.1}$$

with metric components

$$g_{00} = A, \quad g_{11} = -B, \quad g_{22} = -r^2, \quad g_{33} = -r^2\ \sin^2\theta.$$

The metric is diagonal so that $g^{00} = 1/g_{00}$, etc. By using the definition of Equation (5.7) the non-zero components of the metric connection can be obtained:

$$\Gamma^0_{01} = \Gamma^0_{10} = A'/2A$$

$$\Gamma^1_{00} = A'/2B, \qquad \Gamma^1_{11} = B'/2B$$

$$\Gamma^1_{22} = -r/B, \qquad \Gamma^1_{33} = -r\ \sin^2\theta/B$$

doi:10.1088/2514-3433/acc3ffch20

$$\Gamma^2_{12} = \Gamma^2_{21} = \Gamma^3_{13} = \Gamma^3_{31} = 1/r$$

$$\Gamma^2_{33} = -\sin\theta\cos\theta, \qquad \Gamma^3_{32} = \Gamma^3_{23} = \cot\theta.$$

Here the prime denotes differentiation with respect to r. Next using the definition of the Ricci tensor,

$$R_{\beta\delta} = R^{\alpha}_{\beta\alpha\delta}$$

and Equation (6.3) for the Riemann curvature tensor, we obtain

$$R_{00} = \frac{A''}{2B} - \frac{A'}{4B}\left(\frac{A'}{A} + \frac{B'}{B}\right) + \frac{A'}{rB} \tag{C.2}$$

$$R_{11} = -\frac{A''}{2A} + \frac{A'}{4A}\left(\frac{A'}{A} + \frac{B'}{B}\right) + \frac{B'}{rB} \tag{C.3}$$

$$R_{22} = 1 - \frac{r}{2B}\left(\frac{A'}{A} - \frac{B'}{B}\right) - \frac{1}{B} \tag{C.4}$$

$$R_{33} = R_{22}\sin^2\theta \tag{C.5}$$

Using Equations (C.2) and (C.3) gives

$$\frac{R_{00}}{A} + \frac{R_{11}}{B} = \frac{(A'/A + B'/B)}{rB}. \tag{C.6}$$

Now in empty space Einstein's equation reduces to

$$R_{\mu\nu} = 0. \tag{C.7}$$

Therefore Equation (C.6) becomes

$$A'/A + B'/B = 0,$$

whence

$$AB = \text{constant}.$$

At points remote from the source both A and B tend to unity, so that the constant is also unity. Therefore

$$B = 1/A.$$

Substituting this value for B in Equation (C.4) and making use of Equation (C.7) again gives

$$1 - rA' - A = 0.$$

Integrating this with respect to r gives

$$rA - r = \text{constant}.$$

However experiments on the gravitational redshift fix the asymptotic form of A to be $1 - 2GM/rc^2$. Therefore the constant is $-2GM/c^2$, and so the general form for A is

$$A = 1 - \frac{2GM}{rc^2},$$

which is identical with the asymptotic form. Then, since $B = 1/A$, the complete metric equation is

$$ds^2 = \left(1 - \frac{2GM}{rc^2}\right)c^2\,dt^2 - \frac{dr^2}{1 - 2GM/rc^2} - r^2\,d\Omega^2. \tag{C.8}$$

Next we evaluate tensors for the Schwarzschild metric. For convenience we put

$$m = 2GM/c^2, \qquad Z = 1 - m/r.$$

Then

$$g_{00} = Z, \quad g_{11} = -1/Z, \quad g_{22} = -r^2, \quad g_{33} = -r^2\sin^2\theta.$$

The non-zero connections are

$$\Gamma^1_{00} = mZ/2r^2, \qquad \Gamma^0_{01} = \Gamma^0_{10} = m/2r^2Z$$

$$\Gamma^2_{12} = \Gamma^2_{21} = 1/r, \qquad \Gamma^1_{22} = -rZ$$

$$\Gamma^1_{11} = -m/2r^2Z, \qquad \Gamma^1_{33} = -rZ\sin^2\theta$$

$$\Gamma^3_{31} = \Gamma^3_{13} = 1/r, \qquad \Gamma^2_{33} = -\sin\theta\cos\theta$$

$$\Gamma^3_{32} = \Gamma^3_{23} = \frac{\cos\theta}{\sin\theta}.$$

The non-zero elements of the Riemann curvature tensor are

$$R^3_{131} = -m/2r^3Z, \qquad R^1_{010} = -mZ/r^3$$

$$R^1_{212} = -m/2r, \qquad R^2_{323} = m\sin^2\theta/r$$

$$R^0_{303} = -m\sin^2\theta/2r, \qquad R^0_{202} = -m/2r$$

plus components that are related by symmetry operations, for example:

$$-R_{3113} = -R_{1331} = R_{1313} = R_{3131},$$

where

$$R_{3131} = g_{33}R^3_{131}, \text{ etc.}$$

The components of the Ricci tensor

$$R_{\mu\nu} = R^{\alpha}_{\mu\alpha\nu}$$

all vanish, as they must if the metric satisfies Einstein's equation in empty spacetime. Finally we obtain the curvature of two-dimensional surfaces in spacetime. Take the surface defined by geodesics, which at the origin lie in directions such that the polar angle $\theta = \pi/2$ and have t constant. The Gaussian curvature of this surface is

$$K_{r\varphi} = -\frac{R_{3131}}{g_{11}g_{33}} = -\frac{R^{3}_{131}}{g_{11}} = -\frac{m}{2r^3},$$

that is

$$K_{r\varphi} = -\frac{GM}{r^3c^2}.$$

This checks the consistency of Equation (3.43).

Appendix D

Energy Flow in Gravitational Waves

In the linearized approximation appropriate to regions of spacetime where gravitational curvature is small, the metric can be written as in Equation (9.2)

$$g_{\alpha\beta} = \eta_{\alpha\beta} + h_{\alpha\beta}; \qquad h_{\alpha\beta} \ll 1.$$

The stress–energy tensor for gravitational waves was shown in Section 9.2 (Equation (9.14))

$$t_{\alpha\beta} = -\frac{c^4}{8\pi G} G^{(2)}_{\alpha\beta}, \tag{D.1}$$

where only terms quadratic in $h_{\alpha\beta}$ are retained on the right-hand side. This stress–energy tensor will now be calculated for a plane transverse-traceless wave with (+) polarization using Cartesian coordinates. Thus

$$h_+ = h_{11} = -h_{22} = h\cos[k(x^0 - x^3)]. \tag{D.2}$$

The only components of the metric that do not vanish are

$$g_{11} = -1 + h_+, \quad g^{11} = -1 - h_+,$$

$$g_{22} = -1 - h_+, \quad g^{22} = -1 + h_+,$$

$$g_{00} = g^{00} = 1, \qquad g_{33} = g^{33} = -1.$$

Using Equation (5.7) we obtain

$$2\Gamma^1_{01} = 2g^{11}\Gamma_{101} = g^{11}g_{11,\,0} = -h_{+,0} - h_+h_{+,0},$$

where the product

$$h_+h_{+,0} \propto \sin[2k(x^0 - x^3)].$$

doi:10.1088/2514-3433/acc3ffch21

One later step in calculating energy flow requires us to take the average of the stress–energy tensor over a region of spacetime covering several complete waves. With this averaging procedure any contribution from the above product vanishes:

$$\langle h_+ h_{+,0} \rangle = 0. \tag{D.3}$$

Therefore we can ignore such terms from here on. Then

$$\Gamma^1_{10} = \Gamma^1_{01} = \Gamma^0_{11} = -h_{+,0}/2,$$

and

$$\Gamma^1_{13} = \Gamma^1_{31} = -\Gamma^3_{11} = -h_{+,3}/2 = +h_{+,0}/2.$$

Similar formulae hold with the suffix 1 replaced everywhere by 2. Referring to the expression for the Riemann tensor (Equation (6.3)) shows that if only the terms quadratic in h are retained we have

$$R^{(2)\alpha}{}_{\beta\gamma\delta} = \Gamma^\alpha_{\sigma\gamma}\Gamma^\sigma_{\beta\delta} - \Gamma^\alpha_{\sigma\delta}\Gamma^\sigma_{\beta\gamma}.$$

Relevant non-zero components of this Riemann tensor are

$$-R^{(2)1}{}_{010} = -R^{(2)2}{}_{020} = h^2_{+,0}/4,$$

$$-R^{(2)0}{}_{101} = R^{(2)3}{}_{131} = h^2_{+,0}/4,$$

$$R^{(2)1}{}_{013} = R^{(2)1}{}_{310} = -h^2_{+,0}/4,$$

$$-R^{(2)1}{}_{313} = -R^{(2)2}{}_{323} = h^2_{+,0}/4, \tag{D.4}$$

whence it follows that the second-order components of the Ricci tensor are

$$-R^{(2)}_{00} = R^{(2)}_{30} = R^{(2)}_{03} = -R^{(2)}_{33} = h^2_{+,0}/2, \tag{D.5}$$

while for the Ricci scalar

$$R^{(2)} = 0.$$

Therefore the Einstein tensor components are identical with the Ricci tensor components:

$$-G^{(2)}_{00} = G^{(2)}_{30} = G^{(2)}_{03} = -G^{(2)}_{33} = h^2_{+,0}/2. \tag{D.6}$$

Finally, substituting these results in Equation (D.1) we have the components of the stress–energy tensor of the gravitational wave, for example

$$t_{00} = \frac{c^4}{16\pi G} h^2_{+,0}.$$

Now the energy in a wave cannot be localized because only relative displacements have physical significance. It is also not clear whether the energy is located in a peak or trough. Hence it is necessary to average over several complete cycles giving

$$t_{00} = \frac{c^4}{16\pi G}\left\langle h^2_{+,\,0}\right\rangle,$$

where the angular brackets refer to the expectation values. Converting to a time derivative, we obtain

$$t_{00} = \left(\frac{c^2}{16\pi G}\right)\left\langle \dot{h}^2_{+}\right\rangle.$$

If both polarizations are present this result generalizes to

$$t_{00} = \frac{c^2}{16\pi G}\left\langle \dot{h}^2_{+} + \dot{h}^2_{\times}\right\rangle. \tag{D.7}$$

It follows that the energy flux is

$$F = ct_{00} = \frac{c^3}{16\pi G}\left\langle \dot{h}^2_{+} + \dot{h}^2_{\times}\right\rangle, \tag{D.8}$$

meaning the energy crossing unit area in unit time (kg s^{-3}). Equation (D.8) can be rewritten as

$$F = \frac{c^3}{32\pi G}\left\langle \dot{h}^2_{11} + \dot{h}^2_{12} + \dot{h}^2_{21} + \dot{h}^2_{22}\right\rangle,$$

that is

$$F = \frac{c^3}{32\pi G}\left\langle \dot{h}^{\mathrm{TT}}_{ij}\dot{h}^{\mathrm{TT}}_{ij}\right\rangle, \tag{D.9}$$

where we make it explicit, through the superscript TT, that we are working in the transverse-traceless gauge.

AAS | IOP Astronomy

Introduction to General Relativity and Cosmology (Second Edition)

Ian R Kenyon

Appendix E

Radiation from a Nearly Newtonian Source

A Newtonian source is one in which the curvature and strain of spacetime are small, and in which the motion of matter is non-relativistic. Then the metric tensor can be written approximately as:

$$g_{\alpha\beta} = \eta_{\alpha\beta} + h_{\alpha\beta}, \qquad h_{\alpha\beta} \ll 1.$$

When a transverse-traceless gauge is chosen, Einstein's equation reduces to

$$\frac{\partial^2 h_{\alpha\beta}}{\partial x^\mu \partial x_\mu} = -\frac{16\pi G T_{\alpha\beta}}{c^4}, \tag{E.1}$$

where $T_{\alpha\beta}$ is the stress–energy tensor, which vanishes outside the source. The details of the calculation follow the steps leading to Equation (D.9). Far from the source the solution of Equation (E.1) is

$$h_{\alpha\beta}(t) = -\frac{4G}{c^4}\int T_{\alpha\beta}\left(t - \frac{r}{c}\right)\frac{\mathrm{d}V}{r},$$

where the integral is performed over the source and r is the distance from the source to the point where $h_{\alpha\beta}$ is required. This solution closely resembles the retarded solution for the electromagnetic field of a remote source. There is a time delay r/c because the wave has to travel out a distance r. Let y_μ be the coordinates of a representative point in the source. Then

$$h_{\alpha\beta}(t) = -\frac{4G}{rc^4}\int T_{\alpha\beta}\left(t - \frac{r}{c}\right)\mathrm{d}^3y,$$

provided that $y \ll r$. For the assumed non-relativistic source

$$T_{00} \approx \rho_0 c^2 + \rho_0 v^2/2,$$
$$T_{ij} \approx \rho_0 v_i v_j,$$

doi:10.1088/2514-3433/acc3ffch22

where ρ_0 is the rest density and v is the velocity ($\ll c$). Thus

$$\frac{\partial^2}{\partial t^2}\left(\int T_{00}\ y_i y_j\ \mathrm{d}^3 y\right) \approx 2c^2 \int T_{ij}\ \mathrm{d}^3 y,$$

where we have made use of the fact that all accelerations are small. Therefore

$$h_{ij} = -\frac{2G}{rc^4}\frac{\partial^2}{\partial t^2}\left(\int \rho_0 y_i y_j\ \mathrm{d}^3 y\right).$$

We drop the time arguments in order to simplify the presentation. The integrand in brackets is a quadrupole mass moment of the source:

$$I_{ij} = \int \rho_0 y_i y_j\ \mathrm{d}^3 y.$$

The choice we make for h_{ij} is that it is transverse and traceless, and so it follows that equally the quadrupole moment must similarly be rendered transverse and traceless. A conventional starting point is to take the *traceless or reduced* quadrupole moment

$$\mathcal{I}_{ij} = \int \rho_0 (y_i y_j - \delta_{ij} y_k y_k / n)\ \mathrm{d}^3 y, \tag{E.2}$$

for n (2 or 3) dimensional motion. The Kroneker delta δ_{ij} has the value $+1$ when $i = j$ and zero otherwise. Now the component of a vector $\mathbf{d}$ transverse to the unit vector $\mathbf{n}$ is given by

$$d_j^{\mathrm{T}} = P_{jl} d_l,$$

where $P_{jl} = (\delta_{jl} - n_j n_l)$. In the present case $\mathbf{n}$ is a unit vector $\mathbf{r}/r$ pointing away from the source. Similarly a transverse version of $\mathcal{I}_{ij}$ is

$$I_{ij}^{\mathrm{T}} = P_{il} P_{jm} \mathcal{I}_{lm}.$$

However, the trace of this moment does not vanish:

$$\begin{aligned} I_{jj}^{\mathrm{T}} &= P_{jl} P_{jm} \mathcal{I}_{lm} = (\delta_{lm} - n_l n_m)\mathcal{I}_{lm} \\ &= P_{lm} \mathcal{I}_{lm} \\ &= P_{ij} P_{lm} \mathcal{I}_{lm}/2, \end{aligned}$$

where we have used the identity

$$P_{jj} = \delta_{jj} - n_j n_j = 2.$$

The appropriate transverse-traceless quadrupole moment of the source is therefore

$$I_{ij}^{\mathrm{TT}} = P_{il} P_{jm} \mathcal{I}_{lm} - P_{ij} P_{lm} \mathcal{I}_{lm}/2. \tag{E.3}$$

Thus

$$h_{ij} = \frac{2G}{rc^4} \ddot{I}_{ij}^{\mathrm{TT}}. \tag{E.4}$$

Using earlier results from Appendix D we can calculate the energy flux at points distant from the source and also the luminosity of the source. Equation (D.9) gives the flux

$$F = \frac{c^3}{32\pi G}\langle \dot{h}_{ij}\dot{h}_{ij}\rangle \tag{E.5}$$

$$= \frac{G}{8\pi r^2 c^5}\left\langle \dddot{I}_{ij}^{\,\mathrm{TT}}\dddot{I}_{ij}^{\,\mathrm{TT}}\right\rangle. \tag{E.6}$$

Integrating this over a sphere at a distance r from the source gives the total flux, and hence the luminosity of the source is

$$L = r^2 \int F \; \mathrm{d}\Omega.$$

Writing the product of transverse-traceless quadrupole moments in terms of the reduced moments gives

$$F = \frac{G}{16\pi r^2 c^5}\left\langle 2\dddot{\mathcal{I}}_{ij}\dddot{\mathcal{I}}_{ij} - 4n_j n_k \dddot{\mathcal{I}}_{ij}\dddot{\mathcal{I}}_{ki} + n_i n_j n_k n_l \dddot{\mathcal{I}}_{ij}\dddot{\mathcal{I}}_{kl}\right\rangle.$$

It is easy to see that the integral

$$L_2(j, k) = \int n_j n_k \; \mathrm{d}\Omega$$

vanishes if $i \neq j$. Figure E.1 shows a slice through the sphere of integration at constant x_l, where l is the label of the remaining space coordinate ($l \neq k$, $l \neq j$). The value of n_j at P′ is opposite to its value at P but n_k is the same; consequently the contributions near P and P′ cancel. Similarly all contributions to the integral cancel. When $j = k$, this does not happen. For example, if $j = k = 3$ we have

$$L_2(3, 3) = \int_{-\pi}^{+\pi}\int_0^{2\pi} \cos^2\theta \, \sin\theta \; \mathrm{d}\theta \; \mathrm{d}\varphi = 4\pi/3.$$

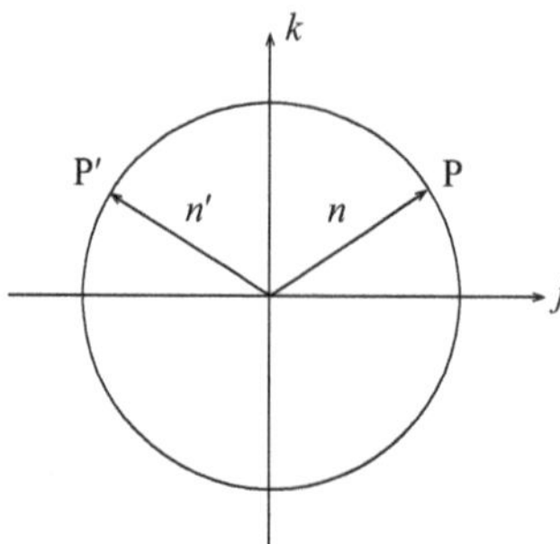

Figure E.1. Schematic diagram of slice through the sphere of integration at constant x_l

In summary $L_2(j, k) = 4\pi\delta_{jk}/3$. A similar analysis shows that

$$\int n_i n_j n_k n_l \, \mathrm{d}\Omega = \frac{4\pi}{15}(\delta_{ij}\delta_{kl} + \delta_{ik}\delta_{jl} + \delta_{il}\delta_{jk}).$$

Finally, using these results in the integral over energy flux gives

$$L = (G/5c^5)\left\langle \dddot{I}_{ij}\dddot{I}_{ij} \right\rangle. \tag{E.7}$$

Ian R Kenyon

Appendix F

The Friedmann Equations

The parameters of supreme interest to the physicist are the sign of curvature and the magnitude of the scale parameter. It emerges that the universe is, if not exactly flat, very close to being so. The scale parameter is then arbitrary and chosen to be unity at present. Interest shifts to enquiring the way this has changed throughout the life of the universe. Equivalently we ask how the Hubble's constant has changed with time.

The universe is taken, for simplicity, to be a perfect fluid with stress–energy tensor

$$T_{\mu\nu} = (p/c^2 + \rho)v_\mu v_\nu - pg_{\mu\nu}, \tag{F.1}$$

where p is the pressure, ρ is the rest density, and v is the fluid velocity. We now set down the components of Einstein's equation. Using the FRW metric of Equation (10.2),

$$g_{00} = g^{00} = 1 \qquad g_{11} = \frac{1}{g^{11}} = \frac{-R^2}{1 - k\sigma^2}$$

$$g_{22} = \frac{1}{g^{22}} = -R^2\sigma^2 \qquad g_{33} = \frac{1}{g^{33}} = -R^2\sigma^2 \sin^2\theta.$$

In a comoving frame $v = (c, 0, 0, 0)$ and

$$T_{00} = \rho c^2 \qquad T_{11} = \frac{pR^2}{1 - k\sigma^2}$$

$$T_{22} = pR^2\sigma^2 \qquad T_{33} = pR^2\sigma^2 \sin^2\theta.$$

The metric connections are evaluated with the help of Equation (5.7): for example,

$$\Gamma^1_{01} = \Gamma^1_{10} = g^{11}\Gamma_{110} = \frac{1}{2}g^{11}g_{11,0}$$

$$= \frac{1}{2}\left(-\frac{1 - k\sigma^2}{R^2}\right)\left\{-\frac{2R\dot{R}}{c(1 - k\sigma^2)}\right\} = \frac{\dot{R}}{Rc}.$$

Equation (6.3) gives the components of the Riemann tensor: for example;

$$R^{1}_{010} = -\Gamma^{1}_{01,0} - \Gamma^{1}_{10}\Gamma^{1}_{01}$$
$$= \left(\frac{\dot{R}^2}{R^2c^2} - \frac{\ddot{R}}{c^2R}\right) - \frac{\dot{R}^2}{c^2R^2} = \frac{-\ddot{R}}{c^2R}.$$

Similarly $R^{2}_{020} = R^{3}_{030} = -\ddot{R}/c^2R$, while R^{0}_{000} is identically zero. Using Equation (6.16) the 00 component of the Ricci tensor is

$$R_{00} = R^{0}_{000} + R^{1}_{010} + R^{2}_{020} + R^{3}_{030} = -3\ddot{R}/c^2R.$$

Its only other non-zero components are

$$R_{11} = \frac{T}{1 - k\sigma^2} \qquad R_{22} = T\sigma^2 \qquad R_{33} = R_{22}\sin^2\theta,$$

where

$$T = 2k + R\ddot{R}/c^2 + 2\dot{R}^2/c^2.$$

Thus the Ricci scalar is given by

$$R(\text{Ricci scalar}) = g^{\mu\nu}R_{\mu\nu} = -6S/R^2,$$

where

$$S = k + R\ddot{R}/c^2 + \dot{R}^2/c^2.$$

Finally the Einstein tensor (Equation (6.17)) is given by

$$G_{\mu\nu} = R_{\mu\nu} - \frac{1}{2}g_{\mu\nu}R(\text{Ricci scalar}),$$

for which the two non-vanishing components are

$$G_{00} = 3\dot{R}^2/R^2c^2 + 3k/R^2,$$

$$G_{11} = -\frac{k + 2R\ddot{R}/c^2 + \dot{R}^2/c^2}{1 - k\sigma^2}.$$

The corresponding 00 and 11 components of the Einstein Equation (6.20) are thus

$$3\dot{R}^2/R^2 + 3kc^2/R^2 - c^2\Lambda = 8\pi G\rho, \tag{F.2}$$

which is generally known as the *Friedmann equation*, and

$$-2\ddot{R}/R - \dot{R}^2/R^2 - kc^2/R^2 + c^2\Lambda = 8\pi Gp/c^2, \tag{F.3}$$

where Λ is the cosmological constant. Other components of the Einstein equation duplicate these results and add no further dynamical information. Equations (F.2) and (F.3) were derived by Friedmann in 1922 (Friedman, 1922) in the period of turmoil following the Russian revolution. He considered the case that the pressure is zero, which is close to being true in the present state of the universe. The equations were derived independently in 1927 by a Belgian priest, Georges Le Maître (Le Maître, 1927) to include pressure.

These equations are manipulated to give forms that are more readily interpreted. First adding Equation (F.2) and 3 times Equation (F.3) gives

$$-6\ddot{R}/R + 2c^2\Lambda = 8\pi G\rho + 24\pi Gp/c^2, \tag{F.4}$$

which can be rewritten as

$$\ddot{R}/R = -\frac{4\pi G}{3c^2}[\rho c^2 + 3p] + c^2\Lambda/3, \tag{F.5}$$

which is known as the *acceleration equation.* A final equation is obtained by first differentiating Equation (F.2)

$$6\dot{R}\ddot{R}/R^2 - 6\dot{R}^3/R^3 - 6kc^2\dot{R}/R^3 = 8\pi G\dot{\rho}. \tag{F.6}$$

To this we add $3\dot{R}/R$ times [Equation (F.2) + Equation (F.3)] to give

$$0 = \frac{8\pi G}{c^2}[\dot{\rho}c^2 + 3\dot{R}p/R + 3\rho c^2\dot{R}/R], \tag{F.7}$$

whence

$$\dot{\rho}c^2 + (3\dot{R}/R)[p + \rho c^2] = 0, \tag{F.8}$$

which is known as the *fluid equation.* The Friedmann, acceleration, and fluid equations can all be simplified by setting

$$R(t) = Ra(t),$$

where R becomes a constant with the dimension of length and $a(t)$ a dimensionless *scale factor*. This change is needed in order to tie the notation to that in general use in cosmology. Then in an expanding universe we set $a = 1$ in the current epoch. Then we have the Friedmann equation

$$3\dot{a}^2/a^2 + 3\kappa c^2/a^2 - c^2\Lambda = 8\pi G\rho; \tag{F.9}$$

where $\kappa/a^2 = k/[Ra]^2$ is the curvature of spacetime, the acceleration equation

$$\ddot{a}/a = -\frac{4\pi G}{3c^2}[\rho c^2 + 3p] + c^2\Lambda/3; \tag{F.10}$$

and the fluid equation

$$\dot{\rho}c^2 + (3\dot{a}/a)[p + \rho c^2] = 0. \tag{F.11}$$

Notice that in flat spacetime $k = 0$ so that the only term in R vanishes.

Simpler expressions are obtained if we introduce an equivalent energy density and pressure to replace the cosmological constant, respectively ε_Λ and pressure p_Λ:

$$\Lambda = \frac{8\pi G}{c^4}\varepsilon_\Lambda, \tag{F.12}$$

where, in order to satisfy the fluid equation,

$$\varepsilon_\Lambda = \rho_\Lambda c^2 = -p_\Lambda . \tag{F.13}$$

The minus sign is surprising, because it reveals that the cosmological constant gives rise to a *repulsive* gravitational force. Then all three equations (Friedmann's, acceleration, and fluid equations) can be expressed more simply in terms of the total energy density and pressure including the cosmological constant:

$$3\dot{a}^2/a^2 + 3\kappa c^2/a^2 = 8\pi G\rho; \tag{F.14}$$

and the acceleration equation becomes

$$\ddot{a}/a = -\frac{4\pi G}{3c^2}[\rho c^2 + 3p]; \tag{F.15}$$

while the form of the fluid equation is unchanged because the contributions from the cosmological constant to both left- and right-hand sides vanish

$$\dot{\rho}c^2 + (3\dot{a}/a)[p + \rho c^2] = 0. \tag{F.16}$$

References

Friedman, A. 1922, ZPhy, 10, 377
Le Maître, G. 1927, ASSB,, A47, 49

AAS | IOP Astronomy

Introduction to General Relativity and Cosmology (Second Edition)

Ian R Kenyon

Appendix G

The Virial Theorem

In a dynamical system consisting of masses m_k at positions $\mathbf{r}_k$ the moment of inertia about the origin is

$$I = \sum m_k r_k^2, \tag{G.1}$$

and its rate of change

$$\dot{I} = 2\sum \mathbf{p}_k \cdot \mathbf{r}_k, \tag{G.2}$$

where $\mathbf{p}_k$ is the momentum of the kth mass. Equally

$$\ddot{I} = 2\sum \dot{\mathbf{p}}_k \cdot \mathbf{r}_k + 2\sum \mathbf{p}_k \cdot \dot{\mathbf{r}}_k = 2\sum \mathbf{F}_k \cdot \mathbf{r}_k + 2\sum m_k \mathbf{v}_k^2, \tag{G.3}$$

where $\mathbf{F}_k$ is the force on the kth mass and $\mathbf{v}_k$ its velocity. If the system is in equilibrium and bound then $\ddot{I}$ vanishes, so we have

$$\sum m_k \mathbf{v}_k^2 = -\sum \mathbf{F}_k \cdot \mathbf{r}_k. \tag{G.4}$$

If the force is gravitational then that on particle k due to particle j is

$$\mathbf{F}_{jk} = Gm_j m_k \mathbf{r}_{jk}/r_{jk}^3$$

and similarly

$$\mathbf{F}_{kj} = -Gm_j m_k \mathbf{r}_{jk}/r_{jk}^3$$

where $\mathbf{r}_{jk} = -\mathbf{r}_{kj} = \mathbf{r}_j - \mathbf{r}_k$. Together they contribute to the right-hand side of Equation (G.4)

$$Gm_j m_k \mathbf{r}_{jk}^2/r_{jk}^3 = Gm_j m_k/r_{jk}$$

doi:10.1088/2514-3433/acc3ffch24

which is simply *minus* their mutual potential energy. Hence Equation (G.4) can be re-written

$$2T \text{ (Kinetic energy)} \\ = -U \text{ (Potential energy)} - W \text{ (Work done by external pressure)}. \tag{G.5}$$

This result will hold for matter, dark plus baryonic, provided it has reached equilibrium. If the system is isolated, which is approximately true of galaxies and galaxy clusters, then

$$2T \text{ (Kinetic energy)} = -U \text{ (Potential energy)}. \tag{G.6}$$

AAS | IOP Astronomy

Introduction to General Relativity and Cosmology (Second Edition)

Ian R Kenyon

Appendix H

Scale Invariance

The near scale invariance of the observed perturbations of matter has been discussed in Chapters 12, 13, and 15. This appendix is used to work through the prediction for the power spectrum of the perturbations resulting from slow roll inflation. The scalar inflaton field, expanded around the background value is

$$\phi = \phi_0 + \delta\phi,$$

where $\delta\phi/\phi_0$ must be of the same small size as the CMB fluctuations. Dimensional analysis shows that the correlation between the field perturbations at $\mathbf{r}$ and $\mathbf{r} + \Delta\mathbf{r}$ is

$$\langle \delta(\mathbf{r},\, t)\delta(\mathbf{r} + \delta\mathbf{r},\, t)\rangle = \text{Length}^{-2}. \tag{H.1}$$

The field is scalar and neither it nor its interactions offer a length. Thus the only length relevant is $\Delta\mathbf{r}$, and we have

$$\langle \delta(\mathbf{r},\, t)\delta(\mathbf{r} + \delta\mathbf{r},\, t)\rangle \sim \Delta r^{-2}. \tag{H.2}$$

During inflation the separation between points grows and eventually they lose causal contact when $\Delta r = c/H$. At that time the correlation freezes and remains unchanged until the points again make causal contact at some time after inflation. Thus the correlation that exists on reentering the horizon is

$$\langle \delta(\mathbf{r},\, t)\delta(\mathbf{r} + \delta\mathbf{r},\, t)\rangle \sim (c/H)^{-2}, \tag{H.3}$$

which is clearly *scale invariant*. The wave equation for a scalar field given in Chapter 13 can be refined to take account of the field kinetic energy, as well as the expansion of the universe and the potential $V(\phi)$:

$$\ddot{\phi} - \frac{c^2}{a(t)^2}\nabla^2\phi + 3H\dot{\phi} + V'(\phi) = 0, \tag{H.4}$$

doi:10.1088/2514-3433/acc3ffch25

where the prime symbol indicates taking the derivative with respect to ϕ. The time dependence of the scale factor is made explicit. In slow roll inflation the last term $V'(\phi)$ is smaller the slower the roll. This term will be ignored, so that we consider the limiting case of infinitely prolonged inflation. Then Equation (H.4) becomes the equation for damped harmonic motion in which the angular frequency ω and the wave number k are related by

$$\omega = ck/a(t). \tag{H.5}$$

Note that the angular frequency of the Fourier components of the perturbations are time dependent. In the case, as here, that the perturbations are tiny the Fourier modes effectively do not interact and evolve independently. The mean square fluctuations of an harmonic oscillator are proportional to $1/\omega$, which translates for a Fourier component of the scalar field to

$$\langle|\delta\phi_k|^2\rangle \propto \frac{1}{a^3\omega}, \tag{H.6}$$

where the normalization of the field is taken into account by the a^{-3} factor. On leaving the horizon we have

$$\omega_{\text{horizon}} \sim H \text{ and } a_{\text{horizon}} \sim k/H. \tag{H.7}$$

Thus the correlation freezes with

$$\langle|\delta\phi_k|^2\rangle \propto \frac{1}{a^3_{\text{horizon}}\omega_{\text{horizon}}} \propto \frac{H^2}{k^3}. \tag{H.8}$$

From this field fluctuation we can infer the corresponding curvature perturbations defined by

$$\xi = \frac{\delta a}{a} = \frac{H}{\dot{\phi}}\,\delta\phi. \tag{H.9}$$

The two point correlation of the curvature perturbations is then

$$\langle|\xi_k|^2\rangle = \frac{H^2}{\dot{\phi}^2}\langle|\phi_k|^2\rangle \propto \frac{H^4}{k^3\dot{\phi}^2}, \tag{H.10}$$

which is equally the two-point power spectrum of the curvature fluctuations

$$P_\xi(k) \propto \frac{H^4}{k^3\dot{\phi}^2}. \tag{H.11}$$

A useful quantity can be defined here,

$$\Delta^2(k) = \frac{1}{2\pi^2}k^3P_\xi(k), \tag{H.12}$$

giving the power in the fluctuations per unit log interval of k. Then

$$\Delta^2(k) \propto \frac{H^4}{\dot{\phi}^2}, \tag{H.13}$$

so that the power in each decade of k is the same. This is another expression of *scale invariance*. Finally the resulting fluctuations in matter can be calculated. Poisson's equation provides the link between the fractional curvature/gravitational field fluctuations (ξ) and the fractional matter fluctuations (δ):

$$\nabla^2 \xi = 4\pi G\, \delta, \tag{H.14}$$

giving for Fourier components

$$k^2 \xi_k = 4\pi G\, \delta_k. \tag{H.15}$$

Then using Equations (H.11) and (H.15) the power spectrum of the matter fluctuations

$$P(k) \propto \left[\frac{k^2}{4\pi G}\right]^2 P_\xi(k) = \frac{H^4}{[4\pi G\, \dot{\phi}]^2} k. \tag{H.16}$$

This justifies the connection, made in Chapter 15: namely that limitingly slow roll inflation gives a power spectrum of matter fluctuations that depends linearly on k. This is just the *scale invariant* Harrison–Zeldovich spectrum. Some modification of the analysis is needed to take account of the finite speed and duration of inflation: the predicted power spectrum is then found to be proportional to k^n, with n slightly less than unity.

Printed in the USA
CPSIA information can be obtained
at www.ICGtesting.com
JSHW070725031123
51218JS00003B/100